ASPECTS OF VERTEBRATE HISTORY

Ned Colbert at Coalsack Bluff camp, Antarctica, 1969.
Photograph by Laurence M. Gould.

ASPECTS OF VERTEBRATE HISTORY

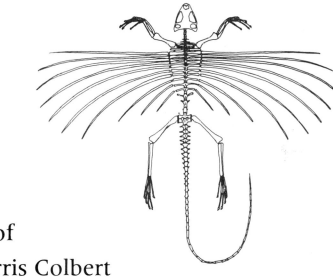

Essays
In Honor of
Edwin Harris Colbert

Louis L. Jacobs, *editor*

MUSEUM OF NORTHERN ARIZONA PRESS

To
Edwin Harris Colbert
on his
seventy-fifth birthday

TABLE OF CONTENTS

LIST OF CONTRIBUTORS

Donald Baird
Museum of Natural History
Princeton University
Princeton, New Jersey 08544

J. F. Bonaparte
Consejo Nacional de Investigaciones
Museo Argentino de Ciencias Naturales
Av. Angel Gallardo 470
1405 Buenos Aires, Argentina

Alan J. Charig
British Museum (Natural History)
Cromwell Road
London, SW7 5BD
England

A. W. Crompton
Department of Biology
    and
Museum of Comparative Zoology
Harvard University
Cambridge, Massachusetts 02138

Joseph T. Gregory
Museum of Paleontology
University of California
Berkeley, California 94720

G. Haas
Department of Zoology
The Hebrew University of Jerusalem
Jerusalem
Israel

Louis L. Jacobs
Museum of Northern Arizona
Route 4, Box 720
Flagstaff, Arizona 86001

Sohan L. Jain
Geological Studies Unit
Indian Statistical Institute
Calcutta 700 035
India

Noye M. Johnson
Earth Sciences Department
Dartmouth College
Hanover, New Hampshire 03755

Emil Kuhn-Schnyder
Institute of Paleontology
University of Zurich
CH-8006 Zurich
Switzerland

Everett H. Lindsay
Department of Geosciences
University of Arizona
Tucson, Arizona 85721

Larry G. Marshall
Field Museum of Natural History
Roosevelt Road at Lake Shore Drive
Chicago, Illinois 60605

Malcolm C. McKenna
American Museum of Natural History
New York, New York 10024

Phillip A. Murry
Department of Geological Sciences
Southern Methodist University
Dallas, Texas 75275

Paul E. Olsen
Biology Department
Bingham Laboratories
Peabody Museum
New Haven, Connecticut 06520

Everett C. Olson
Department of Biology
University of California
Los Angeles, California 90024

Neil D. Opdyke
Lamont-Doherty Geological Observatory
Columbia University
Palisades, New York 10964

John H. Ostrom
Department of Geology and Geophysics,
    and the
Peabody Museum of Natural History
Yale University
New Haven, Connecticut 06520

Timothy Rowe
Museum of Northern Arizona
Route 4, Box 720
Flagstaff, Arizona 86001

Dale A. Russell
National Museum of Canada
Ottawa, Ontario K1A OM8
Canada

Bobb Schaeffer
The American Museum of Natural History
New York, New York 10024

George Gaylord Simpson
The Simroe Foundation
5151 East Holmes Street
Tucson, Arizona 85711
        and
Department of Geosciences
University of Arizona
Tucson, Arizona 85721

Kathleen K. Smith
Department of Anatomy
Duke University Medical Center
Durham, North Carolina 27710

Keith Stewart Thomson
Peabody Museum of Natural History
        and
Department of Biology
Yale University
New Haven, Connecticut 06520

Heinz Tobien
Institut für Geowissenschaften
Johnannes-Gutenberg-Universität
6500 Mainz 1
Federal Republic of Germany

Terry A. Vaughan
Department of Biology
Northern Arizona University
Flagstaff, Arizona, 86011

LIST OF ILLUSTRATIONS

LIST OF TABLES

Preface

This volume is a tribute to the professional life of Dr. Edwin Harris Colbert—"Ned"—on his seventy-fifth birthday. Ned has had a remarkably full and diverse career spanning fifty years. Throughout these fifty years, a period of exciting advancement and achievement in vertebrate paleontology, Ned has contributed substantially as both a researcher and teacher.

The vast range of his contributions to vertebrate history is reflected in his bibliography, presented here as the opening chapter of this volume. The bibliography, which is current only through June 1980, continues to grow because Ned remains active. He is now engaged in the study of Antarctic tetrapods.

Ned's bibliography was selected as the first chapter because his work is the element unifying the papers that follow. All of these studies, on subjects ranging from fossil fish to Recent mammals, can be related in one way or another to Ned's contributions.

The essays in this volume were presented by Ned's friends, colleagues, and students, some of whom were with Ned early in his career. Others are continuing studies initiated or influenced by Ned. Many are from countries other than the United States, attesting to Ned's international reputation. Such a suite of contributors, each with his own approach to vertebrate history, and each addressing the topic of his choice, is appropriate for a man of Ned's achievements. Because of their diverse backgrounds, the authors have been allowed as much stylistic freedom as reasonably possible.

A work such as this is a cooperative endeavor involving many people. It could not have been possible without the support of Malcolm C. McKenna. Hermann K. Bleibtreu, former director of the Museum of Northern Arizona, was encouraging at every stage. John F. Stetter, former publisher of the Museum of Northern Arizona Press, Stephen Trimble, current publisher, Stanley and Emily Stillion of Tiger Typographics, and Earl Hatfield of Classic Printers all did an admirable job in production. Thanks go to R. L. Cifelli, W. Downs, B. L. Fine, and N. K. Viola for assistance in reading proof, and to Kathy Flanagan for undertaking the tedious job of indexing. Dorothy House, librarian at the Museum of Northern Arizona, was particularly helpful in general editorial matters. Lawrence J. Flynn, Everett H. Lindsay, George Gaylord Simpson, G. W. Trussell, and Timothy Rowe assisted in numerous ways. Thanks are due the authors and all those who reviewed the manuscripts. As always, my friend and colleague, Will Downs, was cheerfully helpful.

Margaret Matthew Colbert has been exceedingly helpful throughout this endeavor as she is always. All those who participated in this tribute to Ned are fond admirers of Margaret, and it is fitting to record our affection for her.

Finally, I would like to present my personal thanks to all those with whom I have had the opportunity to work in the preparation of this volume. To Dr. and Mrs. Colbert, I express my sincere gratitude. It has been a genuine pleasure to be associated with them at the Museum of Northern Arizona. Editing this volume has been an honor.

Louis L. Jacobs
Museum of Northern Arizona
Flagstaff, Arizona
23 July 1980

BIBLIOGRAPHY
(THROUGH JUNE 1980)
OF
EDWIN HARRIS COLBERT

1931

With Henry Fairfield Osborn. The elephant enamel method of measuring Pleistocene time. Also stages in the succession of fossil man and Stone Age industries. Amer. Phil. Soc., Proc., 70(2):187–91.

1932

Aphelops from the Hawthorn Formation of Florida. Fla. State Geol. Surv., Bull., 10:55–58. Reprinted in 1956.

1933

The skull of *Dissopsalis carnifex* Pilgrim, a Miocene creodont from India. Amer. Mus. Nov., 603:1–8.
The presence of tubulidentates in the Middle Siwalik beds of northern India. Amer. Mus. Nov., 604:1–10.
A new mustelid from the Lower Siwalik beds of northern India. Amer. Mus. Nov., 605:1–3.
The skull and mandible of *Conohyus,* a primitive suid from the Siwalik beds of India. Amer. Mus. Nov., 621:1–12.
The skull and mandible of *Giraffokeryx punjabiensis* Pilgrim. Amer. Mus. Nov., 632:1–14.
Two new rodents from the Lower Siwalik beds of India. Amer. Mus. Nov., 633:1–6.
An upper Tertiary peccary from India. Amer. Mus. Nov., 635:1–9.

1934

An upper Miocene suid from the Gobi Desert. Amer. Mus. Nov., 690:1–7.
A phylogenetic chart of the Artiodactyla. By W. D. Matthew; foreword and comments by Edwin H. Colbert. J. Mammal., 15(3):207–9.
A new rhinoceros from the Siwalik beds of India. Amer. Mus. Nov., 749:1–13.
Chalicotheres from Mongolia and China in the American Museum. Amer. Mus. Natur. Hist., Bull., 67(8):353–87.

1935

Nebraska—fifteen million years ago. Natur. Hist., 35(1):37–46.
Distributional and phylogenetic studies on Indian fossil mammals. I. American Museum collecting localities in northern India. Amer. Mus. Nov., 796:1–20.
Distributional and phylogenetic studies on Indian fossil mammals. II. The correlation of the Siwaliks of India as inferred by the migrations of *Hipparion* and

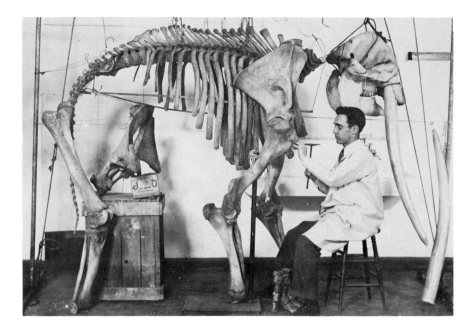

Figure 1.1. Ned Colbert at the University of Nebraska, 1929.

Equus. Amer. Mus. Nov., 797:1–15.

Distributional and phylogenetic studies on Indian fossil mammals. III. A classification of the Chalicotherioidea. Amer., Mus. Nov., 798:1–15.

Distributional and phylogenetic studies on Indian fossil mammals. IV. The phylogeny of the Indian Suidae and origin of the Hippopotamidae. Amer. Mus. Nov., 799:1–24.

Distributional and phylogenetic studies on Indian fossil mammals. V. The classification and phylogeny of the Giraffidae. Amer. Mus. Nov., 800:1–15.

A new fossil peccary, *Prosthennops niobrarensis,* from Brown County, Nebraska. Nebr. Univ. State Mus., Bull., 1(44):419–30.

The beginning of life in North America; fish and reptiles, early rulers of water and land; the Age of Mammals; the first men in North America; Indian hunters; Indians of the Southwest and of Mexico. *In* M. D. Chase (ed.), The story of America. Whitman Publishing Co., Racine, Wisc., pp. 8, 10, 12, 14, 16.

The proper use of the generic name *Nestoritherium.* J. Mammal. 16(3):233–34.

Siwalik mammals in the American Museum of Natural History. Amer. Phil. Soc., Trans., 26:i–x, 1–401.

Abstract of "A critical review of the Mint Canyon mammalian fauna and its correlative significance." Pal. Zentralbl., 7, abs. no. 7347.

Abstract of "A preliminary notice on the Miocene and Pliocene mammalian faunas near Valentine, Nebraska." Pal. Zentralbl., 7, abs. no. 7348.

Abstract of "Pleistocene mammalian fauna from the Carpenteria Asphalt." Pal. Zentralbl., 7, abs. no. 7349.

Abstract of "A rodent fauna from later Cenozoic beds of southwestern Idaho," Pal.

Figure 1.2. Margaret and Ned Colbert in New York, 1934.

Zentralbl., 7, abs. no. 7350.

Abstract of "A Pleistocene bat from Florida." Pal. Zentralbl., 7, abs. no. 7351.

Abstract of "A new antilocaprid and a new cervid from the late Tertiary of Nebraska." Pal. Zentralbl., 7, abs. no. 7352.

Abstract of "Anchitheriine horses from the *Merychippus* zone of the North Coalinga District, California." Pal. Zentrabl., 7, abs. no. 7353.

Abstract of "Preliminary notice of a new species of *Hippohyus* from India." Pal. Zentralbl., 7, abs. no. 7354.

Abstract of "Three oreodont skeletons from the lower Miocene of the Great Plains." Pal. Zentralbl., 7, abs. no. 7355.

Abstract of "Two new species of sheep-like antelope from the Miocene of Mongolia." Pal. Zentralbl., 7, abs. no. 7356.

Abstract of "Preliminary notice of a new genus of lemuroid from the Siwaliks." Pal. Zentralbl., 7, abs. no. 7357.

Abstract of "Preliminary notice of new man-like apes from India." Pal. Zentralbl., 7, abs. no. 7358.

1936

Fossils. Amer. Mus. Natur. Hist., School Nature League Bull., Ser. 6, 5, 3 pp.

Tertiary deer discovered by the American Museum Asiatic Expeditions. Amer. Mus. Nov., 854:1–21.

Palaeotragus in the Tung Gur Formation of Mongolia. Amer. Mus. Nov., 874:1–17.

Was the extinct giraffe (*Sivatherium*) known to the early Sumerians? Amer. Anthropol., 38(4):605–8.

1937

Notice of a new genus and species of artiodactyl from the upper Eocene of Wyoming. Amer. J. Sci., 33:473–74.

A new primate from the upper Eocene Pondaung Formation of Burma. Amer. Mus. Nov., 951:1–18.

The Pleistocene mammals of North America and their relations to Eurasian forms. *In* G. G. MacCurdy (ed.), Early Man. J. B. Lippincott Company, Philadelphia, pp. 173–84.

With Horace Elmer Wood, 2nd. A provincial time-scale for North American continental Tertiary. Geol. Soc. Amer., Abstr., 50th Ann. Mtg., 92.

1938

The giraffe and his living ancestor. Natur. Hist., 41(1):46–50, 78.

The beaver. Amer. Mus. Natur. Hist. School Nature League Bull., Series 8, 6, 3 pp.

The relationships of the Okapi. J. Mammal., 19(1):47–64.

Brachyhyops, a new bunodont artiodactyl from Beaver Divide, Wyoming. Carnegie Mus., Ann., 27:87–108.

The panda: a study in emigration. Natur. Hist., 42(1):33–39.

Pliocene peccaries from the Pacific Coast Region of North America. Carnegie Inst. Washington, Publ,. 487:241–69.

Remarks on the use of the name "Valentine." Amer. J. Sci., 36:212–14.

Fossil mammals from Burma in the American Museum of Natural History. Amer. Mus. Natur. Hist., Bull., 74(6)255–436.

The mastodon and the mammoths of New York. Amer. Mus. Natur. Hist., School Nature League Bull., Series 9, 4, 3 pp.

1939

Wild dogs and tame—past and present. Natur. Hist., 43(2):90–95.

Ancestors of the dog. Science Digest, pp. 56–60. Reprint and condensation of Wild Dogs and Tame—Past and Present.

The origin of the dog. American Museum Guide Leaflet, No. 102. Reprint of Wild Dogs and Tame—Past and Present, in its original form.

A new anchitheriine horse from the Tung Gur Formation of Mongolia. Amer. Mus. Nov., 1019:1–9.

Carnivora of the Tung Gur Formation of Mongolia. Amer. Mus. Natur. Hist., Bull., 76(2):47–81.

The migrations of Cenozoic mammals. N.Y. Acad. Sci., Trans., Ser. 2, 1(6):89–94.

A fossil comes to life. Natur. Hist., 43(5):280–84.

With Robert G. Chaffee. A study of *Tetrameryx* and associated fossils from Papago Spring Cave, Sonoita, Arizona. Amer. Mus. Nov., 1034:1–21.

The tiger in the parlor. Frontiers, 4(1):2–6.

Some studies of adaptations in dentitions of mammals, including man. Amer. J. Orthodontics and Oral Surgery, 25(10):952–68.

Foreword, pp. v–vi, to "Climate and evolution" by William Diller Matthew, 2nd edition. Also annotations throughout the text. N.Y. Acad. Sci., Spec. Publ., 1:i–xii, 1–223.

With William K. Gregory. On certain principles of evolution illustrated in the mammalian orders Perissodactyla and Artiodactyla. Memorial volume to A. N.

Sewertzoff, Acad. Sci., U.S.S.R., vol. 1, pp. 97–116. Russian translation, pp. 117–36.

1940

Some cervid teeth from the Tung Gur Formation of Mongolia, and additional notes on the genera *Stephanocemas* and *Lagomeryx*. Amer. Mus. Nov., 1062:1–6.

Mammoths and men. Natur. Hist., 46(2):96–103.

Mammoths and men. Society for the Preservation of the Fauna of the Empire, Journal, N.S., 41:38–47.

The tar pit tiger. Natur. Hist., 46(5):284–87.

Where the cats came from. Natur. Hist., 46(5):288–89.

Pleistocene mammals from the Ma Kai Valley of northern Yunnan, China. Amer. Mus. Nov., 1099:1–10.

1941

Cats have come a long way. Toronto Star Weekly, Feb. 22, 1941. Reprint of "The tar pit tiger" and "Where the cats came from," 1940.

With Horace E. Wood, 2nd, Ralph W. Chaney, John Clark, Glenn L. Jepsen, John B. Reeside, Jr., and Chester Stock. Nomenclature and correlation of the North American continental Tertiary. Geol. Soc. Amer., Bull., 52:1–48.

The ancestral ursid, *Hemicyon*, in Nebraska. Nebr. Univ. State Mus., Bull., 2(5): 49–57.

The osteology and relationships of *Archaeomeryx*, an ancestral ruminant. Amer. Mus. Nov. 1135:1–24.

A study of *Orycteropus gaudryi* from the island of Samos. Amer. Mus. Natur. Hist., Bull., 78(4):305–51.

Our vanished herds. Frontiers, 6(1):8–10.

Three animals that went to sea. Natur. Hist., 48(2):96–99.

With Robert G. Chaffee. The type of *Clepsysaurus pennsylvanicus* and its bearing upon the genus *Rutiodon*. Notulae Naturae, 90:1–19.

With H. E. Wood and D. Dunkle. Report of Committee on Publication Facilities. The Society of Vertebrate Paleontology News Bulletin, 3:5 11.

An outline of vertebrate evolution. A synopsis of lectures on the evolution of the vertebrates, presented at Bryn Mawr College. 40 pp. (mimeographed)

1942

The wolves and foxes of North America. Amer. Mus. Natur. Hist., School Nature League Bull., Ser. 12, 5, 3 pp.

The geologic succession of the Proboscidea. *In*, Henry Fairfield Osborn, The Proboscidea, Vol. 2, Chapter 22. American Museum Press, New York, pp. 1421–1521.

Circus without spectators. Natur. Hist., 49(5):248–52.

Ice age winter resort. Natur. Hist., 50(1):16–21.

The association of man with extinct mammals in the Western Hemisphere. Eighth Amer. Sci. Cong. Proc., 2:17–29.

Triumph of the mammals. Guide Leaflet Series of the American Museum of Natural History, No. 112, pp. 1–16. Reprint of Circus without Spectators and Ice Age Winter Resort, with some additional material.

An edentate from the Oligocene of Wyoming. Notulae Naturae, 109:1–16.
Notes on the lesser one-horned rhinoceros, *Rhinoceros sondaicus*. 2. The position of *Rhinoceros sondaicus* in the phylogeny of the genus *Rhinoceros*. Amer. Mus. Nov., 1207:1–6.
Finding a new animal in the laboratory. Frontiers, 7(2):41–43.
The study of natural history in war time. Amer. Mus. Natur. Hist., School Nature League Bull., 13, 1, 4 pp.
The Pleistocene faunas of Asia and their relationships to early man. N.Y. Acad. Sci., Trans., Ser. 2, 5(1):1–10.
Vertebrate paleontology. Collier's Year Book for 1941. P. F. Collier & Son, New York.

1943

The native cats of North America. Amer. Mus. Natur. Hist., School Nature League Bull. Ser. 13, 5, 4 pp.
Pleistocene vertebrates collected in Burma by The American Southeast Asiatic Expedition. Amer. Phil. Soc., Trans., N.S., 32(3):95–430.
A Miocene oreodont from Jackson Hole, Wyoming. J. Paleontol., 17(3):298–305.
Report of Committee on Publication Facilities. Society of Vertebrate Paleontology News Bulletin, 9:3–4.
Edgar Billings Howard. Society of Vertebrate Paleontology News Bulletin, 9:5–6.
How do we mount our skeletons? Society of Vertebrate Paleontology News Bulletin 9:18–21.
A lower jaw of *Clepsysaurus* and its bearing upon the relationships of this genus of *Machaeroprosopus*. Notulae Naturae, 124:1–8.
Vertebrate paleontology. Collier's Year Book for 1942. P. F. Collier & Son, New York.

1944

The ancient reptile of Blue Bell. Frontiers, 8(3):67–69.
Nature's greatest bonehead was a dinosaur. Natur. Hist., 53(6):248–51, 284.
A new fossil whale from the Miocene of Peru. Amer. Mus. Natur. Hist., Bull., 83(3):195–216.
Henry C. Raven. Society of Vertebrate Paleontology News Bulletin, 12:25–27.
The beaver. Amer. Mus. Natur. Hist., School Nature League Bull., Ser. 8, 16, 4 pp. (Reprint of 1938 with new illustrations.)
Corrections for a "Check list of the fossil reptiles of New Jersey." J. Paleontol., 18(5):480.

1945

State collecting laws. Society of Vertebrate Paleontology News Bulletin, 14:7.
The distribution of members of the Society. Society of Vertebrate Paleontology News Bulletin, 14:24–25.
A calendar of life. Natur. Hist., 54(6):280–82.
The dinosaur book. The ruling reptiles and their relatives. Amer. Mus. Natur. Hist., Man and Nature Publ., Handbk. 14:1–156.
The wolves and foxes of North America. (Revised edition.) Amer. Mus. Natur. Hist., School Nature League Bull., Series 13, 5, 4 pp.
The hyoid bones in *Protoceratops* and in *Psittacosaurus*. Amer. Mus. Nov., 1301:1–10.

Review of "Vertebrate paleontology" by Alfred Sherwood Romer. Natur. Hist.,
    54(10):443.
Nueva Ballena Fosil del Mioceno del Peru. (Translation and reprint in its entirety
    of "A new fossil whale from the Miocene of Peru," 1944, with illustrations.
    Museo de Historia Natural "Javier Prado," Bol., 9(32, 33):23–60.
Evolution of the horse. Encyclopedia Americana, Americana Corp., Danbury, Conn.,
    pp. 387–91.

1946

Review of "Vertebrate paleontology" by Alfred Sherwood Romer. Science, 103
    (2664):88–89.
Were the dinosaurs failures? Condensation, with minor alterations from The Dino-
    saur Book, 1945. Science Digest, 19(5):62–66.
Hypsognathus, a Triassic reptile from New Jersey. Amer. Mus. Natur. Hist., Bull.,
    87(5):229–74.
With Raymond B. Cowles and Charles M. Bogert. Temperature tolerances in the
    American alligator and their bearing on the habits, evolution, and extinction
    of the dinosaurs. Amer. Mus. Nat. Hist., Bull., 86(7):327–74.
The eustachian tubes in the Crocodilia. Copeia, 1946(1):12–14.
Sebecus, representative of a peculiar suborder of fossil Crocodilia from Patagonia.
    Amer. Mus. Natur. Hist., Bull., 87(4):217–70.

1947

Studies of the phytosaurs *Machaeroprosopus* and *Rutiodon*. Amer. Mus. Natur. Hist.,
    Bull., 88(2):53–96.
A paleontologist's view of the geology curriculum. Geol. Soc. Amer., Interim Proc.,
    1:8–14.
Functions of vertebrate paleontology in the earth sciences. Geol. Soc. Amer., Bull.,
    58:287–91. (Address as Retiring President of Society of Vertebrate Paleontology.)
Were dinosaurs a failure? Toronto Star Weekly, June 28, 1947, pp. 5, 15. (Excerpts
    from The Dinosaur Book.)
With Raymond B. Cowles and Charles M. Bogert. Rates of temperature increase in
    the dinosaurs. Copeia, 1947, (2):141–42.
With Charles L. Camp, Edwin D. McKee and S. P. Welles. A guide to the continen-
    tal Triassic of northern Arizona. Plateau, 20(1):1–8.
The search for fossils in northern New Mexico, 1947. El Palacio, 54(9):209–12.
With James D. Bump. A skull of *Torosaurus* from South Dakota and a revision of
    the genus. Phila., Acad. Natur. Sci., Proc., 99:93–106.
The little dinosaurs of Ghost Ranch. Natur. Hist., 56(9):392–99, 427–28.
With Bobb Schaeffer. Some Mississippian footprints from Indiana. Amer. J. Sci.,
    245:614–23.
Dinosaurs. Amer. Mus. Natur. Hist., Science Guide 70:1–32.
Review of "A census of the determinable genera of the Stegocephalia" by E. C.
    Case. Quart Rev. Biol., 22(3):216–17.

1948

The mammal-like reptile *Lycaenops*. Amer. Mus. Natur. Hist., Bull., 89(6):353–404.
Dinosaurs of Ghost Ranch. The Toronto Star Weekly, March 6, 1948, p. 5. Reprint

and condensation of The Search for Fossils in Northern New Mexico, 1947.

With Norman D. Newell. Paleontologist—biologist or geologist? J. Paleontol., 22(2):264–67.

Review of "Review of the Labyrinthodontia" by Alfred Sherwood Romer. Quart. Rev. Biol., 23(1):50–51.

Triassic life in the southwestern United Staes. N.Y. Acad. Sci., Trans., Ser. 2, 10(7):229–35.

Review of "The corridor of life" by W. E. Swinton. Natur. Hist., 57(6):247.

Pleistocene of the Great Plains—introduction, summary and conclusions. Geol. Soc. Amer., Bull., 59(6):541–42, 627–30.

Evolution of the horned dinosaurs. Evolution, 2(2):145–63.

A hadrosaurian dinosaur from New Jersey. Phila., Acad. Natur. Sci., Proc., 100: 23–37.

1949

With J. T. Gregory. The C-16 (vertebrate paleontology) excursion of the XVIII International Geologic Congress, 1948. Society of Vertebrate Paleontology News Bulletin, 25:6–9.

Albert Thomson. Society of Vertebrate Paleontology News Bulletin, 25:30–32.

The ancestors of mammals. Sci. Amer., 180(3):40–43.

Some paleontological principles significant in human evolution. Studies in Physical Anthropology No. 1, Early Man in the Far East, pp. 103–48.

Progressive adaptations as seen in the fossil record. *In* G. L. Jepsen, G. G. Simpson, and E. Mayr. (eds.), Genetics, paleontology, and evolution. Princeton University Press, Princeton, pp. 390–402.

A new Cretaceous pleisosaur from Venezuela. Amer. Mus. Nov., 1420:1–22.

Evolutionary growth rates in the dinosaurs. Sci. Monthly, 69(2):71–79.

Giants of the animal kingdom. Natur. Hist., 58(9):418–23, 430, 431.

Review of "The meaning of evolution" by George Gaylord Simpson. Natur. Hist., 58(9):391, 431.

The long reign of the dinosaurs. The Northwest Missouri State Teachers College Studies, 13(1):51–82.

1950

The beginning of the Age of Dinosaurs in northern Arizona. Plateau, 22(3):37–43.

Review of "Die Lebensweise der Dinosaurier" by Martin Wilfarth. J. Paleontol., 24(1):116.

Ancient life of northern New Mexico. El Palacio, 57(1)3–9.

Mesozoic vertebrate faunas and formations of northern New Mexico. Guidebook for the Fourth Field Conferences of the Society of Vertebrate Paleontology in Northwestern New Mexico. pp. 56–72.

The fossil vertebrates. *In* E. W. Haury (ed.), The stratigraphy and archaeology of Ventana Cave, Arizona. The University of New Mexico Press, Albuquerque, pp. 126–48.

Review of "Time and its mysteries" by Henry Russell, Adolph Knopf, James T. Shotwell, George P. Luckey. Natur. Hist., 59(10)438–39.

The Fourth Field Conferences of the Society of Vertebrate Paleontology. Society of Vertebrate Paleontology News Bulletin, 30:3–6.

1951

Earth rhythms and evolution. Evolution, 5(1):84. (Review of "Symphony of the earth" by J. H. F. Umbgrove).

With Dirk A. Hooijer. A Note on the Plio-Pleistocene boundary in the Siwalik Series of India and in Java. Amer. J. Sci., 249:533–38.

With Charles Craig Mook. The ancestral crocodilian *Protosuchus*. Amer. Mus. Natur. Hist., Bull., 97(3):143–82.

The dinosaur book (second edition). McGraw-Hill, New York, 156 pp.

Environment and adaptations of certain dinosaurs. Biol. Rev., 20(3):265–84.

With Dirk A. Hooijer. A mastodont tooth from Szechwan, China. Fieldiana, Geol., 10(12):129–34.

Review of "The fall of the sparrow" by Jay Williams. Natur. Hist., 60(10):438–39.

Evolution of the vertebrates. Evolution, 5:416–17. (Review of "Evolution Emerging" by William King Gregory).

1952

The Mesozoic tetrapods of South America. Amer. Mus. Natur. Hist., Bull., 99(3):237–49.

Breathing habits of the sauropod dinosaurs. Ann. Mag. Natur. Hist., Ser. 12, 5:708–10.

A pseudosuchian reptile from Arizona. Amer. Mus. Natur. Hist., Bull., 99(10):561–92.

Review of "Earth song" by Charles L. Camp. Natur. Hist., 61:439.

Darwin's journal and other publications. Evolution, 6:454–55.

1953

Dinosaurs (second edition). Amer. Mus. Natur. Hist., Science Guide 70:1–32.

Explosive evolution. Evolution, 7:89–90. (Review of "Distribution of evolutionary explosions in geologic time. A symposium sponsored jointly by the Paleontological Society, the Society of Vertebrate Paleontology and The Geological Society of America in 1949.)

The origin of the dog. Amer. Mus. Natur. Hist., Science Guide, 102:1–14.

With Dirk Albert Hooijer. Pleistocene mammals from the limestone fissures of Szechwan, China. Amer. Mus. Natur. Hist., Bull., 102(1):1–134.

The Brontosaur Hall of the American Museum of Natural History. The Museums J., 53(7):179–80.

Story of the shells. Saturday Review, 36(45):17–18. (Review of "Man, Time and Fossils" by Ruth Moore.)

The record of climatic changes as revealed by vertebrate paleoecology. *In* H. Shapley (ed.), Climatic Change. Harvard University Press, Cambridge, pp. 249–71.

1954

Paleontology at the Museum of Northern Arizona. Plateau, 26:89–94.

A gigantic crocodile from the Upper Cretaceous beds of Texas. Amer. Mus. Nov., 1688:1–22.

1955

Distillation of Darwin. Saturday Review, 38(5):18. (Review of "Charles Darwin" by Ruth Moore).

Giant dinosaurs. N.Y. Acad. Sci., Trans., Ser. 2, 17:199–209.

Evolution of the vertebrates. John Wiley and Sons, Inc., New York, xiii + 479 pp.

Review of *"Archaeopteryx lithographica.* A study based upon the British Museum specimen," by Gavin de Beer. Sci. Monthly, 80(5):324–25.

Scales in the Permian amphibian, *Trimerorhachis.* Amer. Mus. Nov., 1740:1–17.

Dinosaurs (third edition). Amer. Mus. Natur. Hist., Science Guide 70:1–32.

Review of "Fossil amphibians and reptiles" by W. E. Swinton. Quart. Rev. Biol., 30(4):389–90.

1956

Review of: "Traité de Paléontologie" Vol. V, Amphibiens, Reptiles, Oiseaux. La sortie des eaux, naissance de la tétrapodie, l'exubérance de la view végétative, la conquête de l'air. Masson, Paris, 1955. Science, 123(3197):593.

Rates of erosion in the Chinle Formation. Plateau, 28:73–76.

With John Imbrie. Triassic metoposaurid amphibians. Amer. Mus. Natur. Hist., Bull., 110(6):399–452.

1957

A trip to Philadelphia. Society of Vertebrate Paleontology News Bulletin, 49:24–25.

Review of "On *Millerosaurus* and the early history of the sauropsid reptiles" by D. M. S. Watson. Nature, 180(4577):128.

Battle of the bones. Cope and Marsh, the paleontological antagonists. GeoTimes, 2(4):6–7, 14.

Triassic vertebrates of the Wind River Basin. Wyo. Geol. Assoc., Guidebk., Twelfth Annual Field Confer., pp. 89–93.

Review of "The Permian reptile *Araeoscelis* restudied" by Peter Paul Vaughn. Quart. Rev. Biol., 32(2):179–80.

With Joseph T. Gregory. Correlation of continental Triassic sediments by Vertebrate Fossils. Geol. Soc. Amer., Bull., 68:1456–67.

1958

On being a curator. Curator, 1(1):7–12.

Karl Patterson Schmidt. Curator, 1(1):19.

Origin of the dog. Natur. Hist., 67(2):65–69.

The beginning of the age of dinosaurs. *In* T. S. Westoll (ed.), Studies on fossil vertebrates. Athlone Press, London, pp. 39–58.

With John H. Ostrom. Dinosaur stapes. Amer. Mus. Nov., 1900:1–20.

With Donald Baird. Coelurosaur bone casts from the Connecticut Valley Triassic. Amer. Mus. Nov., 1901:1–11.

Tetrapod extinctions at the end of the Triassic period. Nat. Acad. Sci., Proc., 44(9):973–77.

Animal extinctions are a baffling study. Frontiers, 23(1):26–27. (Excerpts from Tetrapod extinctions at the end of the Triassic period, 1958. Rewritten to some extent.)

Evolution of the vertebrates (second printing). John Wiley and Sons, Inc., New York, xii + 479 pp.

Morphology and behavior. *In* A. Roe and G. G. Simpson (eds.), Behavior and evolution. Yale University Press, New Haven, pp. 27–47.

Where the cats came from. The Illustrated Library of the Natural Sciences. Simon and Schuster, New York, pp. 593–95. (Reprint of article by the same title published in 1940.)

The origin of the dog. The Illustrated Library of the Natural Sciences. Simon and Schuster, New York, pp. 836–43. Reprint of "Wild dogs and tame—past and present," 1939, with an additional figure.

Mammoths and men. The Illustrated Library of the Natural Sciences. Simon and Schuster, New York, pp. 930–41. Reprint of an article by the same title published in 1940.

Giants of the past. The Illustrated Library of the Natural Sciences. Simon and Schuster, New York, pp. 1187–95. Reprint of "Giants of the animal kingdom," 1949, with the title changed and additional illustrations.

Circus without spectators. The Illustrated Library of the Natural Sciences. Simon and Schuster, New York, pp. 1692–99. Reprint of an article by the same title, 1942.

Ice age winter resort. The Illustrated Library of the Natural Sciences. Simon and Schuster, New York, pp. 1700–1707.

The panda. The Illustrated Library of the Natural Sciences. Simon and Schuster, New York, pp. 2144–51. Reprint of an article published in 1938 with abbreviated title.

Millions of years ago. Prehistoric life in North America. Illustrated by Margaret M. Colbert. Thomas Y. Crowell Company, New York, 153 pp.

Chalk murals. Curator, 1(4):10–16.

Relationships of the Triassic Maleri fauna. J. Palaeontol. Soc. India, 3:68–81.

1959

Darwin-Wallace centennial. Science, 129:154–56.

With Katherin Beneker. The Paleozoic museum in Central Park, or the museum that never was. Curator, 2(2):137–50.

1960

With William A. Burns. Digging for dinosaurs. Columbia Record Club, New York, 44 pp.

Roy Chapman Andrews, explorer. Science, 132:21–22.

Review of "The antecedents of Man" by W. E. LeGros Clark. Evolution, 14:392–93.

The museum and geological research. Curator, 3(4):317–26.

Cor Serpentis. Society of Vertebrate Paleontology News Bulletin, 60:28.

Charles J. Lang (Obituary). Society of Vertebrate Paleontology News Bulletin, 60:34–35.

A new Triassic procolophonid from Pennsylvania. Amer. Mus. Nov., 2022:1–19.

Triassic rocks and fossils. N. Mex. Geol. Soc., Guidebk., Eleventh Field Confer., pp. 55–62.

With Chester Tarka. Illustration of fossil vertebrates. Medical and Biological Illustration, 10(4):237–46.

1961

The world of dinosaurs. Home Library Press, New York, 32 pp.

The Triassic reptile, *Poposaurus.* Fieldiana, Geol., 14(4):59–78.

What is a museum? Curator, 4(2):138–46.

Age and origin of species. *In* W. F. Blair (ed.), Vertebrate speciation. University of Texas Press, Austin, pp. 493–97.

Man and geologic time. *In* L. D. Leet and F. J. Leet (eds.), The world of geology. McGraw-Hill Book Company, Inc., New York, pp. 49–55.

Dinosaurs. Their discovery and their world. E. P. Dutton and Co., Inc., New York, xiv + 300 pp.

Evolution of the vertebrates. Science Editions, New York, xiii + 479 pp.

Foreword to "Prehistoric life on Earth" by Kai Petersen. Edited and adapted and supplemented by Georg Zappler. E. P. Dutton and Co., Inc., New York, p. 7.

The career of Charles R. Knight. Curator, 4(4):352–66. (Text; no authorship indicated).

Inexpensive racks for the storage of large specimens. Curator, 4(4):368–70.

1962

The weights of dinosaurs. Amer. Mus. Nov., 2076:1–16.

Some Victorians and the dinosaurs. Natur. Hist., 71(4):48–57. An excerpt (considerably edited and rewritten—not by EHC) from Dinosaurs, 1961.

Dinosaurs—their discovery and their world. Hutchinson, London, 288 pp. (English Edition).

The record of climatic changes as revealed by vertebrate paleoecology. *In* J. J. White (ed.), Study of the Earth—readings in geological science. Prentice-Hall, Inc. Englewood Cliffs, N.J., pp. 239–60. (Article by the same title published in 1953.)

1963

Relationships of the Triassic reptilian faunas of Brazil and South Africa. S. Afr. J. Sci., 59:248–53.

Fossils of the Connecticut Valley—The Age of Dinosaurs begins. Connecticut St. Geol. Natur. Hist. Surv., Bull., 96:i–iv, 1–31.

Phylogeny and the dimension of time. Amer. Natur., 97(896):319–31.

1964

The Triassic dinosaur genera *Podokesaurus* and *Coelophysis.* Amer. Mus. Nov., 2168: 1–12.

The fossils of New Jersey. New Jersey Nature News, 19(1):17–27.

Dinosaurs of the arctic. Natur. Hist., 73:20–23.

Relationships of the saurischian dinosaurs. Amer. Mus. Nov., 2181:1–24.

The ancient life of New Jersey and its future conservation. New Jersey Nature News, 19(4):152–66.

The relevance of palaeontological data concerning evidence of aridity and hot climates in past geologic ages. *In* A. E. M. Nairn (ed.), Problems in palaeoclimatology. Interscience Publishers, London, pp. 378–81.

Climatic zonation and terrestrial faunas. *In* A. E. M. Nairn, Problems in palaeoclimatology. Interscience Publishers, London, pp. 617–39, 641–42.

Adventures in Gondwanaland. Society of Vertebrate Paleontology News Bulletin, 72:46–48.
Ancient reptiles of India. Samvadadhvam, 6(2, 3):116–18.

1965

With Marshall Kay. Stratigraphy and life history. John Wiley and Sons, Inc., New York, 736 pp.
The Age of Reptiles. Weidenfeld and Nicolson, London, xiv + 228 pp.
The Age of Reptiles. W. W. Norton and Co., New York, xii + 228 pp.
The university in the museum. Curator, 8(1):78–85.
Carl Sorensen (Obituary). Society of Vertebrate Paleontology, News Bulletin, 74:56–57.
Die Evolution der Wirbeltiere. Gustav Fischer Verlag, Stuttgart, xvi + 426 pp. (German translation by Gerhard Heberer.)
A phytosaur from North Bergen, New Jersey. Amer. Mus. Nov., 2230:1–25.
With Gordon Reekie. New space, new wings. Curator, 8(3):212–22.
El Libro de los Dinosaurios. Eudeba, Editorial Universitaria de Buenos Aires, Buenos Aires, 181 pp.
Old bones, and what to do about them. Curator, 8(4):302–18.
Remarks *in* Vertebrate Paleontology in Alberta. Report of a conference held at the University of Alberta, August 29 to September 3, 1963. University of Alberta, Edmonton, pp. 21, 66–68, 71, 73.

1966

Rates of erosion in the Chinle Formation—ten years later. Plateau, 38(3):68–74.
A gliding reptile from the Triassic of New Jersey. Amer. Mus. Nov., 2246:1–23.
The appearance of new adaptations in Triassic tetrapods. Israel J. Zool., 14:49–62.
Ancient reptile of Blue Bell. Frontiers, 31:42–44.
The Age of Reptiles. The Norton Library. W. W. Norton and Co., Inc., New York, viii + 228 pp. (Paperback edition.)

1967

Introduction to "Dragons of the air" by H. G. Seeley. Dover Publications, Inc., New York, pp. v–xii.
Adaptations for gliding in the lizard *Draco*. Amer. Mus. Nov., 2283:1–20.
Evolution of the Vertebrates. Tsukiji Shokan Ltd., Vol. 1, xii + 266 pp.; Vol. 2, vi + 304 pp. (Japanese translation by M. Tasumi.)
A new interpretation of *Austropelor*, a supposed Jurassic labyrinthodont amphibian from Queensland. Queensl. Mus., Mem., 15(1):35–42.
New adaptations of Triassic reptiles. Israel Acad. Sci. Human., Section of Sciences, 5:1–13
With D. Merrilees. Cretaceous dinosaur footprints from western Australia. Roy. Soc. West. Austral., J., 50:21–25.
With William A. Burns. Digging for dinosaurs. Childrens Press, Chicago, 62 pp. Reprint of an article by the same title published in 1960, with new illustrations by Robert Borja.

Figure 1.3. Ned Colbert at the American Museum of Natural History, 1968.

1968

Men and dinosaurs. The search in field and laboratory. E. P. Dutton and Company, New York, xviii + 283 pp.

With Peter J. Barrett and Ralph J. Baillie. Triassic amphibian from Antarctica. Science, 161 (3840):460–62.

Memorial to Charles Craig Mook. Geol. Soc. Amer., Proc. vol. for 1966:309–15.

Developments, trends, and outlooks in paleontology—later fossil reptiles. J. Paleontol., 42 (6):1375.

1969

Evolution of the horned dinosaurs. *In* Papers on evolution. Selected by Paul R. Ehrlich, Richard W. Holm and Peter H. Raven. Little, Brown and Company, Inc., Boston, pp. 314–32.

Amphibian fossil. McGraw-Hill Yearbook of Science and Technology, pp. 98–99. (A brief description of the Triassic labyrinthodont found in Antarctica. See Triassic Amphibian from Antarctica, 1968, by Peter J. Barrett, Ralph J. Baillie, and Edwin H. Colbert.)

A Jurassic pterosaur from Cuba. Amer. Mus. Nov., 2370:1–26.

With Dale A. Russell. The small Cretaceous dinosaur *Dromaeosaurus*. Amer. Mus. Nov., 2380:1–49.

Men and dinosaurs. The search in field and laboratory. Evans Brothers Limited, London, xviii + 283 pp.

Evolution of the vertebrates. Second Edition. John Wiley and Sons, Inc., New York, xvi + 535 pp.

Men and dinosaurs. The search in field and laboratory. Hayakawa Shobo and Co., Ltd., Tokyo, 340 pp. (Japanese translation.)

1970

Review of: Handbuch der Palaoherpetologie, Encyclopedia of paleoherpetology, edited by Oskar Kuhn. Part 15, Ornithischia, by Rodney Steel, 1969. Copeia, 1970(1):203–4.

A saurischian dinosaur from the Triassic of Brazil. Amer. Mus. Nov., 2405:1–39.

Gondwanaland and the distribution of Triassic tetrapods. Gondwana Stratigraphy, IUGS Symposium, Buenos Aires, 1967, pp. 355–74.

Arizona and Antarctica. Plateau, 42(4):118–24.

Excavating in Mongolia. Science, 168(3932):719–20. (Review of "Hunting for Dinosaurs" by Zofia Kielan-Jaworowska.)

The fossil tetrapods of Coalsack Bluff. Antarctic J., 5(3):57–61.

What happened in Antarctica. Society of Vertebrate Paleontology News Bulletin, 89:51–54.

Fossils of the Connecticut Valley. The Age of Dinosaurs begins (Revised Edition). Connecticut St. Geol. Natur. Hist. Surv., Bull., 96:i–iv, 1–32.

Paleontological investigations at Coalsack Bluff. Antarctic J., 5(4):86.

Review of "Hunting for dinosaurs" by Zofia Kielan-Jaworowska. Natur. Hist., 79(7):122–23.

With David H. Elliot, William J. Breed, James A. Jensen and Jon. S. Powell. Triassic tetrapods from Antarctica: evidence for continental drift. Science, 169(3951): 1197–1201.

Giant dinosaurs. *In* P. Cloud (ed.), Adventures in Earth History. W. H. Freeman and Co., San Francisco, pp. 716–23. Reprint of an article by the same title published in 1955.

Dinosaurs and Victorians. *In* B. Mears, Jr. (ed.), The nature of geology—contemporary readings. Van Nostrand Reinhold Company, New York, Chapter 11, pp. 133–47. (Reprint of Chapter 2 of Dinosaurs. Their discovery and their world, 1961.)

The Triassic gliding reptile *Icarosaurus*. Amer. Mus. Natur. Hist., Bull., 143(2): 85–142.

Fossil mammals of the North. Encyclopedia Arctica. U. S. Corps of Engineers, St. Paul, Minn.

1971

Men and dinosaurs. Penguin Books, Ltd. (A Pelican Book), Harmondsworth, Middlesex, England, 310 pp.

Evolution of the vertebrates. Wiley Eastern Private Limited, Anand R. Kundaji, New Delhi, xvi + 535 pp. For distribution in India, Ceylon, and Burma.

Antarctic fossil vertebrates and Gondwanaland. Research in the Antarctic—AAAS Special Publication, pp. 685–701.

Tetrapods and continents. Quart. Rev. Biol., 46(3):250–68.

Triassic tetrapods from McGregor Glacier. Antarctic J., 6(5):188–89.

1972

Antarctic fossils and the reconstruction of Gondwanaland. Natur. Hist., 81(1): 66–73.

With James W. Kitching, James W. Collinson, and David H. Elliot. Lystrosaurus Zone (Triassic) fauna from Antarctica. Science, 175(4021):524–27.

Antarctic Gondwana tetrapods. Second Gondwana Symposium, South Africa, 1970, pp. 659–64.

Vertebrates from the Chinle Formation. Mus. North. Ariz., Bull., 47:1–11.

Lystrosaurus and Gondwanaland. *In* T. Dobzhansky, M. K. Hecht, and W. C. Steere (eds.), Evolutionary biology, vol. 6. Appleton-Century-Crofts, New York, pp. 157–177.

Antarctic Triassic tetrapods in the laboratory. Antarctic J., 7(5):141–42.

Antarctic Triassic tetrapods. Scientific Committee on Antarctic Research—International Union of Geological Sciences Symposium on Antarctic Geology and Solid Earth Geophysics. Oslo, 1970, pp. 393–401.

When dinosaurs roamed the Arctic. *In* Marvels and mysteries of the world around us, published by the Readers Digest Association, pp. 29–31.

1973

Early Triassic tetrapods and Gondwanaland. XVII Congress International de Zoologie, Monte Carlo, September 25–30, 1972, pp. 1–27.

Continental drift and the distributions of fossil reptiles. *In* D. H. Tarling and S. K. Runcorn (eds.), Implications of continental drift to the earth sciences, Academic Press, London and New York, pp. 395–412.

Wandering lands and animals. With a foreword by Laurence M. Gould. E. P. Dutton & Co., New York, xxi + 323 pp.

Tetrapods and the Permian-Triassic transition. *In* Logan, A. and L. V. Hills (eds.), The Permian and Triassic Systems and their mutual boundary. Canadian Society of Petroleum Geologists, Memoir 2, pp. 481–92.

Antarctic *Lystrosaurus* defined. Antarctic J., 8(5):273–74.

Further evidence concerning the presence of horse at Ventana Cave. Kiva, 39(1): 25–33.

1974

Antarctic fossils and the reconstruction of Gondwanaland. *In* G. E. Nelson and J. D. Ray (eds.), Contemporary readings in biology. John Wiley and Sons, Inc., New York, pp. 255–64.

Wandering lands and animals. Hutchinson and Co., London, xxi + 323 pp. (British edition of 1973.)

Lystrosaurus from Antarctica. Amer. Mus. Nov., 2535:1–44.

Mesozoic vertebrates of northern Arizona. *In* T. N. V. Karlstrom, G. A. Swann and R. L. Eastwood (eds.), Geology of northern Arizona, Part I. Geological Society of America, Rocky Mountain Section Meeting, Flagstaff, pp. 208–19.

Figure 1.4. Ned Colbert at the Museum of Northern Arizona, 1975.

The rocks and fossils of the Mesozoic. A brief statement. *In* W. J. Breed and E. C. Roat (eds.), Geology of the Grand Canyon. Museum of Northern Arizona and Grand Canyon Natural History Association, Flagstaff, pp. 77–80.

The Triassic paleontology of Ghost Ranch. New Mex. Geol. Soc., Guidebk., 25th Field Confer., Ghost Ranch (Central-Northern N.M.), pp. 175–78.

Labyrinthodont amphibians from Antarctica. Amer. Mus. Nov., 2552.1–30.

P. C. Mahalanobis and the ancient reptiles of India. Samvadadhvam, 10(1–4): 22–25.

The Triassic reptile *Procolophon* in Antarctica. Amer. Mus. Nov., 2566:1–23.

1975

Early Triassic tetrapods and Gondwanaland. Mus. Nat. Hist. Natur., Paris, Mém., Sér. A, 88:202–15.

William King Gregory. May 19, 1876–December 29, 1970. Nat. Acad. Sci., Biographical Mem., 46:91–133.

Little dinosaurs of Ghost Ranch. *In* A. Ternes (ed.), Ants, Indians, and little dinosaurs. Charles Scribner's Sons, New York, pp. 143–53. (Reprint of 1947.)

Mammoths and men. *In* A. Ternes (ed.), Ants, Indians, and little dinosaurs. Charles Scribner's Sons, New York, pp. 180–88. (Reprint of 1940.)

La vita sui continenti alla deriva. *In* Scienza & tecnica 1975. Arnoldo Mondadori Editore—Milano, pp. 289–98.

La vita sui continenti alla deriva. *In* La riscoperta del terra. Ed. Eugenio de Rosa. Arnoldo Mondadori Editore—Milano, pp. 253–62.

Further determinations of Antarctic Triassic tetrapods. Antarctic J., 10(5):250–52.

1976

Triassic tetrapods. Geotimes (April, 1976), pp. 13–14.

Fossils and the drifting of continents. Fossils, 1(1):4–23.

When reptiles ruled. *In* Our continent. A natural history of North America. National Geographic Society, Washington, D. C., pp. 74–97.

The living land. *In* Our continent. A natural history of North America. National Geographic Society, Washington, D. C., pp. 163–67.

With David D. Gillette. Catalogue of type specimens of fossil vertebrates. Academy of Natural Sciences, Philadelphia. Part II: terrestrial mammals. Phila., Acad. Natur. Sci., Proc., 128(2):25–38.

1977

Gondwana vertebrates. Fourth International Gondwana Symposium Section III: Gondwana Fauna (Vertebrates), pp. 1–18.

With James W. Kitching. Triassic cynodont reptiles from Antarctica. Amer. Mus. Nov., 2611:1–30.

The dinosaur world. Stravon Educational Press, New York, 64 pp.

The year of the dinosaur. Charles Scribner's Sons, New York, xiv + 171 pp.

Cynodont reptiles from the Triassic of Antarctica. Antarctic J., 12(4):119–20.

Mesozoic tetrapods and the northward migration of India. J. Palaeontol. Soc. India, 20:138–45. (Published in 1977—for 1975.)

1978

The ancestors of mammals. *In* L. Laporte (ed.), Evolution and the fossil record. W. H. Freeman and Co., San Francisco, pp. 142–45. (Reprint of 1949.)

Evolution of the vertebrates. Second edition. Tsukiji Shokan, Japan, Vol. 1, 314 pp., Vol 2, 307 pp. (Japanese translation.)

With Marc B. Edwards and Rosalind Edwards. Carnosaurian footprints in the Lower Cretaceous of eastern Spitsbergen. J. Paleontol., 52:940–41.

Review of: The distribution of the Karroo vertebrate fauna, by J. W. Kitching. J. Paleontol., 52:948.

It's a hard life. Privately printed, 178 pp.

Animali e continenti alla deriva. Arnoldo Mondadori Editore S.p.A., Milano, 420 pp. (Italian edition of Wandering Lands and Animals).

Roland T. Bird. (Obituary) Society of Vertebrate Paleontology News Bulletin, 114:43–44.

1979

Proto-lizards from the Triassic of Antarctica. Antarctic J., 13(4):20–21.

The enigma of *Sivatherium*. Plateau, 51(1):32.

Gondwana vertebrates. Fourth International Gondwana Symposium, Calcutta, India, 1977, pp. 135–43.

1980

Ancient animals of the Petrified Forest. Plateau, 51(4):24–29.

Evolution of the vertebrates. Third Edition. John Wiley and Sons, Inc., New York, xvi + 491 pp. + index.

REFLECTIONS ON
AGNATHAN-GNATHOSTOME RELATIONSHIPS

by Bobb Schaeffer and Keith Stewart Thomson

Introduction

Until rather recently, the problem of the relationships between the agnathan and gnathostome vertebrates has received little more than passing attention, particularly in regard to the living agnathans or cyclostomes (the petromyzontoids or lampreys plus the myxinoids or hagfishes). Stensiö (1958, 1968) and Jarvik (1964, 1965) have concluded that the gnathostomes could not have arisen from known fossil or living agnathans. Janvier (1978) has argued in favor of a relationship between gnathostomes and osteostracans (one of the major groups of Paleozoic agnathans) and he considers the lampreys to be the sister group of the osteostracans, as originally advocated by Stensiö (1927, 1958). Løvtrup (1977) has assembled a list of characters that are shared by lampreys and gnathostomes but not by hagfishes, from which it may be hypothesized that the lampreys are the sister group of the gnathostomes and the hagfishes are the sister group of these two.

In this paper we wish to explore further the problem of agnathan-gnathostome affinities, with particular emphasis on the role of the living rather than the fossil agnathans in testing hypotheses of relationship. An important aspect of this problem, which we shall consider briefly, is the identification and evaluation of vertebrate synapomorphies in comparisons of the two major and obviously divergent groups of living agnathans.

Patterson and Rosen (1977, q.v. for bibliography) and Løvtrup (1977) have argued that, when possible, hypotheses of relationship should be formulated through outgroup comparison of living taxa, and that fossil groups should be added to or "fitted into" a cladogram of the Recent forms on the basis of shared, derived characters (synapomorphies). This *modus operandi* is based on the fact that living organisms provide more information than fossil ones. Although our knowledge of the fossil agnathans has been greatly amplified and refined, beginning with Stensiö's basic contributions on the osteostracans (e.g., 1927), there is justification for claiming that the relationships of the major fossil agnathan groups to each other, to the living agnathans, and to the gnathostomes remain problematical. As all agnathans and gnathostomes are generally regarded as vertebrates (or craniates), it will be helpful, first of all, to review the vertebrate synapomorphies.

Vertebrate Monophyly

In spite of some arguments to the contrary (Løvtrup, 1977), we regard the cephalochordates as the sister group of the vertebrates. They share a nearly identical, neural induction pattern (Tung, Wu, and Tung, 1962; Deuchar, 1975), and both have a dorsal nerve cord, notochord, and paired rows of somites. Vertebrates, however, have further elaborated certain cellular interactions, mainly during gastrulation, which lead to the "pharyngula stage" (Ballard, 1976) common to all members of this group. Following this stage, the neural crest, brain, cephalic sensory organs, thyroid, neurocranial rudiments, gills, heart, pineal, hypophysis, pancreas, liver, hepatic portal circulation, nephrotomes (or kidney anlagen), and median fins—along with structures shared with the cephalochordates—appear according to a characteristic vertebrate developmental pattern. The neural crest apparently gives rise to, or is involved in, all manifestations of the dermal skeleton (Schaeffer, 1977), the visceral skeleton, part of the nervous system, the median fins, and various other uniquely vertebrate characters. And most importantly in the context of this paper, the vertebrate synapomorphies require for their development a highly specific sequence of tissue interactions observed first with the induction of the neural tube by the archenteron roof and continuing regionally to the final appearance of definitive adult structures.

Newth (1956) has shown experimentally that the lamprey visceral skeleton is derived from neural crest cells, as is the gnathostome visceral skeleton. The cyclostome basicranium is entirely chordal. There is no neural crest contribution under the forebrain to form the trabeculae and no occipital arch (de Beer, 1937). The glossopharyngeal and vagus nerves are thus behind the braincase. As canals for these nerves have been identified in the cephalaspidomorph ostracoderm cranium (Stensiö, 1927; Wänsjö, 1952; Janvier, 1977), the absense of the occipital arch in the cyclostomes may be secondary, or it may mean that the cephalaspidomorphs are more closely allied with the gnathostomes than with cyclostomes.

Lampreys, Hagfishes, and Gnathostomes

With the vertebrate "archetype" in mind, we may now examine the arguments regarding the relationships of the hagfishes, lampreys, and gnathostomes. There are three possible relationships:

1. The lampreys and hagfishes are more closely related to each other than either is to the gnathostomes (Stensiö, 1968, *inter alia*).

2. The hagfishes and gnathostomes are sister groups with respect to the lampreys (no proponents known to us).

3. The lampreys and gnathostomes are sister groups with respect to the hagfishes (Løvtrup, 1977).

Hagfishes and lampreys share several synapomorphies: (a) The gills are at least partly endodermal and are medial to the branchial arches and body wall musculature. (b) The skeleton is uncalcified. (c) There is a single median nostril and nasal capsule, but the olfactory lobes are paired on both groups (Goodrich, 1909, figures 21 and 25). (d) The separate opening of the hypophysial duct outside the mouth is unique to hagfishes and lampreys, although in the former it also opens posteriorly into the pharynx. (e) Both groups of living agnathans have two semicircular canals in the middle ear (Goodrich, 1909). (f) The rasping "tongue" in lampreys has been regarded by some as homologous to the "tongue" of hagfishes (see especially, Romer, 1968), and by others (Strahan, 1963; Jarvik, 1965, and the writers) as independently derived. Which of these six characters are primitive for vertebrates or derived for cyclostomes will be explored in the next section.

A possible synapomorphy for the hagfishes and gnathostomes is the joining of the dorsal and ventral spinal nerve roots, but the nature of the union is unlike that in gnathostomes (Young, 1950, p. 118), which suggests independent acquisition. The dorsal and ventral roots remain separate in lampreys and alternate in position as in the cephalochordates (Romer, 1977, figure 390).

Our evaluation of Løvtrup's hypothesis that the lampreys and gnathostomes are sister groups involves essentially two steps. One, already considered, is an attempt to deal with the hypothesis of vertebrate monophyly. The second step of our approach is a reappraisal of the characters listed by Løvtrup (1977, p. 156) and considered by him to be common to lampreys and gnathostomes but not to hagfishes—and to remark on the nature of these characters (or their absence) in the hagfishes, as follows:

Arcualia. The presence of paired neural arches in the lampreys, and of radials in the caudal fin of hagfishes (Goodrich, 1909, figure 19) implies the presence of active sclerotomic tissue and of neural crest involvement (Balinsky, 1948).

Radial muscles. Radial muscles in the median and caudal fins are absent from myxinoids. There is, however, a muscle in the tail of *Myxine*, the m. cordis caudalis (Marinelli and Stranger, 1956, figure 108) which, like the radial muscles, must arise from the myotomes.

Spiral intestine. Although Løvtrup states that lampreys have a spiral valve in the intestine, as in many gnathostomes, there is uncertainty about this character. The intestine of lampreys is marked by longitudinal folds with a single spiral (Walker, 1980), but without the central "spindle" found in gnathostomes. Hagfishes also have the longitudinal folds but no spiraling.

No persistent pronephros in lampreys and gnathostomes. Although no outgroup comparison is possible, we suspect this to be a secondary condition, not of phylogenetic significance, as probably are the condition of the *mesonephros* and absence of collecting tubules from hagfishes. The condition of the pronephros may possibly be correlated with characteristics of the immune system (see below).

Accessory hearts. This is an interesting characteristic of myxinoids (Johansen, 1963; Bloom, Östland, and Fänge, 1963), but we can see no outgroup comparison that indicates their presence in hagfishes and absence in lampreys and gnathostomes is of primary importance. They could easily be an autapomorphy of hagfishes.

More than one semicircular canal. It has been proposed that the condition of the semicircular canals in hagfishes (apparently single), lampreys (apparently two), and gnathostomes (three) would in some way link lampreys and gnathostomes. The fact that the known agnathans, fossil and Recent, have one or two semicircular canals and the gnathostomes three has suggested to some that one or two is the primitive vertebrate number, but this hypothesis is not testable as the cephalochordates have no canals. However, studies by Lowenstein and Thornhill (1970, pp. 40–41) and Hagelin (1974, e.g., pp. 152, 153, 201) along with an observation of Goodrich (1909) indicate that the lamprey and hagfish conditions are not fundamentally different, and that both have two semicircular canals.

We are unable to comment on the character of the **retinal receptors** in the hagfishes and lampreys, or the **histology of the adenohypophysis,** except to note that these are possibly further instances of the vexing problem of the apparent "degeneracy" of the hagfish. The hagfish eye gives every indication of arrested development. The optic nerve never completely differentiates, as is also the case with the cornea, sclera, and choroid (Weichert, 1965). The retina is folded so as to eliminate the vitreous chamber, and the lens placode begins differentiation, but soon flattens and disappears. In regard to the adenohypophysis (anterior lobe), there is again evidence of incomplete differentiation. The neural and stomodaeal components of the hypophysis remain separated in the adult by a layer of connective tissue, and the pituitary gland is represented by clusters of cells between the infundibulum and the nasopharyngeal pouch (see Falkmer, Thomas, and Boquist, 1974).

Relative blood volume. Differences in blood volume and pituitary structure and function among hagfish, lampreys, and gnathostomes are related to patterns of osmoregulation, and therefore have limited phylogenetic significance.

Hyperosmoregulation. Lampreys and gnathostomes maintain a hyperosmotic blood plasma while hagfish are (more or less) isoosmotic with sea water. However, lampreys and gnathostomes maintain the hyperosmotic condition via different mechanisms, and, further, there is evidence that osmoregulation by urea synthesis and retention is a basic gnathostome adaptation (Thomson, 1980; see also Griffith and Pang, 1979). Lampreys lack this adaptation.

Chondroitin 6-sulphate. This character was thought by Løvtrup to be absent from hagfish cartilage while it is an important constituent of lamprey and gnathostome cartilage. In fact, recent work by Mathews (1975, e.g., table 7) shows that chondroitin 6-sulphate is present in hagfish cranial cartilage, but in small proportions. Other polyanionic glycans, containing the unusual disaccharide periods dis_E and dis_H which are supposedly unique to hagfish, are also possibly found in coelacanth notochord, hog intestinal mucosa, and fibrotic rat liver (Mathews, 1975). It is worth quoting Mathews (1975, p. 140):

> one possible interpretation (of the comparative data on polyanionic glycans of vertebrates) is that the *parent*-type glycan components of cartilage and notochord that are shared by lampreys and non-cyclostome vertebrate embryos represent a primitive condition derived from the common ancestor of both groups. However, in their connective tissue biochemistry . . . hagfish reflect an inheritance from a still more remote chordate ancestor, i.e., one shared by both acraniates and vertebrates. Alternatively, all the major lines of vertebrate evolution . . . may have possessed the full complement of glyosyl transferases and sulfotransferases needed to produce various polyanionic glycans. Subsequent evolution led to selective loss of enzymes in each line. . . .

The weight of evidence so far available suggests that the latter is the more likely interpretation.

Weakly identified immune response in myxinoids. This is in contrast to lampreys and gnathostomes which have a readily observed immune response. However, experimental evidence is limited (see Lundholm, 1972).

Several other biochemical characteristics are listed by Løvtrup in support of the lamprey-gnathostome relationship. We find them (like the polyanionic glycans) difficult to evaluate, particularly because of the problems of outgroup comparison and scanty comparative biochemical and physiological data. They are: **nervous regulation of the heart, absence of a unique dermatan sulphate, amino acid composition of collagen** (a matter of proportions), and **peptide and amino acid composition of hemoglobin.** With respect to this last character, the data presented by

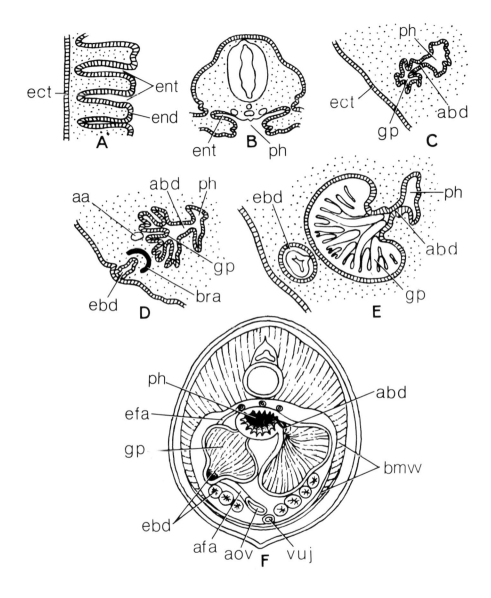

Figure 2.1. Development of branchial pouches and gills in the hagfish. A–E *Bdellostoma stouti.* After Stockard, 1906. F. *Myxine glutinosa.* After Marinelli and Stranger, 1956. Abbreviations: aa, vascular arch; abm, branchial adductor muscle; afa, afferent branchial vessel; agl, anterior gill lamellae; bra, branchial arch; ecm, external constrictor muscle; ect, ectoderm; efa, efferent branchial vessel; end, endoderm; ent, endodermal diverticulum; go, external gill opening; gr, gill ray; lpm, lateral plate mesoderm; pgl, posterior gill lamellae; ph, pharynx; sep, septum.

Løvtrup (table 3-13) show that the proportions of arginine, histidine, and lysine are quite different between *Myxine* and *Petromyzon*. In fact, the *Myxine* data show such a large range of variation that a need for greater precision is indicated. Some additional data are available in Paléus, Liljegvist, and Braunitzer (1972).

Additional points of difference between hagfishes on the one hand and lampreys plus gnathostomes on the other include the apparent absence of **peroxidase activity** in leucocytes of hagfishes (Johansson, 1972) and the absence of **steroid hormone** formation in the ovary (Fernholm, 1972), although again the comparative data are incomplete.

Many of the characteristics listed here that distinguish the hagfishes from all other vertebrates might be the result of incomplete development or interrupted differentiation of systems that are common to all vertebrates. Examples are the eyes and eye muscles (Jollie, 1962), endocrine organs, kidneys, musculo-skeletal system, possibly the immune system and other biochemical characters, and, in addition, the nature of the lateral line system. The term "degenerate" is sometimes used to describe such a condition, and in this case it implies a secondary development from what would have been a generalized vertebrate pattern (shared by the lampreys, gnathostomes, and, presumably, the ancestors of hagfishes). It is noteworthy that the characters under discussion are all ones that in lampreys develop fully only at the metamorphosis between ammocoete larva and adult. Hagfishes lack such a larva and metamorphosis. This fact, coupled with the apparent specialization of hagfishes for a burrowing habit (Strahan, 1963) may then provide a partial explanation for the so-called degeneracy of the group.

Cyclostome Monophyly

Although there are numerous morphological, physiological, and biochemical differences between lampreys and hagfishes, we believe there is one character complex which indicates that the two major groups of cyclostomes share a common ancestor that is not shared with the gnathostomes. In our present state of knowledge, it would appear to compromise any attempt to relate lampreys directly to the gnathostomes. Jarvik (1964) and Stensiö (1968) among others, have noted that the gills of lampreys are situated internally to the branchial arches, while in the gnathostome fishes they are lateral to the branchial arches. The visceral skeleton of the hagfishes is very much reduced, but vestigial branchial cartilages are present near the external gill openings (Figure 2.1). The hagfish gill pouches are situated inside of the branchial musculature as in the lampreys (Figure 2.2), so there is little question that the gills in both are in the same position relative to the body wall and the pharynx. The differences in the form of the pouches, their ducts, and the gills in the lampreys and hagfishes have been well illustrated by Goodrich (1909) and Marinelli and Stranger (1954, 1956). It should be noted that the efferent branchial arteries in the hagfishes enter the gill pouches directly rather than by passing first through interbranchial septa as in lampreys and gnathostomes. This unique situation in the hagfishes may be related to the absence of the septa in these fishes.

In 1901, Goette claimed that the gills and gill sacs in both groups of cycostomes are derived entirely from endoderm while ducts leading to the exterior are ectodermal. For the hagfishes, this was confirmed by Stockard (1906) who observed that the pouch begins as a diverticulum or invagination of the pharyngeal en-

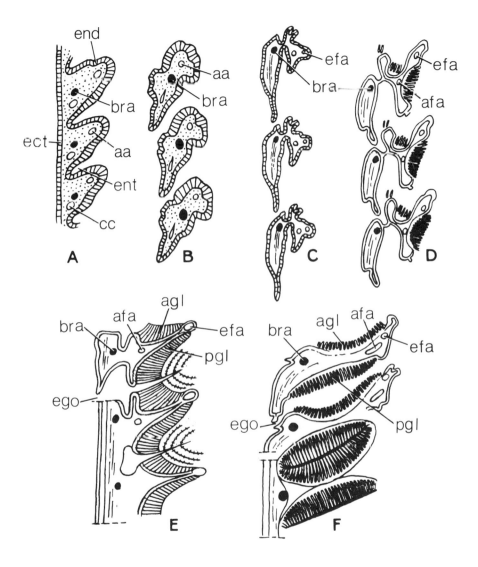

Figure 2.2. Development of branchial pouches and gills in the lamprey. A–E *Petromyzon fluviatilis*. After Goette, 1901; Goodrich, 1930; Daniel, 1934. F. *Lampetra planeri*. After Rauther, 1937. The upper part of E and F at level of external gill openings, lower part below gill openings. Abbreviations as in Figure 2.1.

doderm and reaches its adult form through expansion and folding of its walls (Figure 2.1). The external gill ducts join corresponding ectodermal invaginations shortly before hatching. According to Stockard, the only ectodermal contribution is "a short rim about the external gill opening."

The embryology of the lamprey gill (Figue 2.2), as worked out by Goette (1901) is reviewed by Goodrich (1930, pp. 493–97; 504–5) and Rauther (1937). There is still debate regarding an ectodermal contribution to the gill lamellae, but this can only be settled by studying early developmental stages with marker experiments. Actually, lamprey gills and their pouches form in nearly the same way as in hagfishes, but lamprey gill clefts break through before the pouches begin to form and are incorporated into the pharyngeal wall so that each interbranchial septum contains a branchial arch. The exclusion of direct gill pouch connections with the pharynx during and following metamorphosis is a petromyzontoid autapomorphy.

Regardless of whether lamprey and hagfish gills are entirely endodermal or partly ectodermal, their mode of development and, of course, their adult morphology and location are obviously different from that of the entirely ectodermal gills of gnathostomes (Figure 2.3). One other point of interest in regard to the cyclostome gill position is that it is not compatable with the formation of functional jaws (Jarvik, 1964, p. 27). Accordingly, we are unable to visualize a hypothetical common ancestor for the living agnathans and gnathostomes as an adult that had the gill relationships of either.

We hypothesize an ancestor of agnathans and gnathostomes in which the perforated pharynx was no longer a feeding device. The gill slits were lined with respiratory epithelium but there was no morphological commitment to either the cyclostome or gnathostome type. Conjecturally, one could proceed in two modes, each of which would have different consequences both for the nature of the respiratory surfaces and supporting gill skeleton. The agnathan mode was produced by the elaboration of the endodermal outpocketing of the pharyngeal wall into gill pouches. The gnathostome mode was produced by elaboration of the ectodermal connection between pharynx and exterior into a series of gill lamellae arranged in a pre- and postdermatic pattern (with respect to the gill slit openings).

It is of interest that the embryonic, endodermal, pharyngeal pouches of bony fishes and amphibians induce the ectodermal branchial grooves only if these pouches actually reach the epidermis (Balinsky, 1975, p. 456), indicating a primary role for the pharyngeal endoderm, presumably in agnathans as well as gnathostomes.

Gnathostome Monophyly

At this point in our analysis, it is necessary to consider briefly the bases upon which the gnathostomes are considered monophyletic, in order to place the evidence concerning cyclostomes in context. The gnathostome synapomorphies include the trabeculae cranii, which lengthen the basicranium beneath the enlarged forebrain, segmented visceral arches with the mandibular arch modified into jaws, ectodermal gills situated on the outside of the branchial arches, a third (horizontal) semicircular canal, and controllable, paired, pectoral and pelvic appendages including a dermal plus endoskeletal pectoral girdle and an endoskeletal pelvic girdle.

Because, in our opinion, the internal gills of the cyclostomes could not have given rise to the external gills of the gnathostomes, an unknown eocraniate must have been ancestral to both of these conditions. In other words, the internal gills

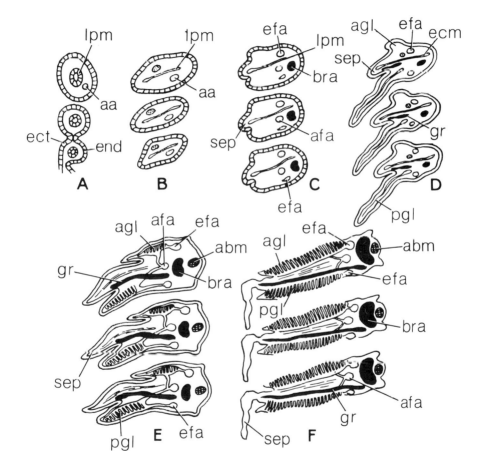

Figure 2.3. Development of gills in the shark. Mostly *Scyliorhinus canicula*. After Goette, 1901 and Goodrich, 1930. Abbreviations as in Figure 2.1.

are synaptomorphic for the cyclostomes and possibly the fossil agnathans, while the external gills are synapomorphic for the gnathostomes. With respect to the other characters discussed, cyclostomes have retained the primitive eocraniate condition, and the gnathostome condition is derived with respect to it.

Note on Fossil Agnathans

In general, the fossil agnathans do not provide much information that can allow us to refine hypotheses of relationships between agnathans and gnathostomes derived from consideration of the living taxa. Our available information on the fossil forms (Osteostraci, Anaspida, Heterostraci, and Thelodonti) has not significantly increased in the last few decades except for some studies by Denison, Janvier, and Ritchie. The interpretation of their detailed structure has relied inevitably but often disconcertingly upon reconstructions that are based on comparisons with living agnathans (particularly the ammocoete larva of lampreys).

Most authors seem to be content with the hypothesis that the Paleozoic osteostracans are related to the lampreys in some way. Both groups have a median dorsal opening for the nasohypophysial duct, as discussed by Stensiö in his classic paper of 1927, and Janvier (e.g., 1974, 1975). An interesting problem with this evidence is that in lampreys the duct is above and behind the expanded "upper lip" (Figure 2.4), while in osteostracans it is above the compact orobranchial chamber. Presumably these differences have their origins in different patterns of head development. In living lampreys, the dorsal nasohypophysial duct opening is formed as a result of the greatly enlarged upper lip, which comes to "push" the nasohypophysial opening and the nasal apparatus to a mid-dorsal position in front of the eyes. The situation in hagfish development is different (Figure 2.4) but the comparison of the two may help to resolve the problem. It may be that an open hypophysial duct (and possibly the common nasohypophysial duct) in the petromyzontoids and the cephalaspidomorphs is not merely the chance product of upper lip enlargement, but is a primitive vertebrate condition. Explanation of the similarity of the lamprey and the cephalaspidomorph arrangements would then become a matter of explaining why the opening is *dorsal* in each. One may question whether or not the dorsal position could have evolved more than once, given a basic arrangement which was probably more like that seen in myxinoids (Figure 2.4). This being the case, Janvier's (1975) conclusion that the nasohypophysial openings of petromyzontoids and osteostracans are not point-for-point homologous might in fact constitute evidence for a hypothesis of convergence.

The osteostracans have a bony exo- and endocranial skeleton, the lampreys only cranial endoskeleton cartilage. The prenasal region of the osteostracans is enlarged, but, as suggested, for a different reason than in the lampreys. There is now good reason for identifying the velar ridge in cephalaspids as the mandibular arch and for regarding its associated foramina as trigeminal nerve exits. Also the evidence in favor of one or more premandibular arches in fossil or living agnathans is inconclusive (reviewed in Schaeffer, 1975).

Reconstructions of soft characters in fossils are frequently suspect unless there is clear supporting comparative evidence from the skeletons of living representatives. Osteostracan endocasts indicate a brain form similar to that of *Petromyzon*, as first noted by Stensiö (1927, also 1963). The presumed vascular canals of the osteostracan head shield show close correspondence with the ammocoete cranial circula-

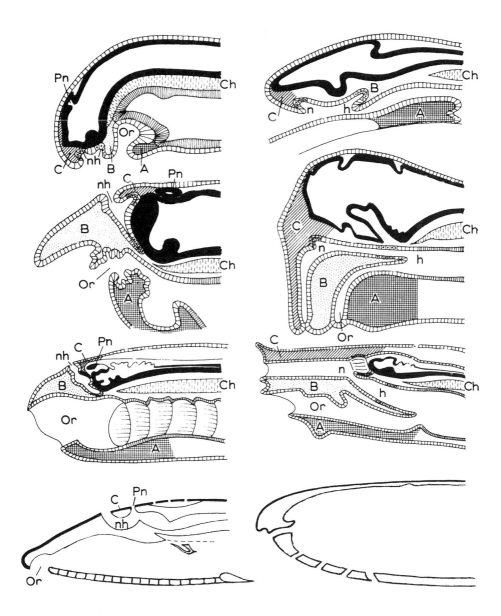

Figure 2.4. Comparison of head development in lamprey (left) and hagfish (right). Midsaggital section of osteostracan head shield in lower left and of heterostracan head armor in lower right with only preserved parts represented. Modified from Heintz, 1963. Abbreviations: A–C, approximately homologous areas related to the mouth, nasal organ and hypophysis; ch, notochord; h, hypothesis; n, nasal organ; nh, naso-hypophysial opening; Or, mouth; Pn, pineal organ.

tion. The lateral line groove pattern on the head shield is much as in the lampreys, but the sensory fields on the osteostracan head shield and their well-developed connections with the lateralis system of the brain have no counterpart in other fishes. Janvier (1975), in particular, has been recently concerned with reconstructing the cranial circulation of various osteostracans, and he believes that the left duct of Cuvier was absent, as in adult lampreys.

In all known fossil agnathans the position of the gills can be discerned merely as a series of impressions on the underside of the cranial shield. These cavities or bulges do not directly indicate the gill pouches or the gill skeleton as both were presumably medial to the branchial muscles. There is nothing in the fossil forms to indicate a different pattern from that seen in living forms.

The hypothesis (for example, Stensiö, 1958) that hagfish and heterostracans form a monophyletic group is a particularly difficult problem. It is our opinion that the scanty evidence on detailed internal anatomy simply does not allow the sort of analysis that is needed to propose or refute such an hypothesis (see, for example, Heintz, 1963). In regard to interpretations of the nasohypophysial region in heterostracans (Stensiö, 1958), there is no direct evidence of a palato-subnasal lamina, and thus all reconstructions of the nasohypophysial connections are conjectural. What is clear is that the heterostracans did not have a dorsally opening nasohypophysial duct (Figure 2.4). Most authors assume that the nasal sacs in the heterostracan fishes were paired.

Finally, there is a difficulty in deciding which heterostracan characters might in fact simply be primitive for all vertebrates: for example, paired (?terminal) nasal capsules, presence of pineal and parapineal organs, two semicircular canals, a simple medulla, lateral eyes, a nearly terminal mouth and a lateral line system. There are, however, three characters that are particularly difficult to evaluate in terms of polarity. One is acellular vs. cellular bone, another is a dermal head shield composed of tesserae vs. continuous bone, and the third is caudal fin form. Although Heintz, Denison, and Halstead Tarlo have considerably improved our knowledge of the heterostracans, in- and outgroup relationships of these fossils remain obscure.

Janvier (1978) has published a new description of paired pectoral fin flaps in a Devonian osteostracan agnathan. The presence of such fins has been known for some time although pelvic fins are apparently absent from all fossil agnathans. Janvier's account is significant because he attempts to show that his osteostracan has pectoral fins that are in major respects homologous with those of gnathostomes in the arrangement of the musculature and the conjectured presence of an endoskeleton. If these reconstructions of the soft anatomy (based on blood vessel and nerve foramina and rather undefined "articular facets") are correct, Janvier believes that it must follow that the Osteostraci are a sister group with the gnathostomes. By extension of the hypothesis of a lamprey-osteostracan sister group relationship, Janvier's arguments could be taken as a further elaboration of the lamprey-gnathostome sister group hypothesis. The problem remains, however, that the direct fossil evidence favoring these hypotheses is extremely tenuous. It is scarcely logical to admit evidence about fossil relatives of lampreys without being confident about the availability of comparable fossil evidence concerning the hagfishes and their putative fossil relatives. Not only are we unsure that the osteostracans described by Janvier actually had a gnathostome-type of pectoral fin, but we also have no knowledge of the paired appendages in other fossil agnathans beyond the lateral fin-folds of anaspids (Ritchie, 1964) and the obscure horizontal caudal flaps described in cephalaspids by Heintz (1967). And, of course, we still have little basis for formulating an hypothesis of relationship among fossil and living agnathans.

Summary and Conclusions

Testing any hypothesis concerning agnathan-gnathostome relationships is restricted by the limitations of outgroup comparison and the availability of ontogenetic data. We propose the cephalochordates as the sister group of the vertebrates, and the living agnathans (cyclostomes) as the sister group of the gnathostomes within the vertebrates. The fossil agnathans, in our opinion, provide minimal information for testing either of these hypotheses.

The characters here regarded as adult vertebrate synapomorphies are the ontogenetic end products of successive, interrelated tissue interactions that ordinarily allow little deviation during the course of normal development. We consider this point to be of primary importance in evaluating the characters of the hagfishes, several of which demonstrate interference with normal morphogenesis.

Most of the characters that have been cited by Løvtrup (1977, p. 156) as relating the lampreys to the gnathostomes should, we believe, be reexamined on the basis that they may be equivocal, primitive vertebrate characters shared with the hagfishes and/or incapable of phylogenetic resolution on the basis of available data.

In attempting to test a different but frequently mentioned hypothesis—that the living agnathans or cyclostomes and the gnathostomes are sister groups—we have found only one character that is synapomorphous for each group. This is the developmental pattern plus the adult morphology and position of the branchial gills in each. We are unable to regard one as the primitive state of the other, and propose that both arose from some unknown condition in the protovertebrates (eocraniates).

The usage of the term Agnatha has become increasingly ambiguous as the known agnathans are frequently defined in terms of negative characters such as absence of jaws and absence of paired controllable fins. Obviously the first vertebrates, including the ancestors of the several ostracoderm groups and the cyclostomes, lacked jaws. Most of the characters frequently used to relate the major ostracoderm groups to each other and to the lampreys and hagfishes have, we believe, a dubious significance as synapomorphies (e.g., the position of the nasohypophysial duct in the osteostracans, anaspids, and lampreys).

Although we have discussed cyclostome monophyly in this paper mostly on the basis of gill structure and position, we nevertheless propose that the hagfishes and lampreys have a common ancestor that is not shared with the gnathostomes. On the basis of present evidence, it is apparent to us that no known agnathan, fossil or Recent, can be regarded as the sister group of the gnathostomes. Furthermore, the difficulties of outgroup comparison have made the analysis of interrelationships among the agnathans nearly impossible. But the fact remains that all known agnathans are vertebrates (or craniates) and that a strong case for vertebrate monophyly demands an ultimate common ancestor for both the known agnathans and the gnathostomes.

Acknowledgments

We are grateful to William Kohlberger, John Maisey, Amy McCune, Stan Rachootin, Donn E. Rosen, and Richard Vari for valuable advice and criticism during the preparation of this paper. Illustrations were drawn by Juan Barbaris. Study supported in part by grant DEB–77–08412 of the National Science Foundation (to KST).

Literature Cited

Balinsky, B. L. 1948. Korrelationen in der Entwicklung der Mund und Kiemenregion und des Darmkanals bei Amphibien. Roux Arch., 143:365–95.

———. 1975. An introduction to embryology. W. B. Saunders Co., Philadelphia, 648 pp.

Ballard, W. W. 1976. Problems of gastrulation: real and verbal. Bioscience, 26:36–39.

Bloom, G.; Östland, E.; and Fänge, R. 1963. Functional aspects of the cyclostome hearts in relation to recent structural findings. *In*, A. Brodal and R. Fänge (eds.), The biology of *Myxine*. Universitetsforlaget Oslo, pp. 397–39.

Daniel, J. Frank. 1934. The circulation of blood in ammocoetes. Calif., Univ. Publ., Zool., 39:311–40.

de Beer, G. R. 1937. The development of the vertebrate skull. Oxford University Press, Oxford, 552 pp.

Deuchar, E. M. 1975. Cellular interactions in animal development. Chapman and Hall, London.

Falkmer, S.; Thomas, N. W.; and Boquist, L. 1974. Endocrinology of the Cyclostomata. *In*, M. Florkin and B. Y. Scheer (eds.), Chemical zoology, v. VII, Deuterostomians, cyclostomes, and fishes. Academic Press, New York, pp. 195–260.

Fernholm, B. 1972. Is there any steroid formation in the ovary of *Myxine?* Acta Reg. Roy. Sci. et Litt. Gothoburgensis, Zool., 8:33–34.

Goette, A. 1901. Über die Kiemen der Fische. Zeitsch, Wissen. Zool., 69:533–77.

Goodrich, E. S. 1909. Vertebrata craniata: treatise on zoology. R. Lankester (ed.), Part 9. Black, London, 518 pp.

———. 1930. Studies on the structure and development of vertebrates. Macmillan, London, 837 pp.

Griffith, R. W., and Pang, P. K. T. 1979. Mechanisms of osmoregulation in the coelacanth: evolutionary implications. Calif. Acad. Sci., Occ. Pap., 134:79–93.

Hagelin, L. O. 1974. Development of the membranous labyrinth in lampreys. Acta Zool., supplement 1974:1–215.

Heintz, A. 1963. Phylogenetic aspects of myxinoids. *In*, A. Brodal and R. Fänge (eds.), The Biology of *Myxine*. Universitetsforlaget Oslo, pp. 9–21.

———. 1967. Some remarks about the structure of the tail in cephalaspids. Coll. Int. CNRS, Paris, 163:21–35.

Janvier, P. 1974. The structure of the naso-hypophyseal complex and the mouth in fossil and extant cyclostomes, with remarks on amphiaspiforms. Zool. Scripta, 3:192–200.

———. 1975. Remarques sur l'orifice naso-hypophysaire des Cephalaspidomorphes. Ann. Paléontol., Vertébrés, 61:3–16.

———. 1977. Contribution a la connaissance de la systematique et l'anatomie due genre *Boreaspis* Stensiö (Agnatha, Céphalaspidomorphi, Osteostraci) du Dévonien inférieur de Spitsberg. Ann. Paléontol., Vertébrés, 63:1–32.

———. 1978. Les nageoires paires des Ostéostracés et la position systematique des Céphalaspidomorphes. Ann. Paléontol., Vertébrés, 64:113–42.

Jarvik. E. 1964. Specialisation in early vertebrates. Soc. Roy. Zool. de Belgigue, Ann., 94:11–95.

———. 1965. Die raspelzunge der Cyclostomen und die pentadactyle Extremität der Tetrapoden als Beweise für monophyletische Herkunft. Zool. Anz., 175:8–143.

Johansen, K. 1963. The cardio-vascular system of *Myxine glutinosa* L. *In*, A. Brodal and R. Fänge (eds.), The biology of *Myxine*. Universitersforlaget Oslo, pp. 289–316.

Johannson, M. L. 1972. Peroxidase in blood cells of fishes and cyclostomes. Acta Reg. Soc. Sci. et Litt. Gothoburgensis, Zool., 8:53–56.

Jollie, M. 1962. Chordate morphology. Reinhold, New York, 478 pp.

Løvtrup, S. 1977. The phylogeny of Vertebrata. Wiley, New York, 330 pp.

Lowenstein, O., and Thornhill, R. A. 1970. The labyrinth of *Myxine*: anatomy, ultrastructure and electrophysiology. Roy. Soc. London Proc. B, 176:21–42.

Lundholm, M. 1972. An attempt to immunise *Myxine*. Acta Reg. Soc. Sci. et Litt. Gothoburgensis, Zool., 8:65–66.

Marinelli and Stranger, A. 1954. Wergleichende Anatomie und Morphologie der Wirbeltiere. I Leiferung. *Lampetra fluviatilis*. Franz Deuticke, Wien. pp. 1–80.

———. 1956. Vergleichende Anatomie und Morphologie der Wirbeltiere II. Lieferung *Myxine glutinosa*. Franz Deuticke, Wien, pp. 81–172.

Mathews, M. B. 1975. Connective tissue: macromolecular structure and function. Springer-Verlag, New York, 318 pp.

Newth, D. R. 1956. On the neural crest of the lamprey embryo. Embryol. Exp. Morph., 4:358–75.

Paléus, S.; Liljeqvist, G.; and Braunitzer, G. 1972. A study of some hemoproteins of *Myxine glutinosa* L., with special reference to the structure of hemoglobin III. Acta Reg. Soc. Sci. et Litt. Gothoburgensis, Zool., 8:42–45.

Patterson, C., and Rosen, D. E. 1977. Review of ichthyodectiform and other Mesozoic fishes and the theory and practice of classifying fossils. Am. Mus. Natur. Hist., Bull., 158(2): 83–172.

Rauther, M. 1937. Kiemen der Anamnier-Kiemendarmderivate der Cyclostomen and Fische. In Handb. vergleichenden Anatomie der Wirbeltiere, v. III. Urban and Schwarzberg, Berlin und Wien, pp 212–78.

Ritchie, A. 1964. New light on the morphology of the Norwegian anaspida. Skr. Norske VidenskAKad. Oslo I mat. natur. Kl., 14:1–35.

Romer, A. S. 1968. Notes and comments on vertebrate paleontology. University of Chicago Press, Chicago, 304 pp.

———. 1977. The vertebrate body. 5th Edition. W. B. Saunders, Philadelphia, 624 pp.

Schaeffer, B. 1975. Comments on the origin and basic radiation of the gnathostome fishes with particular reference to the feeding mechanisms. Coll. Int. CNRS, Paris, 218:101–9.

———. 1977. The dermal skeleton in fishes. *In*, S. M. Andrews, R. S. Mills, and A. D. Walker (eds.), Problems in vertebrate evolution. Academic Press, London, pp. 25–52.

Stensiö, E. A. 1927. The Downtonian and Devonian vertebrates of Spitzbergen. Part 1. Family Cepalaspidae. Skrifter om Svalbard og Nordishavet, 12:1–391.

———. 1958. Les Cyclostomes fossiles ou Ostracodermes. Traité de Zoology. (P-P Grassé, ed.) Vol. 8, Part 1. Masson, Paris.

———. 1963. The brain and cranial nerves in fossil lower craniate vertebrates. Skr. Norske VidenskAKad. Oslo, I. mat. natur. Kl., 1965:1–120.

———. 1968. The cyclostomes with special reference to the diphyletic origin of the Petromyzontida and Myxinoidea. *In*, T. Ørivg (ed.), Current problems of vertebrate phylogeny. Almqvist and Wiksell, Stockholm, pp. 13–71.

Stockard, C. R. 1906. The development of the mouth and gills in *Bdellostoma stouti*. Amer. Jour.Anat., 5:481–517.

Strahan, R. 1963. The behavior of myxinoids. Acta. Zool., 44:73–102.

Thomson, K. S. 1980. Environmental factors in the evolution of vertebrate endocrine systems. *In*, P. K. T. Pang (ed.), Evolution of vertebrate endocrine systems. Texas Tech. University Press, Lubbock, pp. 17–32.

Tung, T. C; Wu, S. C. and Tung, T. T. F. 1962. Experimental studies on the neural induction in *Amphioxus*. Scientia Sinica, 11:805–20.

Walker, W. F. 1980. Vertebrate dissection. 6th Edition. W. B. Saunders, Philadelphia, 425 pp.

Wänsjö, G. 1952. Morphologic and systematic studies of the Spitzbergen cephalaspids. Skr. om Svalbard og lshavet, 97:1–611.

Weichert, C. K. 1965. Anatomy of the chordates. McGraw-Hill, New York, 758 pp.

Young, J. Z. 1950. The life of vertebrates. Oxford University Press, Oxford. 757 pp.

A COMPARISON OF THE VERTEBRATE ASSEMBLAGES FROM THE NEWARK AND HARTFORD BASINS (EARLY MESOZOIC, NEWARK SUPERGROUP) OF EASTERN NORTH AMERICA

by Paul E. Olsen

Introduction

The sediments of the Hartford and Newark Basins (Figure 3.1) occupy a crucial position in the study of the early Mesozoic Newark Supergroup (Olsen, 1978; Olsen, in press 1). Historically, they were the first strata of the Newark in which faunal remains were found; by 1845, abundant reptile footprints (Hitchcock, 1858) and fossil fish (Redfield, 1845) had been recovered from both areas. Since that time a long series of distinguished workers including E. D. Cope, O. C. Marsh, E. Hitchcock, C. W. Gilmore, F. von Huene, J. T. Gregory, R. S. Lull, D. Baird, B. Schaeffer, and of course, E. H. Colbert, have made major contributions to the study of vertebrates from these beds. Considering the new kinds of stratigraphic data now being obtained from the Newark and Hartford basins (such as radiometric dating, paleomagnetic stratigraphy, palynostratigraphy, and physical microstratigraphy), it is appropriate to review the classically studied and newly uncovered vertebrates from these two basins and to place them in their stratigraphic context, thus following the tradition of Colbert (1946), Colbert and Gregory (1957), and Colbert (1965).

Outline of the Stratigraphy of the Newark and Hartford Basins

It is now agreed that the Newark and Hartford basins, along with other Newark Supergroup deposits, formed in conjunction with the initial phases of drift between North America and Africa (Manspeizer, Puffer, and Cousminer, 1978; Van Houten, 1977). Both basins preserve in excess of 4,000 m of largely detrital fill dominated by red clastics (Figure 3.2). Gray and black siltstone and sandstone beds are common (at least locally) and these, along with tholiitic basalt flows, provide the key to the internal physical stratigraphy of the basin section (Olsen, in press 2 and 3). Newark Basin strata are divided into nine formations (Figure 3.2), of which the lowest two, the Stockton and Lockatong formations, have no lithologic counterparts in the Hartford Basin. The remaining seven formations compare well with the seven formations of the Hartford Basin (Figure 3.2). The sequences of both basins are tantalizingly similar; they consist of a middle series of three multiple basalt flows and two major interbedded sedimentary formations sandwiched between two thick, primarily red, sedimentary units. This rough comparability has suggested to many authors that each of the respective formations can be directly correlated (Russel, 1892; Schuchert and Dunbar, 1941; Sanders, 1963). Until recently there were no data available to test what Jepsen (1948, p. 16), called "this undemonstrated but appealing assumption. . . ." Within the last ten years, however, palynological and geophysical evidence has helped to clarify much of the broader

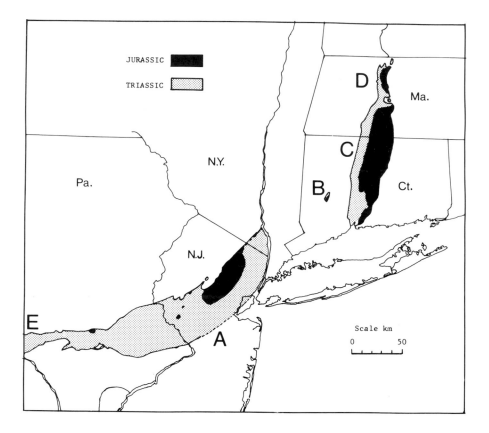

Figure 3.1. Position of the Hartford and Newark basins of the Newark Supergroup: A, Newark Basin; B, Pomperaug Basin; C, Hartford Basin; D, Deerfield Basin; E, northern terminus of Gettysburg Basin.

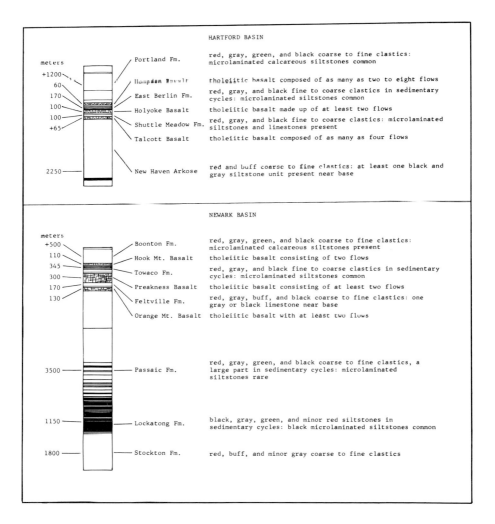

Figure 3.2. Lithologic divisions of the Newark and Hartford basins. Hartford basin thicknesses from Hubert, et al. (1978), and Cornet (1977a). Newark Basin divisions and thicknesses from Olsen (in press 1 and 2). Note that the thicknesses of pre-Orange Mountain Basalt formations are derived from the central Newark Basin while the post-Passaic formation thicknesses are derived from the largest area in which they are preserved, the Watchung Syncline. In the latter area the pre-Orange Mountain Basalt formations are thinner.

picture. As discussed below, the biostratigraphic data roughly confirm the correlation of the larger divisions of the homotaxial sequences, although the details deny any simple one-to-one correlation of basalt flows or sedimentary formations.

The Vertebrate Assemblages: Stockton and Lockatong Formations

The oldest vertebrate fossil from the Newark and Hartford Basins is the lower jaw of a very large amphibian (*Calamops paludosus*) found at the base of the Stockton Formation (Sinclair, 1917). This specimen has been recently prepared and now appears to represent a large capitosaur (W. Seldon, personal communication; Olsen, Baird, Seldon, and Salvia, in preparation). Its stratum lies about 1,700 m below the base of the late Carnian Lockatong Formation, but cannot be readily correlated with any other horizon of the Newark Supergroup because of a lack of relevant biostratigraphic data. The age of *Calamops* is therefore indeterminate, although probably Late Triassic.

The uppermost 100 m of Stockton Formation and lower 500 m of the Lockatong Formation of the Newark Basin have produced a rich and abundant vertebrate assemblage, of which the most characteristic members are fishes (Figure 3.3). The most numerous taxa are the coelacanth *Diplurus newarki*, several species of the palaeoniscoid *Turseodus*, and holosteans of the *"Seminotus brauni* group" (Olsen, McCune, and Thomson, in press). Much less common are the subholosteans *Cionichthys* sp. and *Synorichthys* sp., and the hybodont shark *Carinacanthus jepseni*. These fishes have been found in basins to the south and have proved useful for correlation (Olsen, McCune, and Thomson, in press).

A surprisingly abundant suite of small reptiles is found with the fishes (Figure 3.3). Perhaps the most striking of these is the gliding lizard *Icarosaurus siefkeri* which Colbert (1966, 1970) described from a partially articulated skeleton from the northern Newark Basin. Since the original description a number of very fragmentary specimens have been found, although none of these adds anything to Colbert's original descriptions.

Tanytrachelos (Olsen, 1979) is a small eosuchian found in surprisingly large numbers at a growing list of Lockatong localities (Olsen, in press 3). Originally described from the upper member of the Cow Branch Formation of the Dan River Group in North Carolina and Virginia, it possesses a suite of characters which show that its nearest relative is the long-necked *Tanystropheus* from the Old World Middle Tri-

---

Figure 3.3. The upper Stockton and lower Lockatong vertebrate assemblage: A, *Tanytrachelos* cf. *T. ahynis*, reconstruction; B, *Rutiodon carolinensis*, reconstruction; *Eupelor (Metoposaurus) durus*, reconstruction; D, *Icarosaurus siekeri*, reconstruction; E, "deep-tailed swimmer," tentative reconstruction; F, *Apatopus lineatus*, composite, right manus and pes; G, *Grallator* sp., left pes; H, *Gwynnedichnium minore*, right pes; I, *Rhynchosauroides brunswicki*, right manus and pes; J, *Chirotherium* cf. *evermani*, right manus and pes; K, *Carinacanthus jepseni*, reconstruction; L, *Turseodus* sp., reconstruction; M, *Synorichthys* sp., reconstruction; N, *Cionichthys* sp., reconstruction; O, semionotid of *"Semionotus brauni* group," reconstruction; P, *Diplurus newarki*, reconstruction. Scale 2 cm.

A, original; B, from Colbert, 1965; C, based on *Metoposaurus diagnosticus* from Colbert and Imbrie, 1956; D, modified from Colbert, 1970; E, original; F, from Baird, 1957; G, from Olsen, et al., in prep.; H, original, traced from Bock, 1952; I, from Olsen and Galton, 1977; J, from Olsen, et al., in prep.; K, original, based on PU 13739; L, modified from Schaeffer, 1952b and 1967; M, modified from Schaeffer, 1967; N, modified from Schaeffer, 1967; O, original; P, modified from Schaeffer, 1952a.

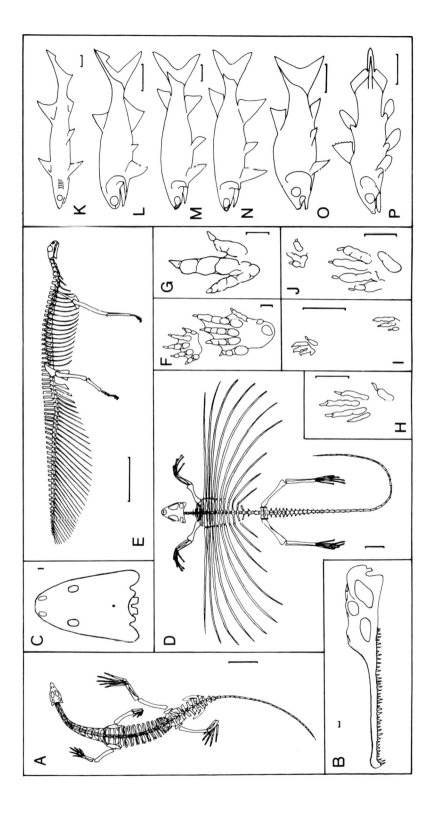

assic. There are two possible senior synonyms of *Tanytrachelos: Rhabdopelix* (Cope, 1870) and *Gwynnedosaurus* (Bock, 1945). Both of these taxa are from the same locality in the Lockatong and consist of disarticulated associations of bones, some individual elements of which cannot be distinguished from *Tanytrachelos*. Huene (1948) felt that *Gwynnedosaurus* was allied with *Macrocnemus* and *Tanystropheus*, a judgment in line with the tanystropheid relationships of *Tanytrachelos*. On the other hand, other elements of *Gwynnedosaurus* and *Rhabdopelix* are very different than *Tanytrachelos* and suggest the specimens represent several mixed taxa; the older two names should, thus, be regarded as *nomina dubia*.

Another small and very peculiar reptile common at several Lockatong localities has come to be known as the "deep-tailed swimmer" (Colbert and Olsen, in preparation; Olsen, in press 3). This reptile's most distinctive character is a deep, ventrally directed tail fin supported by extraordinarily long hemal spines (Figure 3.3). In addition, its front limbs are longer than its hind, and its lower jaw is beak-like and edentulous anteriorly. At this point the "deep-tailed swimmer" cannot be assigned to any reptilian order, although a newly discovered small reptile from the Norian of Italy (R. Zambelli and R. Wild, personal communication) bears some interesting similarities to it.

Lysorocephalus gwynnedensis is the name given to a small skull from the Lockatong of Pennsylvania described as a lysorophid amphibian by Huene and Bock (1954). Although the addition of a lysorophid to the Lockatong assemblage would be intriguing, the overall structure of the skull is very suggestive of a fish skull, particularly *Turseodus* (Baird, personal communication). This taxon is therefore omitted from the list in Table 3.1.

Skeletal remains of larger tetrapods from the upper Stockton and lower Lockatong include the metoposaur amphibian *Eupelor* (*Metoposaurus*) *durus* (Cope, 1866, 1871; Colbert and Imbrie, 1956; Olsen, Baird, Seldon, and Salvia, in preparation) and varied phytosaur material (Figure 3.3). Most of the latter is referable to *Rutiodon* and the rest is indeterminate (Cope, 1871; Huene, 1913; Colbert, 1943, 1965; Olsen, Baird, Seldon, and Salvia, in preparation).

A number of Lockatong reptile taxa, named in the older literature, are based on isolated teeth (see Huene, 1921); most of these teeth are surely phytosaurian but are generically indeterminate. A tooth taxon worth mentioning at this point, however, is "*Thecodontosaurus*" *gibbidens*, which seems to represent an ornithischian dinosaur. Since Huene's (1921) review, the locality for this form has been repeatedly cited as "Lockatong Formation, Phoenixville, York County, Pennsylvania." The correct locality for this important form is New Oxford Formation, near Emiggsville, York County, Pennsylvania (Frazer and Cope, 1886). This locality is in the Gettysburg Basin, not the Newark; Phoenixville is in Chester County, not York County. All tooth taxa from the Lockatong are omitted from Table 3.1.

In addition to these skeletal remains, a suite of footprints has recently been found in upper Stockton beds of New York State (Figure 3.3); these beds appear to be the lateral equivalents of the lower Lockatong (Olsen, in press 3). The forms found so far include the ubiquitous lepidosaurian track *Rhynchosauroides* spp., the thecodont footprint *Chirotherium* cf. *C. eyermani*, the phytosaur track *Apatopus lineatus*, and a variety of small dinosaurian footprints mostly assignable to the theropod track genus *Grallator* (Olsen, Baird, Seldon, and Salvia, in preparation). Added to this list is the peculiar thecodont-like track *Gwyneddichnium* (Bock, 1952) from the

TABLE 3.1

Vertebrates from the Hartford Basin. Comparable data for the Newark Basin
are listed in Olsen (in press 2).

Taxon	Formation	Reference
FISHES		
"Semionotus micropterus group"	Shuttle Meadow and prob-ably East Berlin formations	Olsen, McCune and Thomson, in press
Ptycholepis marshi	Shuttle Meadow Formation	Schaeffer, Dunkle, and McDonald, 1977
Redfieldius spp.	Shuttle Meadow through lower Portland formations	Schaeffer and McDonald, 1978
Diplurus cf. *D. longicaudatus*	Shuttle Meadow and East Berlin formations	Schaeffer, 1948
"Semionotus elegans group"	lower Portland Formation	Olsen, McCune, and Thomson, in press
Acentrophorus chicopensis	middle through upper Portland Formation	Olsen, McCune, and Thomson, in press
REPTILES		
Hypsognathus fenneri	?upper New Haven Arkose	YPM
sphenodontid rhynchocephalian	upper New Haven Arkose	PU 18835
Stegomus arcuatus	lower New Haven Arkose	Marsh, 1896
phytosaur (*Belodon validus*)	upper New Haven Arkose	Lull, 1953
archosaur tooth	Shuttle Meadow Formation	Galton, 1976
Stegomosuchus longipes	middle Portland Formation	Lull, 1953
Anchisaurus polyzelus	middle and upper Portland Formation	Galton, 1967
Ammosaurus major	upper Portland Formation	Galton, 1967
Podokesaurus holyokensis	middle Portland Formation	Lull, 1953
Batrachopus spp.*	Shuttle Meadow, East Berlin, and Portland formations	UM Lull, 1953
Anomoepus spp.*	Shuttle Meadow, East Berlin, and Portland formations	Lull, 1953
Grallator spp.*	Shuttle Meadow, East Berlin, and Portland formations	Lull, 1953
Anchisauripus spp.*	Shuttle Meadow, East Berlin, and Portland formations	Lull, 1953
*Eubrontes giganteus**	Shuttle Meadow, East Berlin, and Portland formations	Lull, 1953

*Indicates a footprint taxon.

lower Lockatong. The majority of other footprint taxa described by Bock (1952) are either completely indeterminate or have prior synonyms (Baird, 1957; Baird, personal communication).

The age of the upper Stockton and lower Lockatong appears to be late Carnian, on the basis of palynomorph assemblages recovered and studied by Cornet (1977a,b). According to Cornet (1977a), the boundary between the Lockatong and Passaic formations appears to coincide with the late Carnian-Norian boundary as palynologically recognized for the central part of the Newark Basin. Since the lower New Haven Arkose of the Hartford Basin has produced palynomorph assemblages of early Norian age (Cornet, 1977a), the entire Stockton-Lockatong sequence and its faunal assemblage predates the entire Hartford Basin column. In addition, most vertebrate assemblages from the Passaic Formation and the New Haven Arkose differ greatly from the Stockton-Lockatong assemblage. This difference cannot be taken literally, however, since the Lockatong is so clearly biased towards small aquatic organisms (compare Figures 3.3 and 3.4).

New Haven Arkose and Passaic Formation

Skeletal assemblages from the New Haven Arkose of the Hartford Basin and the Passaic Formation of the Newark Basin (Figure 3.2) are very similar, both taxonomically and in mode of occurrence. The lower Passaic Formation has produced two partial skeletons of aetosaurian thecodonts both referable to *Stegomus arcuatus* (Figure 3.4) (Jepsen, 1948; Baird, personal communication), a form originally described from the New Haven Arkose (Marsh, 1896; Lull, 1953). The more recently discovered New Jersey specimen (PU 21750) has a skull with a long tapering snout more like that of *Aetosaurus* from the German Stubensandstein than *Stagonolepis* from the Elgin beds of Scotland. Interestingly, scutes identical to those of *Stegomus* and distinct from both *Aetosaurus* and *Stagonolepis* have recently turned up in the same Yale collection of bones from the Stubensandstein from which Gregory (1953) identified a scute of another aetosaur, *Typothorax*. The recently discovered specimen of *Stegomus* from New Jersey comes from beds yielding a Norian palynoflora (near member F, Cornet, 1977a), and the stratigraphic position of both the specimen described by Jepsen (1948) and the New Haven Arkose specimen described by Marsh (1896) are commensurate with this age.

The middle Passaic Formation has produced fragmentary remains of phytosaurs. These include the type material of *Clepsysaurus pennsylvanicus* (Lea, 1852, Colbert and Chaffee, 1941) and a partial maxilla from a different locality. Both of these specimens are Rhaetian based on palynomorph assemblages from stratigraphically close beds (i.e., the Perkasie Member and its lateral equivalents) (Cornet, 1977a). In the Hartford Basin a phytosaur scapula described as *Belodon validus* by Marsh (1893) was found in the upper New Haven Arkose about 1,500 m above and 40 km to the north of the horizon producing the above mentioned early Norian palynomorph assemblage (Figure 3.2). All these phytosaur remains are generically indeterminate (Colbert, 1965).

From sandstones of the upper Passaic Formation and the upper New Haven Arkose come skeletons of the procolophonid *Hypsognathus fenneri*. The structure of the skull shows that *Hypsognathus* was far more specialized than any other known procolophonid. The type specimen was described by Gilmore (1928) and this was folowed by Colbert's (1946) detailed study based on additional, more complete material. One more skull and a skeleton have more recently been collected (Baird, personal communication, and Colbert, personal communication). All of this material was found in the northern Newark Basin, in an interval about 800 m above an early Rhaetian palynomorph-bearing unit (Cornet, 1977a). In 1967 an excellent skull and skeleton of *Hypsognathus* was found in a block in a stone wall in Meriden, Connecticut (Ostrom, 1967, 1969). A combination of information on the heavy mineral suite present in the matrix (Griggs, 1973) and on the coarseness of the sandstone strongly suggest that the upper division of the New Haven Arkose (Krynine, 1950) was the source of the block. This rough lithologic determination corresponds with the stratigraphic range of the New Jersey material.

Two other skeletal forms from the New Haven Arkose Passaic Formation tetrapod assemblage require note. One is from the middle Passaic Formation (probably Norian) and consists of a partial skull and postcranial skeleton, named *Sphodrosaurus pennsylvanicus* by Colbert (1960). Colbert assigned *Sphodrosaurus* to the procolophonids primarily on the basis of very primitive "reptilian" features. The occiput of *Sphodrosaurus* lies far anterior to the rear of the skull, and this, in combination with the sculptured surface of the visible skull bones (Colbert, 1960), indicates that *Sphodrosaurus* is only distantly related to *Hypsognathus,* if it is a procolophonid at all. The other form is from the upper division of the New Haven Arkose of Meriden, Connecticut (Case, 1972) and is a nearly complete skull (PU 18835) of an undescribed small spenodontid rhynchocephalian surprisingly similar to *Sphenodon* (Baird, personal communication) (Figure 3.4). This is the first definite osteological record of a sphenodontid in the North American Triassic, although Baird (1957) identified their presence on the basis of the presence of the footprint taxon *Rhynchosauroides* from the Passaic Formation.

In addition to tetrapods, the lower Passaic Formation has produced some very fragmentary fish, including cf. *Diplurus* sp., *Synorichthys* sp., and *Semionotus* sp. These remains are too rare and too poorly preserved to be very useful in correlation.

While skeletal remains have so far been rare in the Passaic, reptile footprints are abundant throughout the formation. These tracks are very important in providing a link with the footprints from the younger parts of the Newark and Hartford basins. The Passaic footprints can be divided into two broad assemblages on the basis of stratigraphic position (Figure 3.4). The lower assemblage occurs in the portion of the Passaic dated palynologically as Norian (Cornet, 1977a), and is dominated by *Rhynchosauroides,* several species of *Chirotherium,* small species of *Grallator,* and a very unusual form which has been called *"Anchisauripus" milfordensis* and *"Grallator" sulcatus* (Baird, 1957). The latter two ichnospecies clearly belong in their own genus (Baird, personal communication). This taxon (Figure 3.4) is characterized by a *Grallator*-like pes in which, for the size of the track, digit III is unusually short. The metatarsal-phalangeal pads of both digits II and IV are deeply impressed, which results in a tulip-shaped outline of the pes impression. The most distinctive feature, however, is a small three-toed manus impression present in nearly every trackway. This form is known from scores of specimens, from several horizons, and from three Newark Supergroup Basins: the Fundy Basin, Newark Basin, and Gettysburg Basin. Despite its apparent abundance, it corresponds to no known group of reptiles, although its *Gestalt* is dinosaurian. Other footprint taxa present in this lower Passaic assemblage are the phytosaur track *Apatopus lineatus* and the thecodont-like *Gwynnedichnium* (Olsen, in press, 2).

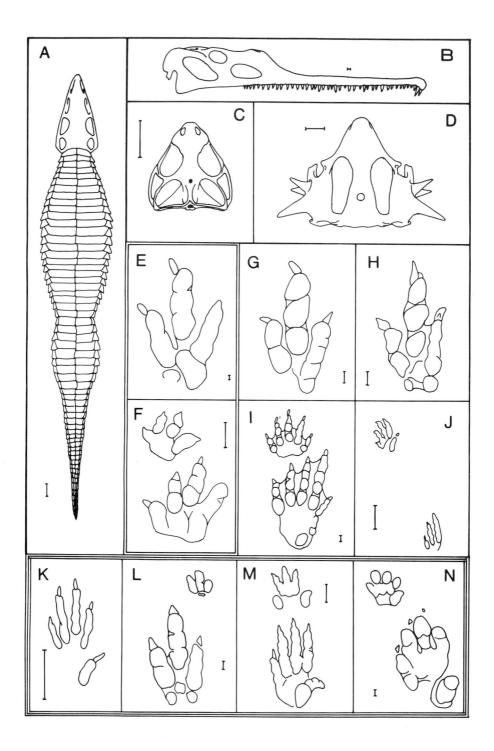

It is important to note that all the footprint taxa found in the Stockton-Lockatong footprint assemblage are also found in the lower Passaic assemblage.

The upper Passaic assemblage of footprints is known from the upper 100 m of the formation. In marked contrast to the lower assemblage, small and large grallatorid footprints are dominant; these include forms traditionally known as *Anchisauripus sillimani* and *Anchisauripus minusculus.* Also common are the earliest Newark occurrences of the probable crocodiliomorph track *Batrachopus.* Present in small numbers are *Apatopus lineatus* and *Rhynchosauroides;* conspicuous by their absence are species of *Chirotherium.* On the basis of the relative position of these beds with respect to the overlying Orange Mountain Basalt (Figure 3.2), the upper Passaic assemblage appears to correlate with late Rhaetian or earliest Jurassic palynomorph-bearing sequences in the southern Newark Basin (Cornet, 1977a). The transitional assemblage between these lower and upper Passaic assemblages would be expected to occur throughout the more than 1,000 m of Passaic Formation of Rhaetian age. The few footprints found in these beds are not very good. They do indicate the presence of more large grallatorids than the lower Passaic assemblage, yet *Chirotherium* is still present. Unfortunately, no New Haven Arkose footprints have found their way into museum repositories, although some have been found (Lull, 1953). Footprints are common, however, in the Hartford Basin sedimentary rocks above the New Haven Arkose (Figure 3.2) and this assemblage has its counterpart in the post-Passaic formations of the Newark Basin. The Passaic Formation footprints thus provide a transitional assemblage between the thecodont-lepidosaur dominated Carnian and Norian Newark Basin beds and the dinosaur-crocodiliomorph dominated assemblages of the Jurassic formations of both the Newark and Hartford basins.

Jurassic Formations of the Newark and Hartford Basins

The Feltville Formation (Figure 3.2) rests on the Orange Mountain Basalt which, in turn, rests on the Passaic Formation and is the oldest Newark Basin formation which appears to be Early Jurassic on the basis of palynomorph correlation (Cornet, 1977a). The Orange Mountain Basalt has yielded radiometric ages of 193–136 Ma,

---

Figure 3.4. The Passaic Formation and New Haven Arkose vertebrate assemblage: A, *Stegomus arcuatus,* reconstruction of skull and dorsal armor; B, a phytosaur skull, reconstruction; C, new sphenodontid rhynchocephalian, reconstruction—temporal arches largely hypothetical; D, *Hypsognathus fenneri,* reconstruction; E, *Anchisauripus minusculus,* right pes; F, *Batrachopus deweyi,* right manus and pes; G, *Anchisauripus* sp., right pes; H, *Grallator* sp., right pes; I, *Apatopus lineatus,* composite, right manus and pes; J, *Rhynchosauroides brunswicki,* right manus and pes; K, *Gwynnedichnium minore,* right pes; L, new genus formerly called *Anchisauripus milfordensis* and *Grallator sulcatus,* right manus and pes; M, *Chirotherium lulli,* composite, right manus and pes; N, *Chirotherium parvum,* composite, right manus and pes. Scale 1 cm. Note that E and F (surrounded by double line) are known only from the upper Passaic assemblage; K, L, M, and N (surrounded by triple line) are restricted to the lower assemblage of the Passaic; and G, H, I, and J occur through the Passaic.

A, original, based on Jepsen (1948), Marsh (1896), Walker (1961), and PU 21750; B, based on *Rutiodon carolinensis* from Colbert (1965); C, original based on PU 18835; D, original based on YPM specimen; E, original, based on AC 16/1; F, from Lull (1953); G, original, based on AC 9/14; H, from Baird (1957); I, from Baird (1957); J, from Olsen and Galton (1977); K, original, traced from Bock (1952); L, original, based on CM specimen; M, from Baird (1954); N, from Baird (1957).

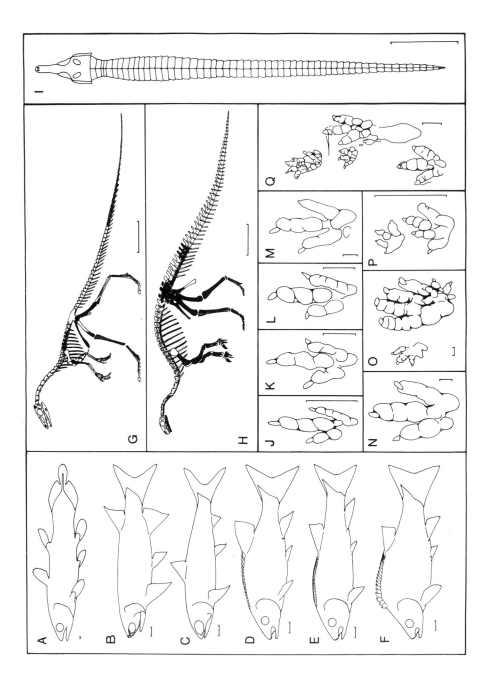

which also suggest an Early Jurassic age (Sutter and Smith, 1979). The Feltville Formation footprint assemblage(Figure 3.5), like that of the upper Passaic assemblage, is dominated by large as well as small grallatorids, including *Grallator* sp., *Anchisauripus sillimani, Anchisauripus minusculus,* and *Eubrontes* sp. As in the upper Passaic assemblage, *Batrachopus* sp. is present. The oldest Newark Basin occurrence of the probable ornithopod footprint *Anomoepus* spp. is in the Feltville, and it is the presence of this genus which distinguishes the Feltville from the upper Passaic assemblage.

The Feltville-type assemblage is also characteristic of the overlying Towaco and Boonton Formations in the Newark Basin and all post-New Haven Arkose sedimentary formations of the Hartford Basin (Figure 3.2). In the Newark Basin, the Towaco Formation has produced the bulk of the Jurassic footprints, while in the Hartford basin this role is taken by the Portland Formation. As far as this analysis goes, these assemblages are indistinguishable. The Portland has, however, very rare examples of the giant ?crocodiliomorph track *Otozoum* (Figure 3.5), and I believe its absence from the Newark Basin could be due to inadequate sampling.

The footprint taxa characteristic of the Feltville Formation have what has been traditionally called a "Connecticut Valley" aspect (Lull, 1953; Baird, 1957). The contrast between this "Connecticut Valley"-type footprint assemblage and the lower Passaic assemblage is, I believe, the most fundamental division of Newark Supergroup assemblages; they share only small *Grallator* species. This contrast was pointed out by Lull (1953), Colbert (1946, 1965), Colbert and Baird (1958), Colbert and Gregory (1957), and Olsen and Galton (1977). It is important to note that, like the Feltville Formation, all Newark and Hartford Basin formations which preserve a "Connecticut Valley"-type of footprint assemblage have also yielded palynomorph assemblages of Early Jurassic age (Hettangian through Toarcian, Cornet, 1977a); also, the interbedded basalt flows have produced Early Jurassic dates (Armstrong and Besancon, 1970; Sutter and Smith, 1980). This is true of the Newark Supergroup as a whole (Olsen, McCune, and Thomson, in press; Cornet, 1977a).

The relative homogeneity of the "Connecticut Valley"-type assemblage through all the post-Passaic and post–New Haven Arkose Formations is not reflected in the

---

Figure 3.5 Jurassic vertebrate assemblage of Newark and Hartford basins: A, *Diplurus longicaudatus,* reconstruction; B, *Redfieldius* spp., reconstruction; C, *Ptycholepis marshi,* reconstruction; D, member of "*Semionotus elegans* group," reconstruction; E, member of "*Semionotus micropterus* group," reconstruction; F, member of "*Semionotus tenuiceps* group," reconstruction; G, *Podokesaurus holyokensis,* reconstruction—only black portions actually preserved; H, *Ammosaurus major,* reconstruction—only black portions preserved; I, *Stegomosuchus longipes,* reconstruction; J, *Grallator cursorius,* right pes; K, *Anchisauripus hitchcocki,* right pes; L, *Anchisauripus* sp.; M, *Anchisauripus minusculus,* right pes; N, *Eubrontes* cf. *E. giganteus,* right pes; O, *Otozoum moodii,* left pes and left and right manus; P, *Batrachopus deweyi,* right manus and pes; Q, *Anomoepus crassus,* right manus (impressed three times) and right pes (impressed twice—once with tarsus). Scale of A-F, 8 cm; scale of G-I, 10 cm; and scale of J-Q, 5 cm.

A-C from Olsen, McCune, and Thomson (in press); D, original based on YPM 6567; E, original, based on YPM 0000; F, original, based on YPM 8162; G, original, data from Lull (1953) and reconstructed partly on the basis of *Coelophysis;* H, based on data in Galton (1976); I, original, based on Walker (1968) and Lull (1953); J, original, based on AC 4/1; K, from Lull (1953); L, original, based on AC 9/14; M, original, based on AC 16/1; N, original, based on AC 45/1; O, original, based on AC 15/14; P, from Lull (1953); Q, original, based on RUGM 50:4:1.

many fish assemblages. The vertical changes in fish assemblage composition allow for their use in correlation within the Newark Supergroup (Olsen, McCune, and Thomson, in press). The most abundant fishes through the Newark and Hartford basin Jurassic are members of the genus *Semionotus*. It is crucial to note that the old concept of *Semionotus* as a Triassic "index fossil" must be abandoned because it is due to long standing taxonomic artifacts; specimens of *Semionotus* spp. are, in fact, common through the Jurassic and Early Cretaceous (Olsen, in preparation). Even though at the generic level *Semionotus* is useless in biostratigraphy, at lower taxonomic levels, such as species and species-groups, it may prove valuable for correlation (Olsen, McCune and Thomson, in press).

Fishes from the Feltville Formation (Figure 3.5) differ completely from those present in the underlying Lockatong (Figure 3.3). Semionotid fishes of the "*Semionotus tenuiceps* group" are overwhelmingly abundant. The only other genus so far found is the subholostean *Ptycholepis*. The Towaco Formation has produced only semionotids and these seem allied to those of the Feltville (Olsen, McCune and Thomson, in press). Semionotids of the Shuttle Meadow Formation of the Hartford Basin, on the other hand, are very different from those of the Feltville and Towaco. Semionotids of the "*Semionotus micropterus* group" and the subholostean *Redfieldius* are the dominant taxa of the Shuttle Meadow; smaller numbers of the coelacanth *Diplurus* cf. *D. longicaudatus* and *Ptycholepis marshi* are also present. The lesser-known East Berlin Formation assemblage seems essentially the same as the Shuttle Meadow. Interestingly, an assemblage like that of the Feltville and Towaco occurs in the Deerfield Basin, to the north of the Hartford Basin, and an assemblage like those of the Shuttle Meadow and East Berlin formations occurs in the Culpeper Basin 500 km to the south of the Hartford Basin. Therefore, simple endemism in individual basins can be ruled out as an explanation for the differences in the fish assemblages. My colleagues and I argue that the simplest hypothesis is that the Feltville and Towaco formations are older than the Shuttle Meadow and East Berlin formations, as depicted in figure 3.6 (Olsen, McCune, and Thomson, in press; Cornet, 1977a).

In contrast to the older Jurassic formations, the upper Boonton Formation of the Newark Basin and lower Portland Formation of the Hartford Basin (Figure 3.2) have essentially the same fish assemblages. Semionotids of the "*Semionotus elegans* group" and *Redfieldius gracilus* are the most common taxa in both assemblages. *Diplurus longicaudatus* and *Ptycholepis* sp. are known from the Boonton Formation but not from the lower Portland, probably because of the very small sample size of the Portland assemblage. Palynomorph assemblages from these beds in both basins show them to be Sinemurian (Cornet, 1977a).

The middle and upper Portland Formation contain a fish assemblage dominated by a form which Newberry (1888) called *Acentrophorus chicopensis*, as well as some *Semionotus* species. Unfortunately, the fish are very poorly preserved (although abundant) and *A. chicopensis* must be regarded as an indeterminate holostean. The beds producing this assemblage appear to be Toarcian in age (Cornet, 1977a), and thus this is the youngest Newark Supergroup fish assemblage.

Apart from footprints and one tooth (Galton, 1971), the only Jurassic tetrapod remains in the two basins have been found in the Portland (Lull, 1953). *Podokesaurus holyokensis* is a small, lightly built theropod found in a glacial boulder at Mt. Holyoke College in Massachusetts. Its probable provenance is the Portland Formation (Lull, 1953), although this, along with its true systematic position, will never be known with certainty, since it was destroyed in a fire. Colbert (1964) synonymized *Podokesaurus* and *Coelophysis*. This was the most economical approach when the two genera were thought to be the same age (i.e., Triassic). If, however, *Podokesaurus* is from the Portland Formation, it is probably about 15 million years younger than *Coelophysis*. The characters cited by Colbert (1964) and Colbert and Baird (1958) as shared between *Coelophysis* and *Podokesaurus* are also shared with other small theropods such as the upper Stormberg *Syntarsus* (Raath, 1969). To avoid the possible artifact of overextending the stratigraphic range of *Coelophysis* it is appropriate to conserve *Podokesaurus* as a separate genus and to regard it as an indeterminate small theropod. This same argument applies to a sandstone natural cast of the impression of parts of a hind limb and pelvis of a small theropod, also from the Portland Formation (Colbert and Baird, 1958). I would prefer to regard this specimen as *incertae sedis* among the theropods rather than to refer it either to *Coelophysis* (Colbert and Baird, 1958) or to *Podokesaurus*.

A very important Portland reptile is *Stegomosuchus longipes*, formerly thought to be a psuedosuchian (Lull, 1953) and now thought to be a crocodile very similar to *Protosuchus* (Walker, 1968) (Figure 3.5). Also present in the Portland are the prosauropod dinosaurs *Ammosaurus* and *Anchisaurus* (Huene, 1906; Lull, 1953; Galton, 1976). The prosauropods and *Stegomosuchus* are from portions of the Portland which are Toarcian in age or younger (Cornet, 1977a). As Galton (1971) pointed out, this skeletal assemblage has its affinities with that of the Navajo Sandstone of the Glen Canyon Group of the southwestern United States. The lower beds of the Glen Canyon Group (Moenave Formation), like the lower beds of the Portland Formation, have produced palynomorph assemblages of Sinemurian or Pliensbachian age (Cornet, 1977a; Olsen and Galton, 1977; Baird, this volume). It can thus be concluded, on the basis of fish and palynomorph correlation that at least the upper Boonton Formation and lower Portland Formation correlate, and that beds equivalent to the upper (?Pliensbachian-Toarcian and younger) beds of the Portland Formation have not been preserved in the Newark Basin (Figure 3.6).

Summary

In a broad way, the major homotaxial lithologic divisions of the Newark and Hartford basins correlate (Figure 3.6). The New Haven Arkose and the Passaic Formation correlate; the basalt flow formations and interbedded sedimentary formations are roughly contemporaneous; and parts of the Boonton and Portland formations correlate. However, the sequences in the two basins differ in many details. Strata equivalent in time to the Stockton and Lockatong are entirely absent in the Hartford Basin; both the Feltville and Towaco formations appear to be older than the Shuttle Meadow and East Berlin formations; and the entire upper two-thirds of the Portland Formation is younger than the youngest beds of the Boonton Formation (Figure 3.6). These correlations are, of course preliminary; their main purpose is to organize the biostratigraphic data in the simplest manner. The conclusions are limited not only by the incompleteness of the paleontological data, but also by realities of biological temporal and spacial heterogeneity, realities necessarily deemphasized in the assumptions of biostratigraphic procedure. Only with the accumulation of additional biostratigraphic and presumably independent geophysical data will these correlations be tested and then improved.

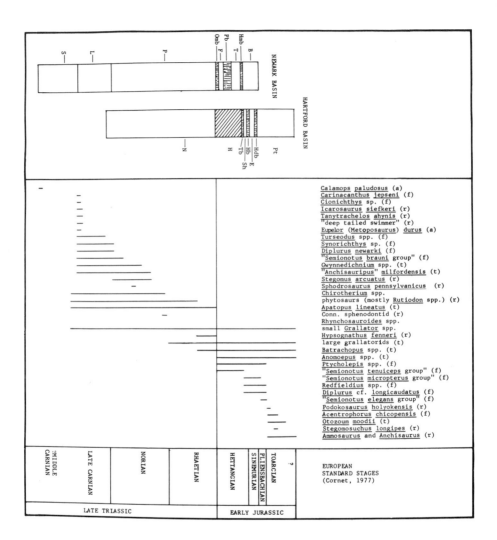

Figure 3.6. Correlation of the formations of the Newark and Hartford Basins. Abbreviations of lithologic divisions as follows: B, Boonton Formation; E, East Berlin Formation; F, Feltville Formation; H, inferred hiatus; Hb, Holyoke Basalt; Hdb, Hampden Basalt; Hmb, Hook Mountain Basalt; L, Lockatong Formation; N, New Haven Arkose; Omb, Orange Mountain Basalt; P, Passaic Formation; Pb, Preakness Basalt; Pt, Portland Formation; S, Stockton Formation; Sh, Shuttle Meadow Formation; and T, Towaco Formation.

All thicknesses of columns proportional to those in Figure 2 except for the New Haven Arkose which is drawn disproportionately thick.

Letters next to taxa denote the nature of the fossil as follows: a, amphibian skeletal remains; b, fish skeletal remains; r, reptile skeletal remains; t, reptile track taxon. Data for taxon distributions in Olsen (in press 2)and Table 3.1.

Acknowledgments

First, I thank Edwin H. Colbert for inspiration through his writing and personal encouragement at a time when I was a young and impressionable amateur fossil collector. In addition, I thank Donald Baird and W. Bruce Cornet for use of many of their unpublished data and ideas. Finally, I thank Kevin Padian and Amy Litt for reading the manuscript and suggesting changes which substantially improved it. This work was supported by National Science Foundation grant number DEB-7921746 to Keith S. Thomson. Any errors of omission or commission are, of course, my own.

Literature Cited

Armstrong. R. L., and Besancon, J. 1970. A Triassic time scale dilemma: K-Ar dating of Upper Triassic igneous rocks, eastern U.S.A. and Canada and post-Triassic plutons, western Idaho, U.S.A. Eclogae Geol. Helv., 63:15–28.

Baird, D. 1957. Triassic reptile footprint faunules from Milford, New Jersey. Harvard Univ., Mus. Comp. Zool., Bull., 117:449–520.

Bock, W. 1945. A new small reptile from the Triassic of Pennsylvania. Notulae Naturae, 154:1–8.

———. 1952. Triassic reptilian tracks and trends of locomotor evolution. J. Paleontol., 26: 395–433.

Case, G. R. 1972. Handbook of fossil collecting. Author, New York, 65 pp.

Colbert, E. H. 1943. A lower jaw of *Clepsysaurus* and its bearing upon the relationships of this genus to *Machaeroprosopus*. Notulae Naturae,124:1–8.

———. 1946. *Hypsognathus*, a Triassic reptile from New Jersey. Amer. Mus. Natur. Hist., Bull., 86:225–74.

———. 1960. A new Triassic procolophonid from Pennsylvania. Amer. Mus. Nov., 29:1–19.

———. 1964. The Triassic genera *Podokesaurus* and *Coelophysis*. Amer. Mus. Nov., 2168:1–12.

———. 1965. A phytosaur from North Bergen, New Jersey. Amer. Mus. Nov., 2230:1–25.

———. 1966. A gliding reptile from the Triassic of New Jersey. Amer. Mus. Nov., 2246:1–23.

———. 1970. The Triassic gliding reptile *Icarosaurus*. Amer. Mus. Natur. Hist., Bull., 143(2): 85–142.

Colbert, E. H., and Baird, D. 1958. Coelurosaur bone casts from the Connecticut Valley Triassic. Amer. Mus. Nov., 1901:1–11.

Colbert, E. H., and Chaffee, R. G. 1941. The type of *Clepsysaurus pennsylvanicus* and its bearing upon the genus *Rutiodon*. Notulae Naturae, 90:1–19.

Colbert, E. H., and Gregory, J. T. 1957. Correlation of continental Triassic sediments by vertebrate fossils. Geol. Soc. Amer., Bull., 68:1456–67.

Colbert, E. H., and Imbrie, J. 1956. Triassic metoposaurid amphibians. Amer. Mus. Natur. Hist., Bull., 110(6):399–452.

Cope, E. D. 1866. On vertebrates of the Mesozoic red sandstone from Phoenixville, Chester Co., Pa. Phila., Acad. Natur. Sci., Proc., 1866:249–50.

———. 1870. On the Reptilia of the Triassic formations of the Atlantic region. Amer. Phil. Soc., Proc., 11:444–46.

———. 1871. Synopsis of the extinct Batrachia, Reptilia, and Aves of North America. Amer. Phil. Soc., Trans., 14:1–252.

Cornet, B. 1977a. The palynostratigraphy and age of the Newark Supergroup. Unpublished Ph.D. Thesis, Pennsylvania State University, 506 pp.

———. 1977b. Preliminary investigation of two Late Triassic conifers from York County, Pennsylvania. *In* R. C. Romans, ed., Geobotany. Plenum Press, New York, pp. 165–72.

Frazer, P., and Cope, E. D. 1886. General notes—sketch on the geology of York County. Amer. Phil. Soc., Proc., 1886:391–410.

Galton, P. M. 1971. The prosauropod dinosaur *Ammosaurus*, the crocodile *Protosuchus*, and their bearing on the age of the Navajo Sandstone of northeastern Arizona. J. Paleontol., 45:781–95.

———. 1976. Prosauropod dinosaurs (Reptilia: Saurischia) of North America. Postilla, 169: 1–98.

Gilmore, C. W. 1928. A new fossil reptile from the Triassic of New Jersey. U. S. Nat. Mus., Proc., 73(7):1–8.

Gregory, J. T. 1953. *Typothorax* scutes from Germany. Postilla, 15:1–6.

Griggs, E. A., Jr. 1973. The origin of *Hypsognathus* in the Triassic sediments of Connecticut. Unpublished senior paper, Yale University senior paper series, 5 pp.

Hitchcock, E. 1858. Ichnology of New England. A report on the sandstone of the Connecticut Valley, especially its fossil footmarks. William White, Boston, 220 pp.

Hubert, J. F.; Reed, A. A.; Dowdall, W. L.; and Gilchrist, J. M. 1978. Guide to the redbeds of central Connecticut: 1978 Field Trip, Eastern Section of S.E.P.M., Univ. Mass., Amherst, Dept. Geol. Geog., Contr. 32, 129 pp.

Huene, F. v. 1906. Ueber die Dinosaurier der aussereuropaeischen Trias. Geol. Pal., Abh. Jena, (N.F.), 8(12):97–156.

———. 1913. A new phytosaur from the Palisades near New York. Amer. Mus. Natur. Hist., Bull., 32(15):275–83.

———. 1921. Reptilian and stegocephalian remains from the Triassic of Pennsylvania in Cope Collection. Amer. Mus. Natur. Hist., Bull., 44(19):561–74.

———. 1948. Notes on *Gwynnedosaurus*. Amer. J. Sci., 246:208–12.

Huene, F. v., and Bock, W. 1954. A small amphibian skull from the upper Triassic of Pennsylvania. Wagner Free Inst. Sci., Bull., 29:27–34.

Jepsen, G. L. 1948. A Triassic armored reptile from New Jersey. N. J. State Dept. Conserv., Misc. Geol. Pap., 1948, 19 pp.

Krynine, P. D. 1950. Petrology, stratigraphy and origin of the Triassic sedimentary rocks of Connecticut. Conn. State Natur. Hist. Surv., Bull., 73:1–239.

Lea, I. 1852. Description of a fossil saurian of the New Red sandstone formation of Pennsylvania—with some account of that formation. Phila., Acad. Nat. Sci., J., (2)2:185–202.

Lull, R. S. 1953. Triassic life of the Connecticut valley. Conn. State Geol. Natur. Hist. Surv., Bull., 81:1–336.

Manspeizer, W.; Puffer, J. H.; and Cousminer, H. L. 1978. Separation of Morocco and eastern North America: A Triassic-Liassic stratigraphic record. Geol. Soc. Amer., Bull., 89: 901–20.

Marsh, O. C. 1893. Restoration of *Anchisaurus*. Amer. J. Sci., 45:169–70.

———. 1896. A new belodont reptile (*Stegomus*) from the Connecticut River Sandstone. Amer. J. Sci., 2:59–62.

Newberry, J. S. 1888. Fossil fishes and fossil plants of the Triassic rocks of New Jersey and the Connecticut Valley. U. S. Geol. Surv., Monogr., 14:1–152.

Olsen, P. E. 1978. On the use of the term Newark for Triassic and Early Jurassic rocks of eastern North America. Newsl. Stratigr., 7:90–95.

———. 1979. A new aquatic eosuchian from the Newark Supergroup (Late Triassic–Early Jurassic) of North Carolina and Virginia. Postilla, 176:1–14.

———. In press 1. The latest Triassic and Early Jurassic formations of the Newark Basin (eastern North America, Newark Supergroup): stratigraphy, structure, and correlation. N. J. Acad. Sci., Bull.

———. In press 2. Triassic and Jurassic Formations of the Newark Basin. Fieldtrip guidebook, N. Y. State Geol. Assoc., Ann. Mtg.

———. In press 3. Comparative paleolimnology of the early Mesozoic Newark Supergroup in New Jersey. N. Y. State Geol. Assoc., Ann. Mtg., Fieldtrip Guidebook.

Olsen, P. E., and Galton, P. M. 1977. Triassic-Jurassic tetrapod extinctions: Are they real? Science, 197:983–86.

Olsen, P. E.; McCune, R. R.; and Thomson, K. S. In press. Correlation of the early Mesozoic Newark Supergroup (eastern North America) by vertebrates, especially fishes. Amer. J. Sci.

Ostrom, J. H. 1967. (On the discovery of *Hypsognathus* in Connecticut.) Discovery, 3:59.

———. 1969. (On preparing *Hypsognathus* from Connecticut.) Discovery, 5:126.

Raath, M. A. 1969. A new coelurosaurian dinosaur from the Forest Sandstone of Rhodesia. Arnoldia (Rhodesia), 28:1–25.

Redfield, J. H. 1845. Catalogue of the fossil fish of the United States as far as known, with descriptions of those found in the New-Red-Sand-Stone. Manuscript in Archives, Peabody Museum (Yale).

Russel, I. C. 1892. The Newark System, correlation papers. U. S. Geol. Surv., Bull., 85: 1–344.

Sanders, J. E. 1963. Late Triassic tectonic history of northeastern United States. Amer. J. Sci., 261:501 24.

Shaeffer, B. 1948. A study of *Diplurus longicaudatus* with notes on the body form and locomotion of the Coelacanthini. Amer. Mus. Nov. 1378:1–32.

———. 1952a. The Triassic coelacanth fish *Diplurus* with observations on the evolution of the Coelacanthini. Amer. Mus. Natur. Hist., Bull., 99(2):29-78.

———. 1952b. The palaeoniscoid fish *Turseodus* from the upper Triassic Newark Group. Amer. Mus. Nov.,1581:1–23.

———. 1967. Late Triassic fishes from the western United States. Amer. Mus. Natur. Hist., Bull., 135(6):287–342.

Schaeffer, B.; Dunkle, D. H.; and McDonald, N. G., 1975. *Ptycholepis marshi,* Newberry, a chondrostean fish from the Newark Group of eastern North America. Fieldiana, Geol., 33:205–33.

Schaeffer, B., and McDonald, N. G. 1978. Redfieldiid fishes from the Triassic-Liassic Newark Supergroup of eastern North America. Amer. Mus. Natur. Hist., Bull., 159(4):129–74.

Schuchert, C., and Dunbar, C. 1941. A textbook of geology: Pt II—Historical geology. John Wiley and Sons, Inc., New York, 544 pp.

Sinclair, W. J. 1917. A new labyrinthodont from the Triassic of Pennsylvania. Amer. J. Sci., 43:319–21.

Sutter, J. F., and Smith, T. E. 1979. ^{40}Ar/^{39}Ar ages of diabase intrusions from Newark trend basins in Connecticut and Maryland: Initiation of central Atlantic rifting. Amer. J. Sci., 279:808–31.

Van Houten, F. B. 1977. Triassic-Liassic deposits of Morocco and eastern North America: comparison. Amer. Assoc. Petrol. Geol., Bull., 61:79–99.

Walker, A. D. 1961. Triassic reptiles from the Elgin area: *Stagonolepis, Dasygnathus* and their allies. Roy. Soc. London Phil. Trans., Ser. B, 709:103–204.

———. 1968. *Protosuchus, Proterochampsa,* and the origin of phytosaurs and crocodiles. Geol. Mag., 105:1–14.

Abbreviations for Specimen Repositories

AC—Pratt Museum, Amherst College, Amherst, Massachusetts
AMNH—American Museum of Natural History, New York, New York
CM—Carnegie Museum of Natural History, Pittsburgh, Pennsylvania
PU—Vertebrate Paleontology collection, Guyot Hall, Princeton University Museum of Natural History, Princeton, New Jersey
UM—University of Massachusetts, Geology Department, Amherst, Massachusetts
WU—Geology Department Museum, Wesleyan University, Middletown, Connecticut
YPM—Peabody Museum, Yale University, New Haven, Connecticut

THE VERTEBRATE COMMUNITY
OF THE TRIASSIC CHINLE FORMATION
NEAR ST. JOHNS, ARIZONA

by Louis L. Jacobs and Phillip A. Murry

Introduction

The Chinle Formation of the southwestern United States is one of the major sources of Late Triassic vertebrate fossils in the western hemisphere. Numerous specimens have been recovered from extensive exposures in Arizona, New Mexico, Utah, and Colorado as the result of over one hundred years of collecting. The fauna from the Chinle is characterized by abundant phytosaurs and metoposaurs. Added to this is a reasonably extensive list of fish and reptiles, all of moderate to large size. The smallest terrestrial vertebrate previously described from the Chinle is the thecodont reptile *Hesperosuchus,* described by Colbert (1952), which was about the size of a fox or slightly larger.

Most vertebrate taxa from the Chinle are known from fragmentary specimens from several localities. However, more complete specimens and large samples of taxa uncommon elsewhere in the Chinle are found in occasional rich pockets.

The size distribution of known taxa from the Chinle Formation strongly suggests that the previously known assemblage is a biased sample of the true Chinle vertebrate community. This was mentioned by Edwin H. Colbert in several of his publications on Chinle vertebrates (e.g., 1952, 1972). There were indications that the Chinle fauna could be expanded by the collection of smaller vetebrates because some small specimens (still undescribed) were found with the skeleton of *Hesperosuchus* near Cameron, Arizona, in 1929 and 1930. These specimens, recovered by dry screening, were reported by Colbert (1952) as "ganoid" scales, small stereospondyl vertebrae, and small amphibian and reptile teeth. An additional sample of small bones (undescribed) was found in 1949 northeast of Winslow, Arizona.

Beginning in 1978, field crews from the Museum of Northern Arizona (MNA) have applied wet screening methods to the Chinle Formation. As a result, thousands of small, albeit fragmentary, specimens have been recovered that represent young individuals of common Chinle forms and a large number of undescribed taxa. The known vertebrate fauna of the Chinle Formation has been nearly doubled. This work, and similar work in the Triassic Dockum Formation of Texas, demonstrates that screening in rocks of these ages is a fruitful approach, just as it has been in younger Mesozoic and Cenozoic deposits.

The major purpose of this paper is to report initial results of screening. Larger specimens removed by quarrying are also reported here in order to present as complete a picture of the Chinle vertebrate fauna as possible. These collections are from two quarries near St. Johns, Arizona. The first is the *Placerias* Quarry, discovered by the University of California (UC). The second quarry, the Downs Quarry, is located east of the *Placerias* Quarry on the same hill. Both quarries apparently sample the same fauna.

The *Placerias* and Downs Quarries

The *Placerias* Quarry is located on a tributary to Big Hollow Wash near Romero Spring about 10.4 km southwest of St. Johns, Arizona. The area is one of grass-covered rolling hills. Exposures are poor by Chinle standards. The exact stratigraphic position of the *Placerias* Quarry relative to the base of the Chinle has not been precisely determined because of the nature of the exposures. Akers (1964) mapped the area of the *Placerias* Quarry as part of the Petrified Forest Member of the Chinle Formation. The Shinarump Member of the Chinle and the underlying Moenkopi Formation are mapped approximately three to four miles east of the quarry (see also Camp and Welles, 1956). Akers (1964) did not include the Mesa Redondo Member of the Chinle (Cooley, 1958; Repenning, Cooley, and Akers, 1969) in his map. The relationships of the Petrified Forest Member and the Mesa Redondo Member in the area of the *Placerias* Quarry are obscure, but it is apparent that the *Placerias* Quarry is in the Petrified Forest Member, low in the Chinle.

The *Placerias* Quarry was worked originally by field crews headed by C. L. Camp of the University of California in the years 1930 to 1934. During that time, thousands of large specimens were meticulously mapped and recorded as they were collected. Camp and Welles (1956) provide a detailed description of the *Placerias* Quarry. They state (p. 259):

> The quarry included two fossiliferous levels separated by 2 feet of barren sediments. The upper level contained masses of coprolites and a few fragmentary phytosaur and stegocephalian bones. Particles of carbon were here abundant in the grayish bentonite. In the lower level were about three thousand identifiable bones of reptiles uncommon elsewhere in the Chinle. The most abundant elements were the scutes of the pseudosuchian, *Typothorax*. The related genus *Episcoposaurus* (= *Desmatosuchus*) with much thicker scutes and long shoulder spines was also common. Skull fragments and phytosaur-like calcanei of two types probably belong to these pseudosuchians, as well as two types of vertebrae with short broadly expanded neural arches, short squarish centra, and elongate peglike diapophyses. Other slender vertebrae, skeletal elements, and fragmentary skulls seem referable to a small dinosaur. A large undetermined carnivorous reptile, possibly a pseudosuchian (or dinosaur), with a short stout maxillary and thecodont, phytosaur-like teeth, was the rarest element in the fauna. The common Chinle phytosaurs and amphibians were meagerly represented by a few fragmentary bones. No *Ceratodus* teeth or shells of *Unio* were found although elsewhere in the vicinity they are common.
>
> More than sixteen hundred dissociated bones of *Placerias* were mixed with five other kinds of reptiles on the quarry floor.
>
> The matrix encasing the bones was largely a soft, brownish-yellow, bentonitic clay which became in places the usual blocky grayish bentonite. Here it was more earthy and less consolidated than elsewhere in the Lower Chinle bentonites. This clay contained small chunks of dark brown and blackish carbonaceous matter ("coal") and lumpy masses of hard calcareous concretions from a few millimeters to 30 cm in diameter. Concretions were evidently formed in place since they enclosed many of the bones. Bones were

sometimes encased in crystalline gypsum and even the nodules were so encased, indicating secondary deposition of the gypsum.

. . . .

At the low northeast corner of the quarry was a 10-inch layer of dirty, dendritic gypsum. Elsewhere a thin layer of clear crystalline gypsum lay about 4 inches below the lower bone level. The floor of the quarry seems to have been the bottom of a shallow depression which sloped up around its edges to where little or no bone was found.

The heavy concentration of lime, beyond anything normal in the Chinle, would indicate a pool, marsh, or spring bog, subject to evaporation and receiving sediments intermittently. Patches of free sulfur and the abundant gypsum may have been formed by chemosynthetic bacterial action in the highly organic sediments. Concentration of small masses of carbonized vegetation, together with the yellow earth which may be an old pond mud, seems to harmonize with this view. The bones are sometimes partly rotted and broken, with skull elements almost entirely macerated along suture lines and widely separated. Some were pressed into the mud, probably by trampling; some flat dermal plates were found in a vertical position as if disturbed by rooting.

Most of the bones were scattered to form a pavement on the irregular quarry floor. Some few were in clusters and some had evidently been rotted dorsally and were better preserved where they had been pressed into the mud. Coprolites were numerous.

From this evidence we regard the bonebed as having been a soft, wet, vegetated pond bottom where reptiles congregated and fed on vegetation and probably on each other. There the skeletons underwent maceration even to the separation of tightly sutured skull elements, and most of the bones were widely scattered.

MNA field crews opened the unexcavated area adjacent to the northeastern limit of the UC quarry. The geology and occurrence of bones were essentially as described by Camp and Welles (1956); however, *Placerias* was less common in the area worked by MNA. Large bone was concentrated in two levels, but was reasonably abundant throughout the section, including below the gypsum layer underlying the lower bone level. Thin sand stringers rich in bones and coprolites comprised the upper bone level. Small bone was abundant throughout the sediments, but particularly rich in the sand stringers. Approximately 5.8 metric tons of sediment from the *Placerias* Quarry were wet-screened.

The *Placerias* Quarry is located in the northwestern extent of a low grass-covered hill or table. Bone is weathering out at several places on the table, particularly near the edges. Lag gravel and vegetation obscure most of the top of the table. The Downs Quarry was opened approximately 72 m east of the *Placerias* Quarry. Large bone was found on the surface down to about 80 cm below the surface (measured from the datum stake at the southwest corner of the excavation). Small bone was found for an additional 30 cm below the large bone. A carbonate layer is present about 20 cm below the lowest bone collected. This carbonate can be traced roughly to the *Placerias* Quarry. It overlies the upper bone level of the *Placerias* Quarry by approximately 3 m.

The microstratigraphy of the Downs Quarry is largely obscured by a deeply weathered Quaternary soil zone. Below the soil is gray bentonite underlain by 20 cm of yellow marly bentonite resting on the carbonate layer. Many of the bones from the Downs Quarry are encased in gray carbonate concretions which have an odor of hydrogen sulfide when broken. When treated with acetic acid, the concretions yield beautifully preserved plant fragments, invertebrates, and vertebrates. The plants do not appear to be mineralized. They are generally small fragments, some of which are pubescent. Cell walls can be seen in most specimens. The invertebrates include extremely fragile ?branchiopod crustaceans covered with delicate hair-like processes. The vertebrates from the Downs Quarry are represented by isolated elements. Small vertebrates occur in the concretions and additional small specimens have been recovered by screening approximately 70 kg of sediment. The scattered occurrence of bone in several places on the table near the *Placerias* and Downs quarries suggests the presence of numerous significant concentrations of bones.

Vertebrates of the *Placerias* and Downs Quarries

The fauna from the *Placerias* and Downs quarries provides one of the most complete pictures of the North American Upper Triassic vertebrate fauna. In this paper, we consider the fauna from the *Placerias* and Downs quarries to represent the vertebrate community that lived in the area at the time of deposition. According to E. C. Olson (1980, pp. 10–11):

> A *community* is defined in the ecological sense as a *group of organisms living together within a definite locality.* Further, living together is assumed to imply interaction of the constituent individuals either directly or by influence of organisms upon each other through their impacts on the common environment. The integrated structure of interactions of the elements of the system (e.g., trophic structure) defines the limits of the system both geographically and temporally. When the structure becomes seriously altered, the community as such ceases to exist. Although the last aspect, as well as the areal and compositional limits, is necessarily imprecise, the definition as a whole serves reasonably well as a working base for studies of fossil communities. . . .

The sediments of the *Placerias* and Downs quarries are characteristic of relatively low energy depositional environments. We agree with the interpretation of Camp and Welles (1956) that the *Placerias* Quarry represents a pond, marsh, or bog. There is no evidence that bones were washed into the deposits from appreciable distances.

Many of the bones show tooth marks. Coprolites, some of which contain small bone, are common. The association of taxa at the *Placerias* and Downs quarries, coupled with the positive evidence of tooth marks and coprolites, implies that the assemblage represents a community. Similar assemblages have been collected from the Chinle in Petrified Forest National Park, Arizona, (Jacobs, in press) and from the Dockum Group, Texas (Murry, in preparation).

The discussion below presents the individual members of the fauna in terms of their inferred roles in the community. The inferences presented here are tentative and preliminary; they can be tested by studies of functional anatomy and paleoecology. Several taxa discussed herein are also known from the Newark Group of eastern North America. Many of these taxa are illustrated by P. E. Olsen (this volume). Specimens forming the subject of this paper will be deposited in the

Museum of Paleontology, University of California, Berkeley. A faunal list for both quarries is given in Table 4.1.

Aquatic Omnivores

Included in this category are taxa assumed to be facultative in dietary habits. Their diets probably included a variety of tiny invertebrates and plants. Aquatic omnivores comprise hybodont sharks, palaeoniscoid fish, and lungfish.

A single tooth, several dorsal fin spines, and dermal denticles from the *Placerias* Quarry are assigned to the genus *Lonchidion*, a freshwater hybodont shark previously known from the Cretaceous of North America and Great Britain, and from the Upper Triassic Dockum Formation of Crosby County, Texas (Cappetta and Case, 1975; Estes, 1964; Patterson, 1966; Thurmond, 1971). These small sharks, with their batteries of crushing teeth, probably fed on aquatic vegetation, small mollusks, and crustaceans common in the Chinle environment. A small tooth recovered from the *Placerias* Quarry is similar to conservative hybodonts such as *Lonchidion breve breve*, *Lonchidion babulskii*, and the Dockum species. However, it is much smaller than the Cretaceous teeth, and may be distinct from the Dockum species because of its better developed labial and occlusal cusps.

A number of redfieldiid rostral bones recovered from the *Placerias* Quarry are similar to those found in the genera *Synorichthys, Lasalichthys, Cionichthys,* and *Redfieldius*. According to Schaeffer (1967), these distinctive, tuberculated rostral elements may have served as a point of insertion for a fleshy protrusible upper lip, useful for feeding upon detritus of the Chinle lakes and streams.

Specimens of *Cionichthys* have been found in the upper part of the Chinle Formation in San Juan County, Utah, and in Montrose and possibly San Miguel County, Colorado. Another species has been described from the Dockum Group near Otis Chalk, Texas (Schaeffer, 1967). In the fault basins of the eastern Triassic, *Cionichthys* has been found in the "Lower Barren Beds" of the Chatham Group, in the Cow Branch Formation, Dan River Group, and in the Lockatong Formation of the Newark Basin.

Synorichthys has been found in the upper Chinle Formation of San Juan County, Utah, and Montrose County, Colorado, and in the eastern Triassic basins from the Pekin and Cumnock Formations, Chatham Group, in the Cow Branch Formation, Dan River Group, from the New Oxford Formation of the Gettysburg Basin, and in the Lockatong and Passaic Formations of the Newark Basin (Schaeffer and McDonald, 1978; Olsen, McCune, and Thomson, in press). Partial skulls and skeletons recovered from the Dockum Group in Howard County, Texas, belong to either *Lasalichthys* or *Synorichthys* (Schaeffer, 1967). *Synorichthys* is found in beds of middle to late Carnian age, although its range may extend into the Norian (Olsen, McCune, and Thomson, in press).

The genus *Redfieldius* seems to be restricted to the Liassic (Hettangian-Sinemurian) of the East Coast (Schaeffer and McDonald, 1978). This genus is characterized by posteriorly denticulated scales, the most ventral denticle generally of much larger size than the others. No scales of this type were recovered from the *Placerias* and Downs quarries and it is assumed that the redfieldiids present represent genera other than *Redfieldius*.

A peculiar toothplate with mammillated striated teeth similar to the Triassic genus *Colobodus* was found in the *Placerias* Quarry. This genus is known from Europe,

TABLE 4.1
Fauna of the *Placerias* and Downs quarries
(asterisk indicates previously unknown from the Chinle fauna).

	Placerias Quarry	Downs Quarry
Class Chondrichthyes		
Subclass Elasmobranchii		
Order Xenacanthodii		
Family Xenacanthidae		
Xenacanthus moorei	*	*
Order Hybodontiformes		
Family Hybodontidae		
Lonchidion sp.	*	
Class Osteichthyes		
Subclass Actinopterygii		
Order Palaeonisciformes		
Family Palaeoniscidae		
Turseodus sp.	+	+
Family Redfieldiidae		
Cionichthys sp.	+	
Lasalichthys sp. and/or *Synorichthys* sp.	+	
Family Colobodontidae (?)	*	?
Order Semionotiformes		
Family Semionotidae	+	+
Subclass Sarcopterygii		
Order Dipnoi		
Family Ceratodontidae		
Ceratodus sp.	+	+
Class Amphibia		
Subclass Labyrinthodontia		
Order Temnospondyli		
Family Metoposauridae		
Metoposaurus sp.	+	+
Class Reptilia		
Subclass Cotylosauria		
Order Procolophonomorpha		
Family Procolophonidae	*	
Subclass Lepidosauria		
Order Eosuchia		
Family Tanystropheidae		
Tanytrachelos sp.	*	*
Subclass Archosauria		
Order Thecodontia		
Family Rauisuchidae	*	*
Family Stagonolepidae		
Typothoras sp. and/or *Desmatosuchus* sp.	+	+
Family Phytosauridae		
Phytosaurus sp.	+	+
?Family Ornithosuchidae		
Hesperosuchus agilis		+
Order Saurischia		
Family Podokesauridae		
Coelophysis sp.	+	+
Order Ornithischia		
Family Fabrosauridae	*	*

	Placerias Quarry	Downs Quarry
Subclass Euryapsida		
Order Protosauria		
Family Trilophosauridae		
Trilophosaurid sp. 1	*	*
Trilophosaurid sp. 2	*	*
Subclass Synapsida		
Order Therapsida		
Family Kannemeyeriidae		
Placerias gigas	+	+
Problematica	*	*

Asia, Africa, and Australia, but no previous indications of Upper Triassic colobo-dontids are known from North America. Other toothplates, similar in morphology but lacking striae, were found in both quarries.

Lungfish toothplates have been recovered sparsely throughout the Triassic of the Southwest. Toothplates from the St. Johns quarries show considerable variation. However, these differences may be ontogenetic rather than phylogenetic. Similarly, many of the "species" of *Ceratodus* described from the Southwest probably represent various stages in the life history of a single species. The transition from a cutting to a crushing dentition may reflect a change in food habits during the ontogeny of the lungfish (Branson and Mehl, 1931; Case, 1921; Varob'yeva, 1967; Warthin, 1928).

Aquatic Carnivores

The aquatic carnivores of the Chinle Formation are represented by four classes of vertebrates, including freshwater sharks, bony fishes, amphibians, and reptiles. Numerous tricusped teeth of the predaceous shark *Xenacanthus* were found in the *Placerias* and Downs quarries. Although xenacanth sharks were probably common in the Upper Triassic of North America, their tiny teeth have been found only in those sediments that have been screen-washed: the Chinle and Dockum formations. The teeth of these sharks are referred to the species *Xenacanthus moorei*, first described from the Keuper of Somersetshire, England, and found in the Gipskeuper of Gaildorf, Germany, and the Maleri Formation of India (Jain, et al., 1964; Woodward, 1889; Seilacher, 1943; Johnson, in press).

At least two species of predaceous actinopterygian fish are present in the Chinle near St. Johns. Semionotids, represented by numerous scales at both quarries, were similar to *Amia* in jaw structure. However, these fish possessed a rather small gape and were probably not as effective in capturing larger prey as modern teleosts (P. E. Olsen, personal communication). Semionotids of Upper Triassic age have been found in the upper Chinle Formation in Utah and Colorado, in the Upper Member of the Cow Branch Formation, Dan River Group, and in the Lockatong and Passaic Formations of the Newark Basin (Olsen, McCune, and Thomson, in press).

Ring-shaped centra and ridged scales, referred to the palaeoniscid genus *Turseodus*, were recovered from both the *Placerias* and Downs quarries. Although the jaw morphology of *Turseodus* was not as complex as that in the semionotids, *Turseodus* was a common and successful predator, as suggested by its numerous remains at many Triassic localities. *Turseodus* has been found in Montrose County, Colorado, in the "Durham Basin Lacustrine Beds" of the Chatham Group, in the Upper Member of the Cow Branch Formation, Dan River Group, in the New Oxford Formation of the Gettysburg Basin, and in the Lockatong Formation of the Newark Basin (Bock, 1959; Leidy, 1857; Olsen, McCune, and Thomson, in press; Schaeffer, 1952, 1967; Schaeffer and McDonald, 1978).

Numerous isolated elements of metoposaurs were found in both quarries, including jaws, dermal armor, and vertebrae. These amphibians, the terminal members of the labyrinthodonts, had large skulls and jaws equipped with several rows of formidable teeth for trapping fish in their huge mouths. Metoposaurs were surely aquatic, as their poorly-ossified tiny limbs would almost certainly be useless for terrestrial locomotion (Colbert, 1948, 1960a, 1974; Gregory, 1972). Colbert and

Imbrie (1956) synonymized most of the nominal genera of metoposaurs with *Eupelor* (now considered *Metoposaurus*, see Roy-Chowdhury, 1965). A tiny clavicle was recovered in screening the *Placerias* Quarry. This specimen probably represents a juvenile. With a length of 1.27 cm, it is much smaller than clavicles reported by Colbert and Imbrie (1956) or Sawin (1945).

Metoposaurs appear to be restricted to Carnian age formations from North America, Europe, India, and North Africa where they are common fossils. Interestingly, they have not been found in South America or Australia (Carroll, 1977; Roy-Chowdhury, 1965; Colbert, 1967; Dutuit, 1976).

A number of small procoelous vertebrae and plow-shaped cervical ribs referred to the eosuchian genus *Tanytrachelos* were found in both the *Placerias* and Downs quarries. *Tanytrachelos* has been reported from the late Carnian Dan River Group of North Carolina and the lower Lockatong Formation of the Newark Basin (Olsen, 1979). The remains from the Chinle Formation represent the first tanystropheids discovered from western North America. *Tanytrachelos* was probably aquatic. It was long-tailed and had long hind limbs. The presence of crescent-shaped heterotopic bones lateral to the anterior caudal vertebrae reported by Olsen (1979) indicates affinities with European *Tanystropheus,* and suggests the presence of hemipenes or similar structures.

Several partial phytosaur skulls were found in the Downs Quarry and numerous phytosaur teeth illustrating marked heterodonty were recovered from both quarries. The heavy rostrum and pronounced rostral crest exhibited by these skull fragments are characteristic of the genus *Phytosaurus* (= *Nicrosaurus, fide* Westphal, 1976), which has previously been found in numerous Chinle localities (Colbert, 1947; Stovall and Savage, 1939; Camp, 1930; Camp and Welles, 1956). These large-snouted forms were probably similar in habits to the modern Nile crocodile and the American alligator. They probably preyed on a variety of fishes, as well as tetrapods that assembled at the Triassic ponds and streams.

Terrestrial Herbivores

Few taxa of terrestrial herbivores have been reported from any given site in the North American Triassic. Screening and quarrying the Chinle Formation has produced a variety of both small and large herbivores from the *Placerias* and Downs quarries.

A small jaw fragment containing two transversely oriented bicusped teeth recovered from the *Placerias* Quarry has been assigned to the Procolophonidae, a family of small herbivorous or possibly omnivorous reptiles. This jaw fragment and jaw fragments and a fragmentary palate with four bicusped teeth from the Dockum of Texas indicate that several types of procolophonids lived in the Southwest during the Upper Triassic. These fragments differ from *Hypsognatus* in having chisel-shaped teeth (Gilmore, 1928; Colbert, 1946). Several types of procolophonids with similar dentitions have been recovered from Nova Scotia, but have not been described (Baird and Take, 1959; Carroll, et al., 1972). Colbert (1960b) described the procolophonid *Sphodrosaurus* from the Brunswick Formation of Pennsylvania. *Sphodrosaurus* is known from the postcranial skeleton and part of the skull, but no dentition.

A fragmentary dentary and numerous isolated teeth indicate the presence of a new species of trilophosaurid. These specimens are similar to teeth found in the Dockum Formation of Texas. The isolated teeth comprise two kinds. The first set has cingula connecting the cusps. The dentary fragment contains portions of five teeth which indicate that the anterior teeth in this species are peglike. An isolated bulbous tooth is possibly the most posterior tooth of this species. This suggests affinities with *Variodens inopinatus* from the Emborough Quarry, Somersetshire, England, of Upper Triassic age (Robinson, 1957). Baird (1963) reported a trilophosaurid similar to *Variodens* from the Wolfville Formation of Nova Scotia.

The second set of trilophosaurid teeth is comparable in morphology to the type specimen of *Trilophosaurus buettneri*. Although some teeth are similar in size to the type of *Trilophosaurus*, many teeth recovered from both the *Placerias* Quarry and from the Dockum of Crosby County, Texas, are smaller than those from quarries near Otis Chalk, Howard County, Texas (Case, 1928a,b; Gregory, 1945).

Desmatosuchus was described from the Dockum of Texas by Case (1920). Since that time, it has been reported from Dickens and Crosby counties, Texas, and from northeastern Arizona (Brady, 1954, 1958; Camp and Welles, 1956). Skeletal material attributed to *Typothorax* has been found both in the Chinle and Dockum formations in Texas, New Mexico, Arizona, and Utah. The relationships of *Typothorax* and *Desmatosuchus* have been debated (see Gregory, 1953). A shoulder spine of *Desmatosuchus* was found in the *Placerias* Quarry. Scutes belonging either to *Typothorax* or *Desmatosuchus* have been found in both quarries.

Teeth tentatively referred to the Ornithischia were found in the *Placerias* and Downs quarries. These teeth are similar to fabrosaurids found in the Upper Triassic and Lower Jurassic of North America, England, Africa, and possibly China (Crompton and Charig, 1962; Galton, 1978; Simmons, 1965; Thulborn, 1974). In North America, remains referrable to Triassic-Liassic ornithischians have been found in the Sanford Basin of North Carolina (P. E. Olsen, personal communication), the Wolfville Formation of Nova Scotia (Carroll, et al., 1972), in the Kayenta Formation of Arizona (Colbert, in press), and from the Emigsville copper mine in the New Oxford Formation, York County, Pennsylvania. Some of the teeth from the St. Johns quarries are similar to those found in the Dockum Group, Texas, and show affinities to "*Thecondontosaurus*" *gibbidens* from the New Oxford Formation. There are several distinct morphologic groups of teeth in the Dockum and Chinle, probably representing several herbivorous taxa.

The *Placerias* Quarry was named for the abundant remains of the ox-like dicynodont *Placerias*, which has been found at both quarries near St. Johns. The genus is rare, but occurs in the Popo Agie Formation of Wyoming, the Chinle Formation of Arizona, and the Pekin Formation of North Carolina (Camp and Welles, 1956; Baird, 1967). It has been suggested that *Placerias* was a large herbivore which used its edentulous beaked jaws in digging for roots and rhizomes (Camp and Welles, 1956). Rowe (1979) has speculated on the intraspecific behavior of *Placerias*.

Terrestrial Carnivores

Both large and small terrestrial carnivores have been found near St. Johns, representing three suborders of Triassic reptiles. A straight and slender radius found in the Downs Quarry is referred to the species *Hesperosuchus agilis*. This small pseudosuchian was previously known only from the Ward Bone Bed near Cameron, Arizona. *Hesperosuchus* was evidently an active little predator, with sharp teeth, strong hind limbs, and a bipedal gait (Colbert, 1952).

Isolated teeth and vertebrae probably belonging to the small theropod dinosaur

Coelophysis were found at both quarries. *Coelophysis* is a well-known and broadly distributed genus with two species recognized by Colbert (1964). *Coelophysis bauri* is found at many localities in the Chinle and Dockum formations of western North America. *Coelophysis (Podokosaurus) holyokensis,* based on a single partial skeleton described by Talbot (1911), was found in a glacial boulder on the campus of Mt. Holyoke College in Holyoke, Massachusetts. Welles (1972) mentions the presence of *Coelophysis* at the *Placerias* Quarry.

A beautifully preserved rauisuchid braincase was recovered from the Downs Quarry. Large, compressed, serrated teeth from the *Placerias* Quarry may belong to the same taxon. Camp and Welles (1956) reported a large carnivorous reptile from the *Placerias* Quarry that may be this rauisuchid. The rauisuchids have been described from rocks of Anisian to Carnian age in South America (Brazil and Argentina), East Africa (Tanzania), Europe (Switzerland), the United States, and possibly China (Dawley, Zawiskie, and Cosgriff, 1979; Sill, 1974). The rauisuchids were large (3–6 m in length) quadrupedal carnivores, which probably preyed on saurischians, ornithischians, and other reptiles during Chinle times.

Problematica

A major disadvantage of collecting fossils by screening is the loss of association of parts. Many specimens from the *Placerias* and Downs quarries are too fragmentary to identify, even tentatively, until more complete material is found and more detailed comparisons are made. Among the indeterminate specimens are some peculiar saddle-shaped vertebrae and numerous other postcranial elements. Most indeterminate specimens, however, are teeth and jaw fragments of fish and reptiles. Some jaw fragments have pleurodont teeth reminiscent of lizards. Others resemble sphenodontids. Still other teeth are pierced by a canal. Some flattened multicuspid teeth generally resemble small therapsid and pterosaur teeth. The sharp trenchant shape of many of the problematical teeth suggests that they may represent small insectivores.

Discussion

Colbert (1972) listed seventeen vertebrate taxa from the Chinle Formation. Not including problematica, this paper lists a total of twenty-one taxa. Only four taxa listed by Colbert have not been identified from the *Placerias* and Downs quarries: the palaeoniscid *Tanaocrossus,* the semionotid *Hemicalypterus,* the coelacanth *Chinlea,* and the phytosaur *Rutiodon.* In addition, Carroll, et al. (1972), mention an undescribed large cynodont from the Chinle which has not been recovered from the *Placerias* or Downs quarries. Nine of the taxa identified here are new to the Chinle. There are several unidentified problematical taxa. We do not feel it is an exaggeration to state that the vertebrate fauna of the Chinle has been essentially doubled as a result of work at the Downs and *Placerias* quarries.

The fauna of the St. Johns quarries presents a more complete picture of the Chinle vertebrate fauna than has previously been realized. Figure 4.1 is an interpretation of the feeding relationships and energy flow in the Chinle community near St. Johns. It is admittedly oversimplified, but it may be realistic at the level of resolution at which it is used here. Taxa grouped together in boxes are to simplify the figure. They are assumed to play similar roles in the community. For example, trilophosaurids and procolophonids are small terrestrial herbivores. Even though

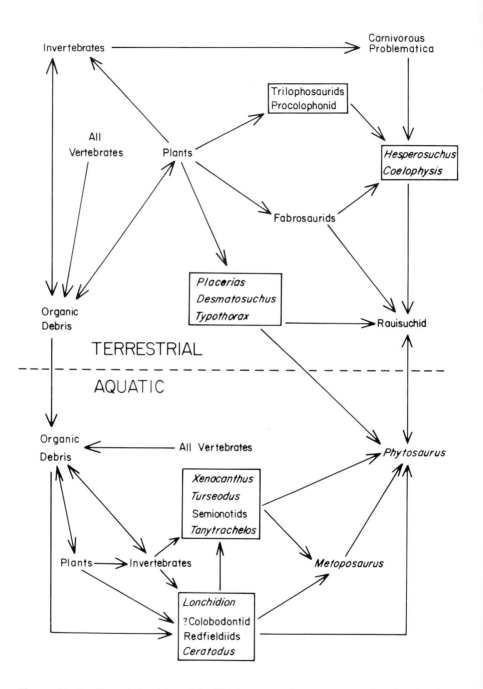

Figure 4.1 Feeding relationships of the Chinle community as represented at the *Placerias* and Downs quarries. Arrows indicate direction of energy flow. Taxa grouped together probably had similar dietary habits.

they are considered terrestrial herbivores here, we realize that they were probably not strictly obligatory in their diets and may have occasionally taken, say, a juicy cockroach.

The carnivores probably ate anything they could catch and subdue, including others of their own species. Thus, size was probably an important factor in determining the preferred prey of a carnivore. For example, *Desmatosuchus* was probably not the preferred prey of *Hesperosuchus,* simply because its size would have made it difficult for *Hesperosuchus* to subdue. However, *Hesperosuchus* may very well have taken young (i.e., small) or eggs of *Desmatosuchus.*

Many of the vertebrates included here in the terrestrial community probably fed on plants or insects along shorelines or in shallows. In such a case, phytosaurs would certainly have preyed on some of them, particularly the larger ones such as *Placerias.* On the other hand, the large rauisuchid may have taken the occasional phytosaur, and *Hesperosuchus* and *Coelophysis* may have preyed opportunistically on small phytosaurs or stranded metoposaurs. Carrion may have been taken by a variety of the vertebrates.

Exchange between the terrestrial and aquatic ecosystems was apparently a significant aspect of community structure. This was accomplished by the transfer of organic debris (detritus, carcasses, feces, etc.) from the terrestrial to the aquatic systems, and by predation. In this context, the role of the dominant aquatic carnivores could be viewed as simply one more way to introduce organic debris into the aquatic ecosystem. Terrestrial and aquatic ecosystems were closely linked, rather than divisible into upland and lowland communities. Thus, ecological factors that influenced energy flow in the community were also important taphonomic factors in the formation and compositon of the fossil deposit.

The contribution of this paper to understanding the Chinle community is in documenting both the dominant terrestrial carnivore and the smaller vertebrates important at lower trophic levels. The latter of these would not have been accomplished without the application of wet-screening techniques.

We make no claim that all smaller vertebrate taxa of the Chinle have been recovered. Certainly much remains to be done. However, we have demonstrated the utility of wet-screening in Chinle sediments. As a result, the biostratigraphy of the Chinle can potentially be refined. Placed in the context of evolving communities and community succession (*fide* E. C. Olson, 1980), we can look forward to a better understanding of the Late Triassic origination and diversification of some major vertebrate groups, particularly teleosts, frogs, lizards, and mammals.

Acknowledgments

We have been assisted in this endeavor by many people; to all we are grateful. Mrs. Dorathalene Tuckness allowed us access to her land, on which the *Placerias* and Downs quarries are located. The field crew consisted of Will Downs, Larry Flynn, Rich Citelli, Yuki Tomida, Alex McCord, Sheila Donovan, Bonnie Fine, Francine Bonnello, Laura O'Hara, Scott Madsen, Edgar Roberts, and Dave Lawler. Special thanks are due Will Downs, foreman of the crew and preparator in charge.

Rupert Wild, Donald Baird, Karl Flessa, Gary Johnson, Paul Olsen, and Hans-Dieter Sues have been particularly helpful in sharing their opinions on identifications, although we are, of course, completely responsible for errors. Paul Olsen, currently studying the fauna and flora of the Newark Supergroup, has openly provided any and all assistance requested, and more. We happily acknowledge his influence on us. M. E. Cooley and C. A. Repenning discussed the stratigraphy of the area with us. Joseph T. Gregory, Donald Baird, Paul Olsen, Will Downs, Rich Cifelli, Bob Slaughter, and Catherine Badgley reviewed the manuscript. Field work was supported by a National Geographic Society grant to LLJ.

Literature Cited

Akers, J. P. 1964. Geology and ground water in the central part of Apache County, Arizona. U. S. Geol. Surv., Water Supply Paper, 1771:1–107.

Baird, D. 1963. Rhynchosaurs in the Late Triassic on Nova Scotia. Geol. Soc. Amer., Spec. Paper, 73:107.

———. 1967. Dicynodont-Archosaur fauna in the Pekin Formation (Upper Triassic) of North Carolina. Geol. Soc. Amer., Abs., p. 11.

Baird, D., and Take, W. F. 1959. Triassic reptiles from Nova Scotia. Geol. Soc. Amer., Bull., 70:1565–66.

Bock, W. 1959. New eastern American Triassic fishes and Triassic correlation. Phila., Acad. Natur. Sci., Geol. Center Res. Ser., 1:1–184.

Brady, L. F. 1954. *Desmatosuchus* in northern Arizona. Plateau, 27(1):19–21.

———. 1958. New occurrence of *Desmatosuchus* in northern Arizona. Plateau, 30(3):61–63.

Branson, E. B., and Mehl, M. G. 1931. Fish remains of the western interior Triassic. Geol. Soc. Amer., Bull., 42:330–31.

Camp, C. L. 1930. A study of the phytosaurs, with description of new material from western North America. Calif., Univ., Mem., 10:1–161.

Camp, C. L., and Welles, S. P. 1956. Triassic dicynodont reptiles. Calif., Univ., Mem., 13(4): 255–348.

Cappetta, H., and Case, G. R. 1975. Contribution a l'étude des Sélachians du Groupe Monmouth (Campanien-Maestrichtien) du New Jersey. Palaeontographica, Abt. A, 151:1–46.

Carroll, R. L. 1977. Patterns of amphibian evolution; an extended example of the incompleteness of the fossil record. *In* A. Hallam (ed.), Patterns of evolution as illustrated by the fossil record. Elsevier, Amsterdam, pp. 405–37.

Carroll, R. L.; Belt, E. S.; Dineley, D. L.; Barid, D.; and McGregor, D. C. 1972. Vertebrate paleontology of eastern Canada. XXIV Int. Geol. Congress, Montreal, Quebec, 113 pp.

Case, E. C. 1920. Preliminary description of a new suborder of phytosaurian reptiles, with a description of a new species of *Phytosaurus*. J. Geol., 28(6):524–35.

———. 1921. A new species of *Ceratodus* from the Upper Triassic of western Texas. Mich., Univ., Mus. Zool., Occ. Pap., 101:1–2.

———. 1928a. A cotylosaur from the Upper Triassic of western Texas. J. Wash. Acad. Sci., 18:177–78.

———. 1928b. Indications of a cotylosaur and of a new form of fish from the Triassic beds of Texas, with remarks on the Shinarump conglomerate. Mich., Univ., Mus. Paleontal., Contrib., 3:1–14.

Colbert, E. H. 1946. *Hypsognathus*, a Triassic reptile from New Jersey. Amer. Mus. Natur. Hist., Bull., 86:225–74.

———. 1947. Studies of the phytosaurs *Machaeroprosopus* and *Rutiodon*. Amer. Mus. Natur. Hist., Bull., 88:53–96.

———. 1948. Triassic life in the southwestern United States. N. Y. Acad. Sci., Trans., ser. 2, 10:229–35.

———. 1952. A pseudosuchian reptile from Arizona. Amer. Mus. Natur. Hist., Bull., 99: 565–92.

———. 1960a. Triassic rocks and fossils. N. Mex. Geol. Soc., Guidebk, 11th Field Conf., pp. 55–62.

———. 1960b. A new Triassic procolophonid from Pennsylvania. Amer. Mus. Nov., 2022:1–19.

———. 1964. The Triassic dinosaur genera *Podokesaurus* and *Coelophysis*. Amer. Mus. Nov., 2168:1–12.

————. 1967. A new interpretation of *Austropelor,* a supposed Jurassic labyrinthodont amphibian from Queensland. Queensl. Mus., Mem., 15(1):35–41.

————. 1972. Vertebrates from the Chinle Formation. Mus. North. Ariz., Bull., 47:1–11.

————. 1974. The Triassic paleontology of Ghost Ranch. New Mex. Geol. Soc., Guidebk., 25th Field Conf., Ghost Ranch, pp. 175–78.

————. In press. A primitive ornithischian dinosaur from the Kayenta Formation of Arizona. Mus. North. Ariz., Bull.

Colbert, E. H., and Imbrie, J. 1956. Triassic metoposaurid amphibians. Amer. Mus. Natur. Hist., Bull., 110:399–452.

Cooley, M. E. 1958. The Mesa Redondo member of the Chinle Formation, Apache and Navajo Counties, Arizona. Plateau, 31(1):7–15.

Crompton, A. W., and Charig, A. J. 1962. A new ornithischian from the Upper Triassic of South Africa. Nature, 196:1074–77.

Dawley, R. M.; Zawiskie, J. W.; and Cosgriff, J. W. 1979. A rauisuchid thecodont from the Upper Triassic Popo Agie Formation of Wyoming. J. Paleontol., 53:1428–31.

Dutuit, J. M. 1976. Introduction à l'étude paléontologique du Trias continental marocain. Description des permiers stegocephales recueillis dans le couloir d'Argana (Atlas Occidental). Mus. Nat. Hist. Natur., Paris, Mem., Sér. C, 36:1–253.

Estes, R. 1964. Fossil vertebrates from the Late Cretaceous Lance Formation, eastern Wyoming. Calif., Univ., Publ. Geol. Sci., 49:1–180.

Galton, P. 1978. Fabrosauridae, the basal family of ornithischian dinosaurs (Reptilia, Ornithopoda). Palaeontol. Z., 52:138–59.

Gilmore, C. W. 1928. A new fossil reptile from the Triassic of New Jersey. U. S. Nat. Mus., Proc., 73(7):1–8.

Gregory, J. T. 1945. Osteology and relationships of *Trilophosaurus.* Tex., Univ., Publ., 4401: 273–359.

————. 1953. *Typothorax* and *Desmatosuchus.* Postilla, 16:1–27.

————. 1972. Vertebrate faunas of the Dockum Group, Triassic, Eastern New Mexico and West Texas. N. Mex. Geol. Soc., Guidebk., 23rd Field Conf., pp. 120–23.

Jacobs, L. L. In press. Additions to the Triassic vertebrate fauna of Petrified Forest National Park, Arizona. J. Ariz. Nev. Acad. Sci.

Jain, S. L.; Robinson, P. L.; and Roy-Chowdhury, T. K. 1964. A new vertebrate fauna from the Triassic of the Deccan, India. Quart. Geol. Soc. London., Quart. J., 120:115–24.

Johnson, G. D. In press. Xenacanthodii (Chondrichthyes) from the Tecovas Formation (Upper Triassic) of west Texas. J. Paleontol.

Leidy, J. 1857. Notices of some remains of extinct fishes. Phila., Acad. Natur. Sci., Proc., 9:167–68.

Olsen, P. E. 1979. A new aquatic eosuchian from the Newark Supergroup (Late Triassic-Early Jurassic) of North Carolina and Virginia. Postilla, 176:1–14.

Olsen, P. E.; McCune, A. R.; and Thomson, K. S. In press. Correlation of the Newark Supergroup by vertebrates, especially fishes. Amer. J. Sci.

Olson, E. C. 1980. Taphonomy: Its history and role in community evolution. *In* A. K. Behrensmeyer and A. Hill (eds.), Fossils in the making; vertebrate taphonomy and paleoecology. University of Chicago Press, Chicago, pp. 5–19.

Patterson, C. 1966. British Wealden sharks. Brit. Mus. (Natur. Hist.), Bull., Geol., 11:283–350.

Repenning, C. A.; Cooley, M. E.; and Akers, J. P. 1969. Stratigraphy of the Chinle and Moenkopi Formations, Navajo and Hopi Indian Reservations, Arizona, New Mexico, and Utah. U. S. Geol. Surv., Prof. Paper, 521-B:1–34.

Robinson, P. L. 1957. An unusual sauropsid dentition. Linn. Soc., Zool. J., 43:283–93.

Rowe, T. 1979. *Placerias,* an unusual reptile from the Chinle Formation. Plateau, 51(4):30–32.

Roy-Chowdhury, T. 1965. A new metoposaurid amphibian from the Upper Triassic Maleri Formation of Central India. Roy. Soc. London, Phil. Trans., Ser. B, 250:1–52.

Sawin, H. J. 1945. Amphibians from the Dockum Triassic of Howard County, Texas. Tex., Univ., Publ., 4401:361–99.

Schaeffer, B. 1952. The palaeoniscoid fish *Turseodus* from the Upper Triassic Newark Group. Amer. Mus. Nov., 1581:1–24.

————. 1967. Late Triassic fishes from the western United States. Bull. Amer. Mus. Natur. Hist., Bull., 135:287–342.

Schaeffer, B., and McDonald, N. G. 1978. Redfieldiid fishes from the Triassic-Liassic Newark Supergroup of eastern North America. Bull. Amer. Mus. Natur. Hist., Bull., 159:133–73.

Seilacher, A. 1943. Elasmobranchier-Reste aus dem oberen Muschelkalk und dem Keuper Wurtemburgs. Neues Jahrb. Geol. Paläotol., Abh., B:256–71, 273–92.

Sill, W. D. 1974. The anatomy of *Saurosuchus galilei* and the relationships of the rauisuchid thecodonts. Harvard Univ., Mus. Comp. Zool., Bull., 146:317–62.

Simmons, D. J. 1965. The non-therapsid reptiles of the Lufeng Basin, Yunnan, China. Fieldiana, Geol., 15:1–93.

Stovall, J. W., and Savage, D. E. 1939. A phytosaur in Union County, New Mexico. J. Geol, 47:759–66.

Talbot, M. 1911. *Podokesaurus holyokensis*, a new dinosaur from the Triassic of the Connecticut Valley. Amer. J. Sci., 31:469–79.

Thulborn, R. A. 1974. A new heterodontosaurid dinosaur (Reptilia: Ornithischia) from the Upper Triassic Red Beds of Lesotho. Linn. Soc. London, Zool. J., 55:151–75.

Thurmond, J. T. 1971. Cartilaginous fishes of the Trinity Group and related rocks (Lower Cretaceous) of North Central Texas. Southeastern Geol., 13:217–18.

Varob'yeva, E. I. 1967. Triassic ceratod from South Pergana and remakes on the systematics and phylogeny of ceratodontids. Paleontol. J., 4:80–87.

Warthin, A. S. 1928. Fossil fishes from the Triassic of Texas. Mich., Univ., Mus. Paleontol., Contrib., 3:15–18.

Welles, S. P. 1972. Fossil-hunting for tetrapods in the Chinle Formation: a brief pictorial history. Mus. North. Ariz., Bull., 47:13–18.

Westphal, F. 1976. Phytosauria, *In* O. Kuhn (ed.), Encyclopedia of paleoherpetology. Gustav Fischer Verlag, New York, pp. 99–120.

Woodward, A. S. 1889. Paleichthyological notes. 2. On *Diplodus moorei*, sp. nov., from the Keuper of Somersetshire. Ann. Mag. Natur. Hist., 6:297–302.

Figure 4.2. Ned Colbert, with badlands of Chinle Formation in background, near St. Johns, Arizona, 1946.

JURASSIC TETRAPODS FROM SOUTH AMERICA AND DISPERSAL ROUTES

by J. F. Bonaparte

Introduction

The collections and knowledge of South American Mesozoic tetrapods have been greatly improved during the last twenty years. Several institutions and men were (and are) involved in this endeavor, notably the Fundación M. Lillo-Universidad Nacional de Tucumán, Argentina; the Museum of Comparative Zoology, Harvard University; and the Instituto de Geociencias, Universidade Federal do Rio Grande do Sul, Brazil. Ned Colbert led an expedition from the American Museum of Natural History to southern Brazil, and published on the oldest known dinosaur (Colbert, 1970) from South America (upper Middle Triassic), collected there by a previous expedition from the Museum of Comparative Zoology.

Improvements in knowledge dealt mainly with Triassic and Cretaceous discoveries, while the Jurassic, so richly documented in North America, remained almost unknown in South America, except for several isolated and fragmentary finds of ichthyosaurs along the southern Andes, a fragmentary cetiosaurid sauropod in Patagonia (Cabrera, 1947), and the ascaphid anuran *Notobatrachus* (Stipanicic and Reig, 1955). However, the Jurassic record became better documented during the sixties and seventies with the discovery of an assemblage of clear footprints of small dinosaurs and mammals (Casamiquela, 1961a), the recent discovery of footprints of tetrapods in northeastern Brazil (Leonardi, in press a), and the discovery of well-preserved remains of marine crocodiles in Argentina (Gasparini and Dellapé, 1976) and Chile (Gasparini and Chong, 1977). Since 1975, the National Geographic Society, Washington, D. C., has sponsored a project on Jurassic and Cretaceous vertebrates under which was discovered a rich Middle Jurassic locality in Patagonia with well-preserved remains of carnosaurs and sauropods (Bonaparte, 1979).

Four assemblages of Jurassic tetrapods are documented in South America. They are formed by two or more species recorded from the same or similar levels of one place, or small area. For these I use the term "local fauna" as recommended by Simpson (1971). A fifth "assemblage" is recorded from marine deposits of Middle and Upper Jurassic age of the Neuquén province. Finally, a few species are represented by isolated finds (Figure 5.1).

Roca Blanca Local Fauna (Middle or Upper Liassic)

This fauna from Santa Cruz province, Patagonia, comprises a specimen of primitive Ascaphidae and a supposed fragmentary lacertilian. Both specimens, and a few undescribed fish remains, were collected by R. Herbst while working on his geological and paleontological thesis.

Vieraella herbstii, an ascaphid anuran, is of particular significance in understand-

Figure 5.1 Map of South America with indications of tetrapod localities. ①, Roca Blanca local fauna; ②, Cerro Cóndor local fauna; ③, Laguna Manantiales local fauna (ichno-fauna); ④, Araraquara local fauna (ichnofauna); ⑤, Jurassic fauna of Neuquén. 1, *Noto-batrachus;* 2, *Amygdalodon;* 3, *Macropterygius;* 4, *"Purranisaurus";* 5, several discoveries of ichthyosaurs; 6, *Metriorhynchus;* 7, *ornithischian footprints;* 8, cetiosaurid *indet.*

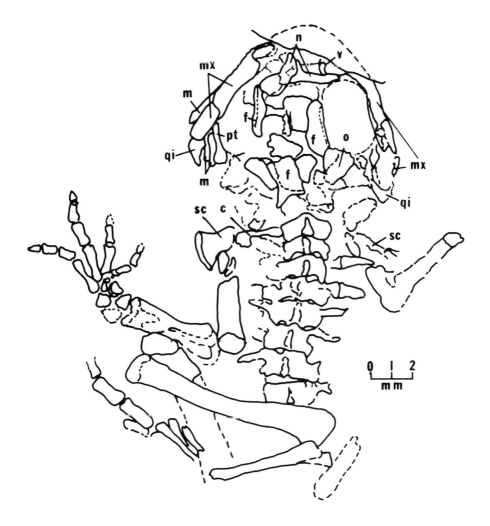

Figure 5.2. *Vieraella herbstii,* a primitive anuran from the Lower Jurassic of Argentina. Mold of the dorsal surface of the type specimen. (After Estes and Reig, 1973.)

ing the early history and distribution of anuran evolution. Actually it is the oldest known species of the Euanura. It is a small individual, 25 mm long, with anatomical features sufficient to understand its systematics and phylogenetic relationships (Figure 5.2). After a brief paper by Reig (1961), Casamiquela and Herbst discovered the counterpart impression of the type of *Vieraella,* and Casamiquela (1965) published a new study on *Vieraella herbstii* based on both impressions of the same individual. Estes and Reig (1973) restudied the species and discussed in detail both morphological and phylogenetic aspects of this primitive Euanura.

Protolacerta patagonica, considered Lacertilia *incertae sedis,* is based on two groups of impressions of small bones supposedly from the same individual in two separate rocks. A group of molds corresponds to some vertebrae, a femur, and fragmentary humerus, radius, and ulna. The vertebrae, 13 mm in height and 5 mm in length, bear flat and wide neural spines in lateral view, with ridges bordering hollows near the transverse processes. My observations of the two specimens did not identify the pelvis remains as indicated and reconstructed by Casamiquela (1975a). The systematic position of these remains is not definite, although they eventually may prove to be lacertilian. At present I would consider *Protolacerta patagonica* as a possible member of the Lacertilia.

Cerro Cóndor Local Fauna (Middle Jurassic, probably Callovian)

This assemblage was recently discovered by the author and associates in Chubut province, Patagonia (Figure 5.1). The fossil-bearing unit corresponds to the Cañadón Asfalto Formation, with an algal flora, conchostracans, gastropods, and pelecypods (Tasch and Volkheimer, 1970), a macroflora (Stipanicic and Bonetti, 1970), and fishes (Pascual, et al., 1969). The tetrapods are represented by several incomplete specimens of two species of cetiosaurid sauropods and one megalosaurid carnosaur.

Patagosaurus fariasi (Figure 5.3) is the most abundant species in this local fauna. It is a huge cetiosaurid with known femora up to 152 cm. The remains of about seven specimens have provided information on the greater part of the postcranial skeleton. *P. fariasi* shows more pronounced "pleurocoels" than the Bathonian species from Patagonia, *Amygdalodon patagonicus* described by Cabrera (1947). It seems to be more primitive than the Morrison species *Haplocanthosaurus priscus* (McIntosh, in press), as shown by the morphology of the dorsal vertebrae, with less pronounced pleurocoels and a shorter distance between the transverse process and the base of the neural arch. *P. fariasi* resembles most closely the Callovian species *Cetiosaurus leedsi* from England, with a few differences in the morphology of the ischium and the shape of the neural spines of the caudals. *Patagosaurus fariasi* shows a large expansion in the neural canal of the sacral vertebrae, involving the second, third (with the biggest cavity), and fourth sacrals. The few isolated teeth of *P. fariasi* are similar to those of *Camarasaurus.*

Volkheimeria chubutensis is a cetiosaurid known from an incomplete skeleton. It has dorsal vertebrae different from those of *P. fariasi,* with lower neural arches and with lateral compressed neural spines (Figure 5.4) as in *Bothriospondylus* sp. from the Bathonian of Madagascar (Oigier, 1975). The ischium is primitive as in *Cetiosaurus leedsi* but with the distal end massive. The ilium is shorter and higher than in *P. fariasi* (see Figure 5.4). This species is undoubtedly a cetiosaurid, probably more

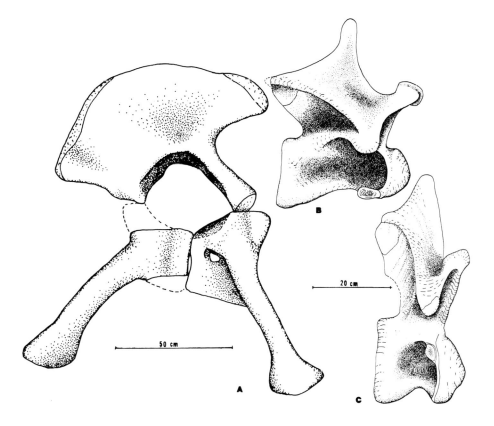

Figure 5.3. *Patagosaurus fariasi,* a cetiosaurid from the Callovian of Argentina. A, pelvis in lateral view; B, a posterior cervical in lateral view; and C, an anterior dorsal in lateral view.

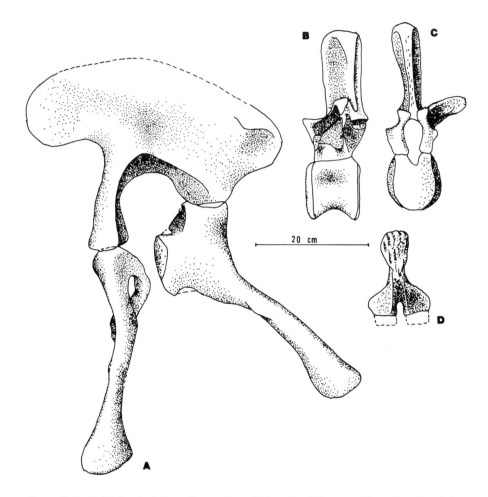

Figure 5.4. *Volkheimeria chubutensis,* a cetiosaurid from the Callovian of Argentina. A, pelvis in lateral view; B and C, a dorsal vertebra in lateral and anterior views; and D, a neural spine of a sacral in anterior view.

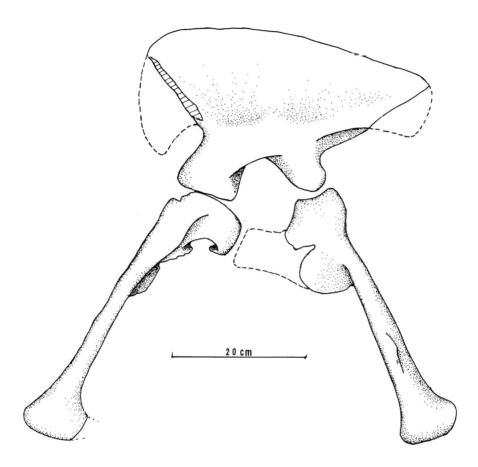

Figure 5.5. *Piatnitzkysaurus floresi,* a megalosaurid from the Callovian of Argentina. Pelvis in lateral view.

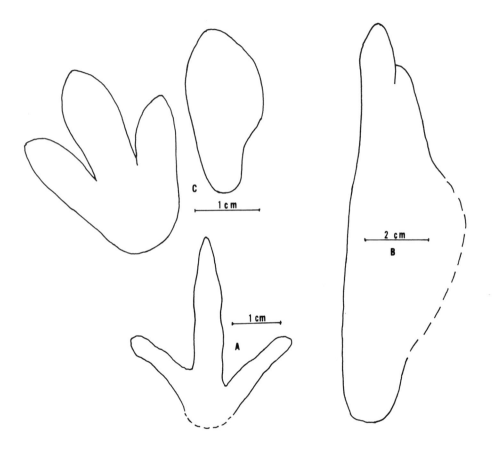

Figure 5.6 Coelurosaurian footprints of the Laguna Manantiales local fauna, Oxfordian of Argentina. A, *Wildeichnus navesi;* B, *Sarmientichnus scagliai;* C, *Delatorrichnus goyenechei.*

primitive than *Cetiosaurus* and *Patagosaurus*, related to but different from *Bothrio-spondylus*, particularly in the morphology of the ischium.

Piatnitzkysaurus floresi is a carnosaur of the family Megalosauridae represented by a good part of a skeleton of modest size (femur 55 cm). It is related to *Allosaurus fragilis* (Madsen, 1976), but is probably more primitive. The "foot" of the pubis is smaller, the obturator is almost completely bordered by bone (Figure 5.5), and the humerus is proportionally longer than in *A. fragilis*. The affinity with *Ceratosaurus* is less certain.

Laguna Manantiales Local Fauna (Ichnofauna, Upper Jurassic)

The assemblage of four different types of small footprints from the Estancia Laguna Manantiales, northern Santa Cruz, was communicated by Casamiquela (1961a, 1961b, 1964) based on detailed tracks of three species of coelurosaurs and one species of a primitive mammal. They were discovered in sandstones of the La Matilde Formation considered Oxfordian by Stipanicic, et al. (1968), and by Stipanicic and Bonetti (1970) based on geological and paleobotanical evidence.

Wildeichnus navesi is a small, biped, digitigrade ichnospecies of coelurosaur (Figure 5.6). It has a large medial digit III and approximately symmetrical lateral IV and internal II digits, the latter a bit longer. The distance from the tip of finger III to the central posterior border of the footprint is 50 mm. In some footprints of this form a posterior small digit of avian type is impressed.

Sarmientichnus scagliai is an ichnospecies of a specialized coelurosaur. It is a small digitigrade form, functionally monodactyl (Figure 5.6). The single impressed finger measures 130 mm and bears an acuminate claw. A second, smaller finger is present and a supposed plantar expansion is weakly marked behind it.

Delatorrichnus goyenechei is an ichnospecies representing a small ?coelurosaur (Figure 5.6) with quadrupedal, digitigrade, tridactyl feet, and ?one functional digit in the hands. The digits of the foot measure (II) 22 mm, (III) 31 mm, and (IV) 24 mm; and the hand impression is 22 mm. Several associated footprints of this species have been collected.

Ameghinichnus patagonicus is an ichnospecies representing a small, pentadactyl, digitigrade or semidigitigrade mammal, with feet and hands of approximately the same size (Figure 5.7). The impressions of the hands are nearer to the axial plane than those of the feet. Digits II, III, and IV are of the same size, and digits I and V a bit shorter. The hand digits are slightly smaller than those of the feet. The footprints of this species are very common and dominate the fossil locality. Some of them show the animal walking, while others show it running in the form of a rabbit (a ricochetal-type gait, Figure 5.7).

Araraquara Local Fauna (Ichnofauna, Upper Jurassic)

The discovery of this ichnofauna in the Botucatu Formation of Sao Paulo State, Brazil, was described by Leonardi (in press a) in a preliminary way and no names were given to the three different forms present. The age of the Botucatu Formation is not definitely established, but an Upper Jurassic–Lower Cretaceous age is commonly accepted (Camargo Mendes and Petri, 1971). The overlying basaltic flows have been dated at over 130 m.y. (Leonardi, in press a). Two types of the footprints of tridactyl feet probably represent species of small-sized bipedal Coelurosauria. The third type corresponds to a specialized quadruped, pentadactyl form,

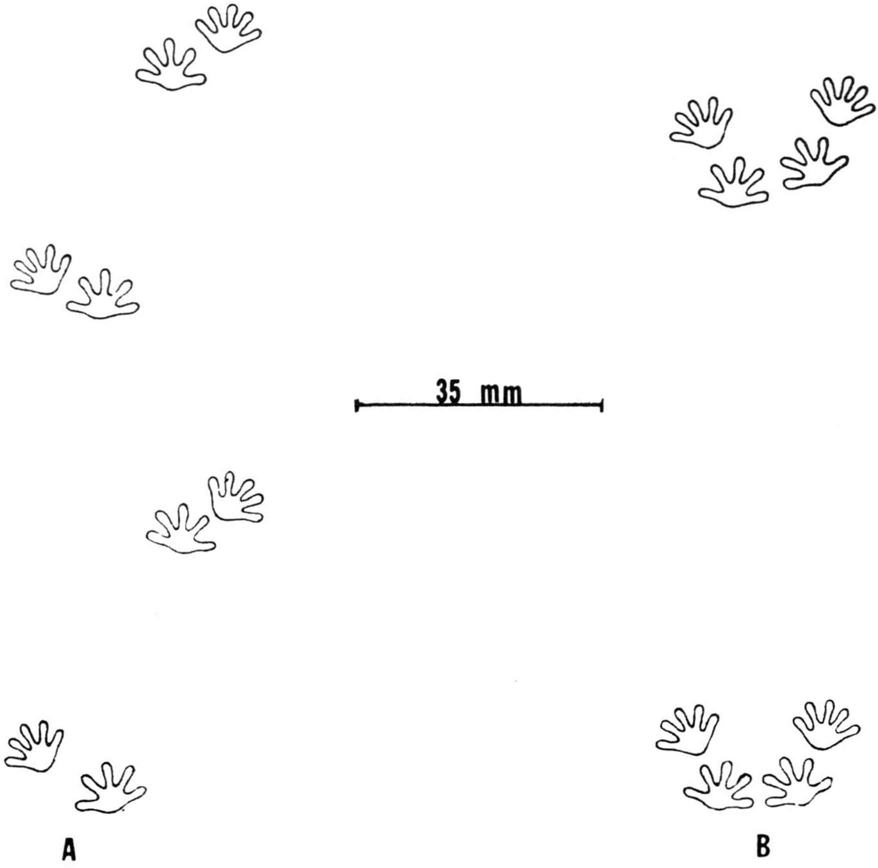

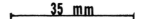

A

B

Figure 5.7. Mammalian footprints of the Laguna Manantiales local fauna, Oxfordian of Argentina. *Ameghinichnus patagonicus;* A, trackways of the animal walking; B, trackway of the animal in ricochetal progression. (After Bonaparte, 1978.)

with short digits and phalangeal formula of 2-3-3-3-3, more related to the mammals than to the reptiles. Gallop and ricochet were observed in the trackways of this form (Leonardi, in press a).

This assemblage of ichnospecies appears to be related to those forms of the Laguna Manatiales local fauna, and tentatively a similar age is accepted for both.

Jurassic Fauna of Neuquén (Callovian-Tithonian)

The Vaca Muerta (Tithonian) and Lotena (Callovian) formations of Neuquén province, Argentina, yielded a marine crocodile (Gasparini and Dellapé, 1976), a chelonian (Wood and Freiberg, 1977) and a pterosaur (Casamiquela, 1975 b, Bonaparte, 1978). These species have been collected in different levels of the Middle and Upper Jurassic in the central region of the Neuquén province. For a better understanding of the Jurassic record all these taxa are considered together.

Herbstosaurus pigmaeus was discovered in the Lotena Formation (Callovian) of southern Neuquén, and represents the only pterosaur known from the Jurassic of South America. The remains include an incomplete sacrum, an ilium, a pre-pubis, and both incomplete femora (Figure 5.8). They were described by Casamiquela (1957b) as a coelurosaur. However, the pterosaurian morphology of each piece is easily seen (Bonaparte, 1978, pp. 419–20). At present the phylogenetic relationships of this pterosaur have not been studied.

Notoemys laticentralis described by Cattoi and Freiberg (1961) from the Vaca Muerta Formation (Tithonian), is represented by a nearly complete carapace and a plastral fragment (Figure 5.9). Wood and Freiberg (1977) restudied the specimen, defined its relationships, and considered it as a member of the Plesiochelydae. Wood refers to the characters by which *Neotemys laticentralis* can be unequivocally distinguished from its European contemporaries. This species, recorded in marine beds, is the oldest chelonian known from South America.

Several specimens of *Geosaurus araucanensis* have been discovered in the Vaca Muerta Formation (Tithonian). The type specimen is a complete skull (Figure 5.10) and jaws associated with postcranial remains. A few distinctive characters of *G. araucanensis* distinguish it from the European species *G. suevicus*, such as the preorbital opening and less-separated teeth in the Argentine species (Gasparini and Dellapé, 1976).

Isolated Finds

Isolated finds include species of anurans, ichthyosaurs, marine crocodiles, sauropods, and footprints of an ornithopod.

Notobatrachus degiustoi (described by Reig in Stipanicic and Reig, 1955), is a rather large species of the primitive anuran family Ascaphidae (Figure 5.11). It is known from three localities in the La Matilde Formation of the Santa Cruz province, Patagonia (Stipanicic and Reig, 1957). This formation was considered upper Middle Jurassic or lower Upper Jurassic by the cited authors, and Oxfordian by Stipanicic, et al. (1968), and Stipanicic and Bonetti (1970). Callovian-Oxfordian age is most reasonable, but there is some doubt.

Several specimens have been collected, some of them complete and very well preserved as molds. After the papers by Reig (1957) and Casamiquela (1961c) on this species, Estes and Reig (1973) published a detailed account of its anatomy and relationships.

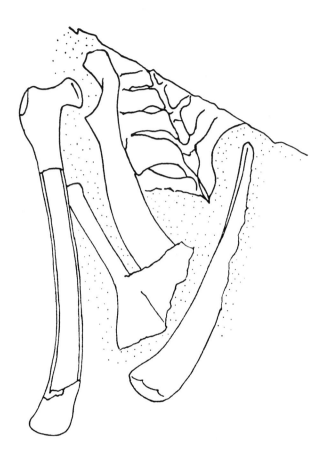

Figure 5.8. *Herbstosaurus pigmaeus,* type specimen with molds and bone fragments of the sacrum, ilium, prepubis, and both femora, from the Callovian of Neuquén, Argentina, X1. (After Bonaparte, 1978.)

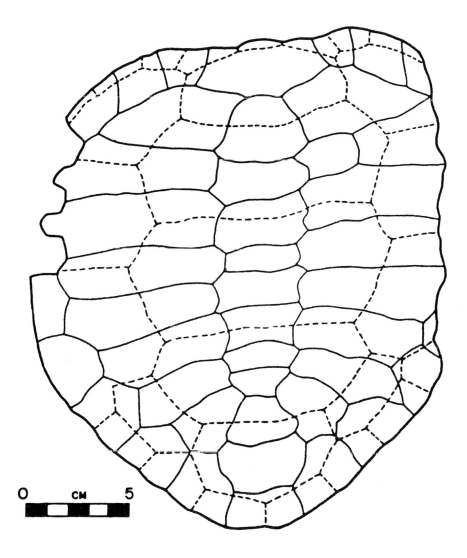

O   CM   5

Figure 5.9. *Notoemys laticentralis,* carapace, from the Tithonian of Neuquén, Argentina. (After Wood and Freiberg, 1977.)

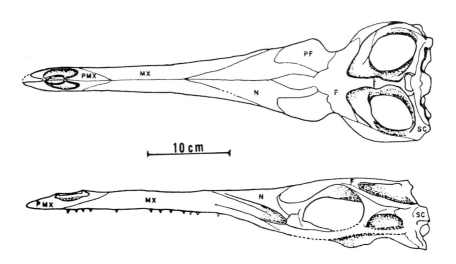

Figure 5.10. *Geosaurus araucanensis*, dorsal and lateral view of the skull, from the Tithonian of Neuquén, Argentina. (After Bonaparte, 1978, based on Gasparini and Dellapé, 1976.)

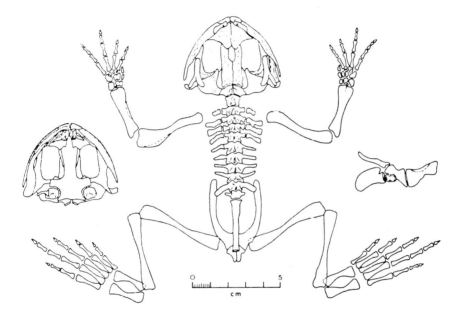

Figure 5.11. *Notobatrachus degiustoi,* restoration of the skeleton in dorsal view; skull and shoulder girdle in ventral view, from the Oxfordian of Santa Cruz, Argentina. (After Estes and Reig, 1973.)

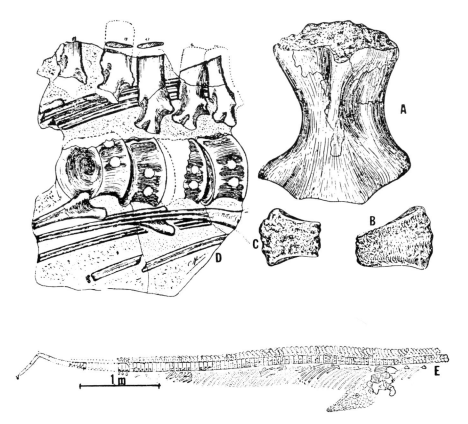

Figure 5.12. *Ancanamunia mendozana,* ichthyosaur from the Upper Jurassic of Mendoza, Argentina. A, humerus; B, C, ulna and radius; D, vertebrae and ribs as preserved; E, restoration of the postcranial skeleton. (After Rusconi, 1948.)

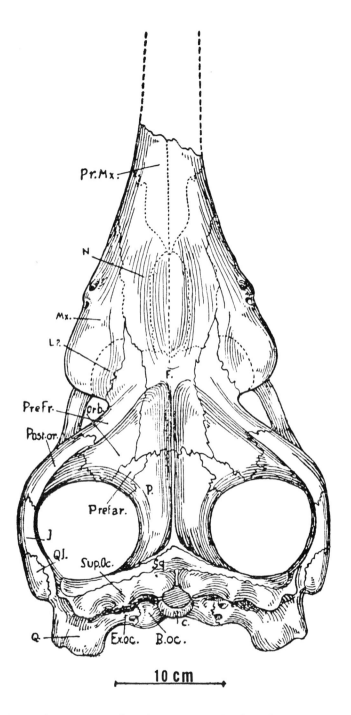

Figure 5.13. *"Purranisaurus" potens,* from the Upper Jurassic of Mendoza, Argentina. Dorsal view of the incomplete skull. (After Rusconi, 1948.)

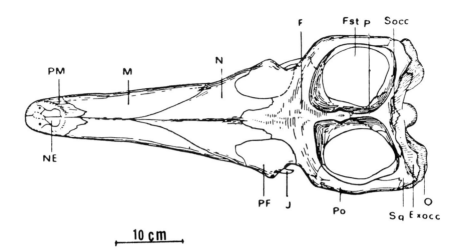

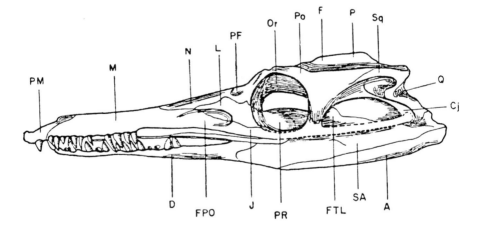

Figure 5.14. *Metriorhynchus casamiquelai,* from the Callovian of northern Chile. The skull and jaws in dorsal and lateral views. (After Gasparini and Chong, 1977.)

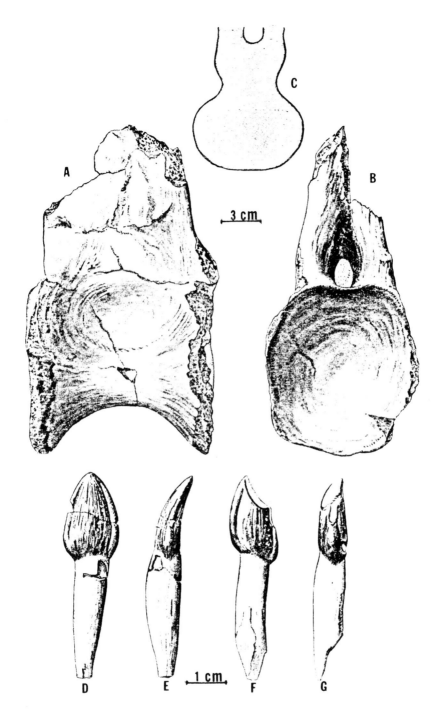

Figure 5.15. *Amygdalodon patagonicus*, from the Bajocian of Chubut, Argentina. Upper group: a dorsal vertebra in (A) lateral, and (B) posterior views; C, a section of the vertebral body. Lower group: two teeth; D, and E, labial and anterior views of one; and F, and G, labial and posterior views of another one. (After Cabrera, 1947.)

Several fragmentary specimens of *Ichthyosaurus* have been described since the end of the last century, and they were recognized as new species; i.e., *Ichthyosaurus inmanis, Ichthyosaurus bodenbenderi, Ichthyosaurus inexpectatus,* and *Ichthyosaurus sanjuanensis.* All of them are recorded from the Jurassic of west central Argentina and their specific distinction is highly questionable.

Ancanamunia mendozana is the most significant specimen of the family Ichthyosauridae recorded in South America (Figure 5.12). The specimen includes about seventy vertebrae, some ribs, incomplete forelimbs, and pectoral girdle. Rusconi (1948) studied this species in some detail and made an incomplete reconstruction (without skull) of it. The probable age of these remains is Upper Jurassic. They were collected in southern Mendoza province, Argentina.

Ancanamunia espinacitensis is a doubtful species based on a single vertebral body from the Jurassic of Espinacito, San Juan province, Argentina.

Macropterygius sp. is based on four vertebrae recovered in the Jurassic of Neuquén (Huene, 1931).

Stenopterygius grandis, a member of the family Stenopterygiidae, is based on a skull fragment with several teeth discovered in beds of the Middle Jurassic (Bayocian) of Neuquén province.

The type specimen of *"Purranisaurus" potens* (Figure 5.13) comprises a skull and jaws without the more anterior part of the muzzle. Its morphology appears to correspond to the variation of the genus *Metriorhynchus* (Romer, 1966; Gasparini, 1973). It was collected in southern Mendoza province, probably from beds of Tithonian age.

Metriorhynchus casamiquelai (Figure 5.14) is based on a complete skull and jaws from Callovian beds of northern Chile. It is distinctive from other species of *Metriorhynchus* in lacking ornamentation on cranial and rostral bones. Gasparini and Chong (1977) suggested that the centers of dispersal of the genus may have included the warm seas on the western margin of South America.

Amygdalodon patagonicus is a cetiosaurid species described by Cabrera (1947) based on a fragmentary and incomplete skeleton collected in beds of the Cerro Carnerero Formation, Bayocian (Stipanicic, et al., 1968), Chubut province, Argentina. The specimen includes several incomplete vertebrae, fragments of cervical and dorsal ribs, an incomplete pubis, a fragment of tibia, and seven teeth, some of them complete (Figure 5.15). Casamiquela (1963) restudied the specimen and discussed the affinities and significance of this primitive sauropod. The primitiveness of this species is apparent because of the relatively slight depressions on the lateral sides of the centrum, which are not truly pleurocoels.

Langston and Durham (1955) described a rather complete anterior dorsal vertebra of a big sauropod discovered in pre-upper Aptian beds, "possibly Jurassic," in northern Colombia. I agree with them in the possibility of a Jurassic age for this sauropod specimen and suspect it may be referred to the Cetiosauridae.

Von Huene (1927) restudied an isolated vertebra from the upper Liassic of Chile, and considered it *Plesiosaurus neogaeus.* It was previously studied by Burmeister and Giebel (1861) and was considered a crocodile, *Teleosaurus neogaeus.* In the same paper von Huene described a vertebra from the upper Liassic of Mendoza, Argentina, and considered it as *Steneosaurus gerhti.* It seems to me that the evidence in both cases is poor and the taxa based on it rather dubious. Chong and Gasparini

(1973) communicated the discovery of four metatarsi of a marine crocodile from Liassic beds of northern Chile.

Footprints of Supposedly Jurassic Age

Leonardi (in press b) reported ornithischian trackways, supposedly of Jurassic age, discovered in the Corda Formation, State of Goiás, Brazil. The footprints correspond to a big, bipedal, tridactyl ornithischian, and resemble those of the iguanodontid *Sousaichnium pricei* (see Leonardi, 1976).

Paleobiogeography and Dispersal Routes

The twenty-four species of Jurassic tetrapods recorded in South America represent eight orders (Salientia, Ichthyosauria, Chelonia, Pterosauria, Saurischia, Crocodilia, Squamata, and ?Eupantotheria) and eleven suborders (Euanura, Longipinnati, Latipinnati, Amphichelydia, Rhamphorhynchoidea, Theropoda, Coelurosauria, Sauropodomorpha, Thalattosuchia, Lacertilia, and suborder indet., see Table 5.1). This record is quite different from that used by Cox (1974) who reported no Jurassic terrestrial vertebrate locality in South America. Charig (1973) discussed the evidence of Jurassic (and Cretaceous) dinosaurs of the world and reported only one species of sauropod known at that time in South America.

Most species recorded from southern South America are below the latitude of 35S. However, the Araraquara local fauna (ichnites) and one of the two known species of marine crocodiles (*Metriorhynchus casamiquelai*) are at the latitude of 25S. The cetiosaurid indet. from northern Colombia is 11N. The known geographic distribution of localities suggests that the range of sauropods may have covered the whole continent; the small coelurosaurs and mammal (footprints) at least occupied a region from eastern Brazil to southern Patagonia; and the marine reptiles were all along the Pacific Chilean coast.

The Continental Species

The frogs from the Lower and Upper Jurassic of Patagonia, both referred to the Ascaphidae, suggest that the origin and radiation of this family occurred in the Southern Hemisphere (Estes and Reig, 1973).

The available information from the Argentine and other Gondwana sauropods (Bajocian and Callovian of Argentina; Lias of Australia and India; and Middle Jurassic of Morocco and Madagascar) strongly suggest that the first radiation of sauropods may have taken place in the Southern Hemisphere. The close similarity between Morrison and Tendaguru sauropods suggests the existence of terrestrial communication between South America–Africa and Euroamerica (Cox, 1974) during this time.

The only South American Jurassic carnosaur is from the Callovian. Accordingly, it shows more primitive features than *Allosaurus* from the Morrison beds which is of later age. The association of this carnosaur and cetiosaurids represents an ancestral assemblage that may or may not be interpreted as ancestral to the Morrison and Tendaguru faunas.

The coelurosaurian footprints of the Oxfordian Laguna Manantiales local fauna prove the existence of advanced biped forms like *Sarmientichnus*, and very small species such as *Wildeichnus* and *Delatorrichnus*, indicating a varied assemblage of small dinosaurs. At least at the family level they probably occurred in the Arara-

TABLE 5.1

Table of genera, orders and suborders of tetrapods recorded in the Jurassic
of South America. The genera with ° are ichnogenera.

Upper Jurassic			
	Notobatrachus	Salientia	Euanura
	Notoemys	Chelonia	Amphichelydia
	Ichthyosaurus	Ichthyosauria	Latipinnati
	Macropterygius	Ichthyosauria	Latipinnati
	Ancanamunia	Ichthyosauria	Latipinnati
	"Purranisaurus"	Crocodilia	Thalattosuchia
	Geosaurus	Crocodilia	Thalattosuchia
	Wildeichnus°	Saurischia	Coelurosauria
	Sarmientichnus°	Saurischia	Coelurosauria
	Delatorrichnus°	Saurischia	Coelurosauria
	Ameghinichnus°	Eupantotheria?	Suborder indet.
Middle Jurassic	*Stenopterygius*	Ichthyosauria	Longipinnati
	Herbstosaurus	Pterosauria	Rhamphorhynchoidea
	Metriorhynchus	Crocodilia	Thalattosuchia
	Piatnitzkysaurus	Saurischia	Theropoda
	Amygdalodon	Saurischia	Sauropodomorpha
	Patagosaurus	Saurischia	Sauropodomorpha
	Volkheimeria	Saurischia	Sauropodomorpha
Lower Jurassic	*Vieraella*	Salientia	Euanura
	Protolacerta	Squamata?	Lacertilia?

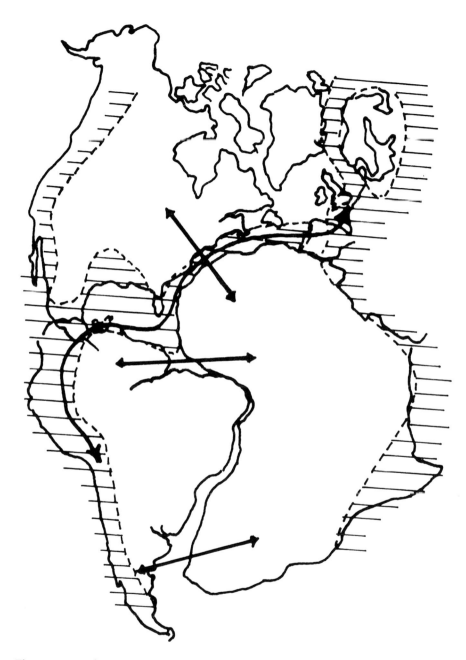

Figure 5.16. Paleogeographic map of the Middle Jurassic. The position of the continents is after Sclater, Hellinger, and Tapscott (1977), and the distribution of the seas and oceans is after Smith, Briden, and Drewry (1973; in Cox, 1974). The assumed terrestrial (straight lines) and marine (sinuous line) tetrapod dispersal routes are indicated for the Middle and Upper Jurassic. Supposedly the connections between Euroamerica and Africa were restricted and not permanent.

quara local fauna of eastern Brazil. It seems to me that the terrestrial tetrapods from the Jurassic of South America suggest faunal relationships with the Gondwana continents and Euroamerica.

The Marine Species

The more significant marine species from the Jurassic of South America probably are the thalattosuchian crocodiles *Metriorhynchus casamiquelai,* Callovian of Chile, and *Geosaurus araucanensis,* Tithonian of Argentina. Both supposedly are good indicators of warm seas along the western coast of South America.

Gasparini and Chong (1977) emphasized the importance of the Callovian *M. casamiquelai,* which is of similar age as the oldest record of the genus in Europe. The cited authors assumed that the dispersal of *Metriorhynchus* probably occurred sometime before the Callovian along the warm waters of the Tethys and the western coast of South America. This dispersal route was probably open until the end of the Jurassic as suggested by the presence in western South America and Europe of the related species *Geosaurus araucanensis* and *Geosaurus suevicus* (Gasparini and Chong, 1977). However, terrestrial connections probably existed from time to time during the Middle and Upper Jurassic for the dispersal of dinosaur faunas between South America–Africa and Euroamerica (Figure 5.16).

The Tithonian Amphichelydia *Notoemys laticentralis* (Wood and Freiberg, 1977) was discovered in marine rocks and considered by Wood (after a long discussion) as a probable marine turtle which followed a similar dispersal route as the above discussed crocodiles.

If it is assumed that the Jurassic pterosaurs were associated with marine environments as was *Herbstosaurus pigmaeus* from the Callovian marine beds of northwestern Patagonia, there is a third line of evidence of a dispersal route along the warm waters and coasts of the European Tethys and western border of South America.

Literature Cited

Bonaparte, J. F. 1978. El Mesozoico de América del Sur y sus Tetrápodos. Opera Lilloana, 26:1–596.

———. 1979. Dinosaurs: a Jurassic assemblage from Patagonia. Science, 205 (4413):1377–79.

Burmeister, H., and Giebel, C. 1861. Die Versteinerungen von Juntas im Thal des Rio Copiapo. Natur. Ges. Halle, Abh., 6:122–32.

Cabrera, A. 1947. Un Saurópodo nuevo del Jurásico de Patagonia. La Plata, Univ. Nac., Mus. Notas, 12:95.

Camargo Mendes, J., and Petri, S. 1971. Geología do Brasil. Enciclopédia Brasileira. Inst. Nac. do Livro, 9.

Casamiquela, R. M. 1961a. El hallazgo del primer elenco (icnológico) Jurásico de vertebrados terrestres de Latinoamérica (Noticia). Rev. Assoc. Geol. Argen., 15:1–2.

———. 1961b. Sobre la presencia de un mamífero en el primer elenco (icnológico) de vertebrados del Jurásico de la Patagonia. Physis, 22(63):225–33.

———. 1961c. Nuevos materiales de *Notobatrachus degiustoi* Reig. La Plata, Mus., Rev., 4:35–69.

———. 1963. Consideraciones acerca de *Amygdalodon* Cabrera (Sauropoda-Cetiosauridae) del Jurásico medio de la Patagonia. Ameghiniana, 3(3):79–95.

———. 1964. Estudios icnologicos. Gob. Prov. Rio Negro, Minist. Asuntos Sociales, Buenos Aires, 229 pp.

———. 1965. Nuevo material de *Vieraella herbstii* Reig. La Plata, Mus., Rev., 4:265–317.

———. 1975a. La presencia de un Sauria (Lacertilia) en el Liásico de la Patagonia. Actas, I Congr. Argent. Paleont. Bioestr., 2, pp. 57–70.

————. 1975b. *Herbstosaurus pigmaeus* (Coeluria, Compsognathidae) n. g. n. sp. del Jurásico medio del Neuquén, (Patagonia Septentrional). Actas, I Congr. Argent. Paleont. Bioestr., 2, pp. 87–104.

Cattoi, N., and Freiberg, M. 1961. Nuevo hallazgo de Chelonia extinguidos en la República Argentina. Physis, 22(63):202.

Charig, A. J. 1973. Jurassic and Cretaceous dinosaurs. *In,* A. Hallam (ed.), Atlas of paleobiogeography. Elsevier, Amsterdam, pp. 339–52.

Chong Díaz, G., and Gasparini, Z. de, 1973. Presencia de Crocodilia marinos en el Jurásico de Chile. Rev. Asoc. Geol. Argent., 27(4).

Colbert, E. H. 1970. A saurischian dinosaur from the Triassic of Brazil. Amer. Mus. Nov., 2405:1–39.

Cox, C. B. 1974. Vertebrate paleodistributional patterns and continental drift. J. Biogeogr., 1:75–94.

Estes, R., and Reig, O. A. 1973. The early fossil record of frogs: A review of the evidence. *In,* J. L. Vial (ed.), Evolutionary biology of the anurans. Univ. Missouri Press, Columbia, pp. 11–63.

Gasparini, Z. de. 1973. Revisión de *"Purranisaurus potens"* Rusconi 1948 (Crocodilia, Thalattosuchia). Los Thalattosuchia como un nuevo infraorden de los Crocodilia. Actas, V Congr. Geol. Argent., 3, pp. 423–31.

Gasparini, Z. de, and Dellapé, D. 1976. Un nuevo Crocodilia marino (Thalattosuchia, Metriorhynchidae) de la Formación Vaca Muerta (Jurásico, Tithoniano) de la Provincia de Neuquén (República Argentina). I Congr. Geol. Chileno.

Gasparini, Z. de, and Chong Díaz, G. 1977. *Metriorhynchus casamiquelai* n. sp. (Crocodilia, Thalattosuchia), a marine crocodile from the Jurassic (Callovian) of Chile, South America. Neues Jahrb. Geol. Palaontol., Abh., 153(3):341–60.

Huene, F. von, 1927. Beitrag zur Kenntnis mariner mesozoischer Wirbeltiere in Argentinien. Centralbl. Min. Geol. Pal., Abt. B, 1:22–29.

————. 1931. Verschiedene Mesozoische Wirbeltierreste aus Südamerika. Neues Jahrb. Min. Geol. Pal., Abt. B, 66:181–98.

Langston, W., and Durham, J. W. 1955. A sauropod dinosaur from Colombia. J. Paleontol., 29(6):1047–51.

Leonardi, G. 1976. Nota preliminar sobre seis pistas de dinossauros Ornithischia da Bacia do Rio do Peixe (Cretáceo inferior) em Sousa, Paraíba, Brasil. Atas, XXIX Congr. Brasil. Geol.

————. In press a. On the discovery of an abundant ichno-fauna (vertebrates and invertebrates) in the Botucatú Formation s.s. in Araraquara, Sao Paulo, Brazil.

————. In press b. Ornithischian trackways of the Corda Formation (Jurassic), Goiás, Brazil. Actas II Congr. Argent. Paleont. Bioestr.

Madsen, J. H. 1976. *Allosaurus fragilis:* a revised osteology. Utah Geol. and Min. Surv., Bull., 109:1–163.

McIntosh, J. In press. A catalog of the dinosaurs in the collection of the Carnegie Museum of Natural History.

Oigier, A. 1975. Etude de noveaux ossement de *Bothriospondylus.* Thesis 3rd. Cycle, Université Paris.

Pascual, R.; Odreman Rivas, O. E.; and Tonni, E. P. 1969. Las unidades estratigráficas del Jurásico de la Argentina portadoras de vertebrados. Correlaciones y edades. Cuart. J. Geol. Argent., 1:469–83.

Reig, O. A. 1957. Los anuros del Matildense. Act. Geol. Lilloana, 2:231–97.

————. 1961. Noticia sobre un nuevo anuro fósil del Jurásico de Santa Cruz (Patagonia). Ameghiniana, 2(5):73–78.

Romer, A. S. 1966. Vertebrate Paleontology. 3rd ed. Univ. Chicago Press, Chicago, 468 pp.

Rusconi, C. 1948. Ictiosaurios del Jurasico de Mendoza (Argentina). Mus. Hist. Natur. Mendoza, Rev. 2(1/2):17–160.

Sclater, J. G.; Hellinger, S.; and Tapscott, C. 1977. The paleobathymetry of the Atlantic Ocean from the Jurassic to the present. J. Geol., 85(5):509–52.

Simpson, G. G. 1971. Clasificación, terminología y nomenclatura provinciales para el Cenozoico mamalífero. Rev. Asoc. Geol. Argent. 26(3):281–97.

Stipanicic, P. N., and Bonetti, M. 1970. Posiciones estratigráficas y edades de las principales floras Jurásicas Argentinas. II. Floras Doggerianas y Malmicas. Ameghiniana, 7(2): 101–18.

Stipanicic, P. N., and Reig, O. A. 1955. Breve noticia sobre el hallazgo de anuros en el denominado "complejo porfírico de la Patagonia extraandina" con consideraciones acerca de la composición geológica del mismo. Rev. Asoc. Geol. Argentina, 10(4): 215–33.

————. 1957. El "Complejo porfírico de la Patagonia Extraandina" y su fauna de anuros. Acta Geol. Lilloana, 1:185–230.

Stipanicic, P. N.; Rodrigo, F.; Baulies, O. L.; and Martinez, C. G. 1968. Las formaciones presenonianas en el denominado Macizo Nordpatagónico y regiones adyacentes. Rev. Asoc. Geol. Argent., 23(2).

Tasch, P., and Volkheimer, W. 1970. Jurassic conchostracans from Patagonia. Univ. Kansas. Paleont. Contrib., Paper 50, 23 pp.

Wood, R., and Freiberg, M. A. 1977. Redescription of *Notoemys laticentralis*, the oldest fossil turtle from South America. Acta Geol. Lilloana, 23(6).

THE CONTINENTAL LOWER JURASSIC FAUNA FROM THE KOTA FORMATION, INDIA

by Sohan L. Jain

Introduction

Although an excellent account of the Jurassic geology of the world by Arkell (1956) and Jurassic environments by Hallam (1975) are available it will be seen that these are largely confined to the marine strata. The continental deposits are so few that they hardly get a mention. This article, to honor Dr. Edwin H. Colbert on his 75th birthday, deals with a continental deposit in India. The Kota Formation in the Pranhita-Godavari valley has been given a Lower Jurassic age on the basis of Liassic fishes occurring therein, yet it is a continental deposit in which, among other animals, are found bones of a large sauropod.

Dr. Colbert has taken keen interest in Indian Gondwana vertebrates for over two decades since his review of the Triassic Maleri fauna (1958). His first visit to India materialized in 1964 when he met vertebrate paleontologists in India and participated in the fieldwork program in the Indian Gondwanas. This included a visit to a few localities of the Kota Formation and the excavation of dinosaur bones (Figure 6.1). A few of his experiences have been included in his volume *Wandering Lands and Animals* published in 1973.

During the Fourth International Gondwana Symposium in Calcutta, January 1977, Dr. Colbert made his second visit to India. The staff of the Geological Studies Unit (GSU) at that time were mounting the skeleton of *Barapasaurus tagorei* in the Geology Museum of the Indian Statistical Institute (ISI). He was immensely pleased with this first attempt of mounting a dinosaur skeleton in India and freely made his comments available. In his keynote address to the symposium he dwelt upon the relationships of the Gondwana vertebrate faunas with the continental relationships as inferred from the theory of plate tectonics. He elaborated his views on the northward migration of India as evidenced by the Mesozoic tetrapods in his contribution to the J. A. Orlov Memorial Number of the *Journal of the Palaeontological Society of India* (1978). The vertebrate fauna from the Kota Formation was emphasized in the discussion of the Indian Jurassic in these contributions by Dr. Colbert (1977, 1978).

In view of the active interest and close participation of Dr. Colbert in the work on Indian continental vertebrates, I have thought it befitting to summarize, and comment when necessary, on our knowledge of the Kota fauna. During the last two decades there has been an enormous increase in our knowledge of this fauna, but the publications are in widely scattered journals. The present article is basically a piecing together of the results of these investigations in a systematic manner. This summary will be useful for an overall view of the Indian Gondwanas to paleontologists abroad who may not have easy access to the scattered literature. It may also be helpful for further research in this area.

Figure 6.1. Dr. Edwin H. Colbert (right) at the excavation site of the Kota dinosaur (*Barapasaurus tagorei*) in 1964; also seen in the photograph is Dr. S. L. Jain.

The exploration of the Kota beds in recent times has been largely at the initiative of Dr. Pamela L. Robinson (University College, London) as guest scientist of the Indian Statistical Institute, other Indian colleagues at ISI, Geological Survey of India (GSI) and Oil & Natural Gas Commission. I have had the opportunity to participate in the field work of ISI since 1958 and owe my sincere thanks to Dr. Robinson for providing leadership in field work, for her untiring efforts in introducing me to paleontological research and guiding me through innumerable discussions. Various observations included here have been drawn from those discussions. Citations have been made as far as possible, but repeated citations have been avoided for the sake of easy reading.

The Kota Formation

The Kota beds were named after the village of Kota situated on the eastern bank of the river Pranhita, being one of the major tributaries of the river Godavari flowing to the eastern seaboard of the Indian peninsula. During the nineteenth century fossil fishes were discovered in one of the limestone bands near Kota. These fishes were recognized to be species of three genera, *Tetragonolepis, Dapedium,* and *Lepidotes*. These also happen to be well known members of the European marine fish faunas of the Lower Jurassic (Liassic). The Kota beds, however, like other members of the Gondwana Group, are continental deposits. This is evidenced by the occurrence of fluviatile sand and clays, and thin limestones representing deposits in inland lakes.

The Kota Formation contains two rather different kinds of sediments according to Robinson (1970). The sandstones are slightly ferruginous arkoses, poorly sorted and with large-scale trough cross-bedding, having coarser lenses containing moderately well-rounded quartz and quartzite pebbles. The limestones are well bedded, often finely laminated in the upper part of the band, and may be interbedded with laminated clays. They are followed in sequence by red and green clay shales, finely laminated or cross-bedded on a very minute scale.

The limestones are largely of inorganic origin, and clastic material is almost wholly absent from the limestones proper, though there may be a trace of iron residue, and thin chert nodules occur in certain layers. Desiccation polygons, usually filled in with limestone, are common, and worm-bored layers occur. There seems little doubt that the Kota limestones are of freshwater origin. Although King (1881), who pioneered work on the Kota beds, thought that there may be three bands of limestone in the Kota Formation, recently Rudra (1972) has suggested that in fact there is only one distinct "zone" of limestone in the Kota. Mapping has revealed that the limestone "zone" is repeated due to a number of faults.

Tasch, et al. (1973), made a detailed examination of the Kota beds during the 1971–72 field season. They have described an outcrop belt of Kota limestone and conchostracan-insect beds at seven different localities, named K-1 to K-7. Comments on the more important aspects of this work are given in a later section. It is sufficient to mention here that the freshwater origin of the Kota limestone is supported and the fauna investigated indicates the presence of small basins or ponds.

Govindan (1975) recently mapped the northeastern part of the Pranhita-Godavari valley, including the Kota Formation. He obtained a faunule of freshwater ostracods which will be discussed in a later section. The results of his investi-

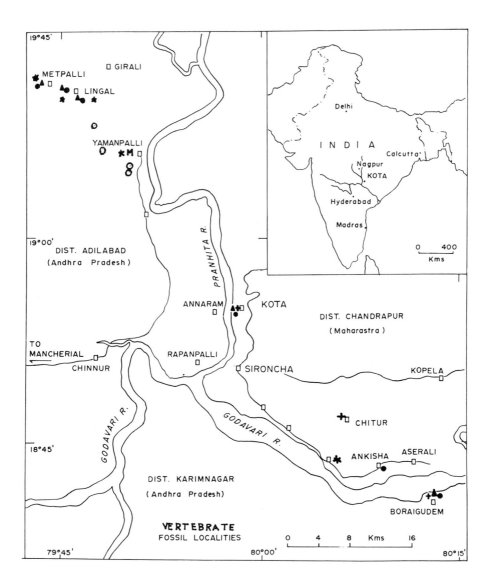

Figure 6.2. Main vertebrate fossil localities in the Kota Formation; *inset,* map of India showing Pranhita-Godavari river system and the location of the village Kota. The semionotid (●),
coelacanth (▲) and pholidophorid fishes (0), dinosaur (*), pterosaur (+), and mammalian
(M) localities are marked.

gation have suggested that the Kota limestone represents a freshwater facies deposited under lacustrine conditions.

The Kota fossil assemblage has received attention for over a century, but it is mainly during the last two decades that there has been a vast increase in knowledge. This includes the first discoveries of land fauna (dinosaurs) in the early sixties, and a flying reptile, a coelacanthid, a pholidophorid, a new semionotid, and early mammals in the seventies. Localities from where vertebrate fossils have been discovered in the Kota Formation are given in Figure 6.2. Detailed geological mapping of the Kota beds has also, in recent years, led to a more precise definition of rock units in the Kota Formation. The Kota fauna encompasses aquatic, terrestrial, and aerial vertebrates and invertebrates, most of them excellently preserved. These merit individual consideration. The following section deals with this assemblage, including the status of various taxa, in a systematic manner.

Fishes

The Kota Formation has yielded well-preserved fauna of fossil fishes. These include members of the families Semionotidae (three genera), Coelacanthidae (one genus), and Pholidophoridae (one genus).

During the second half of the nineteenth century, P. M. G. Egerton designated five species of the semionotid *Lepidotes* (spelled *Lepidotus*), one species of *Dapedium* (spelled *Dapedius*) and three species of *Tetragonolepis*. Thereafter, the Kota beds remained in obscurity for the next eighty years or so, since there were no published reports or descriptions of new collections. During 1958 a fresh collection was made and a part of it exhibited in the Geological Society of London's meeting in 1959, where the collections of the nineteenth century were earlier exhibited (Jain, 1959).

Lepidotes is the most commonly occurring member of the entire piscine fauna. It is found in the laminated beds as well as nodules (Figure 6.3). It appears to have been an active fusiform fish of moderate size with relatively slender teeth suitable for preying on estherians and other soft-bodied invertebrates. The species found in the nineteenth century are *L. deccanensis*, *L. longiceps*, *L. breviceps*, *L. calcaratus*, and *L. pachylepis*. A detailed study of Kota specimens of *Lepidotes* and an examination of the validity of characters used to distinguish species within the genus suggest that all Kota specimens should be placed in a single species, recognizing that within that species variations in certain characters are possible, as it is possible in all other species of the genus. The first named species from the Kota was *L. deccanensis*, and all the specimens of the genus so far obtained from the Kota are now included within this species.

A study of the cheek plates in *Lepidotes* (Jain and Robinson, 1963) reveals that these are variable within a species and sometimes on the two sides of the head of a single individual. Other characters associated with the cheek plate conditions in *Lepidotes* are orbital size, dentition, skull ornamentation, operculum proportions, and symphysis depth. *L. deccanensis* with very slender and acuminate teeth has a low symphysis depth comparable only to *L. elvensis*. Several other features of the two species also indicate close morphological characters. This suggests an upper Liassic age for *L. deccanensis*, since *L. elvensis* is restricted to upper Liassic.

The semionotid *Dapedium* is known from Rhaetic to upper Liassic and its presence in the Kota beds was recognized in the nineteenth century. However, a re-

Figure 6.3. *Lepidotes deccanensis* in a limestone nodule in the Kota Formation: *above*, a portion of the block has been chipped to expose part of the fish; *below*, same block exposing almost complete fish after chipping the entire surface. Note desiccation polygons on the block.

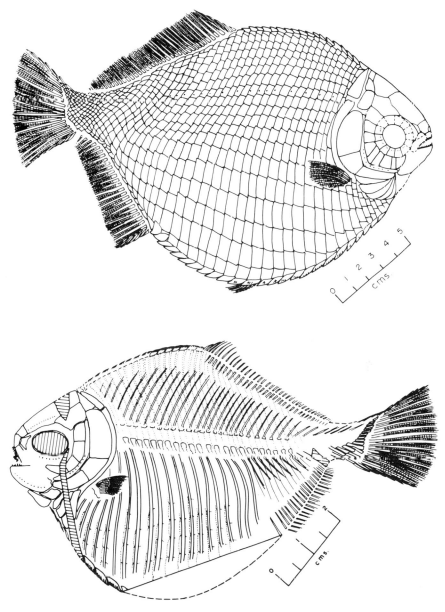

Figure 6.4. The hypsisomid semionotid fishes from the Kota Formation: *above*, a restoration of *Paradapedium egertoni*; *below*, a restoration of *Tetragonolepis oldhami* (both from Jain, 1973).

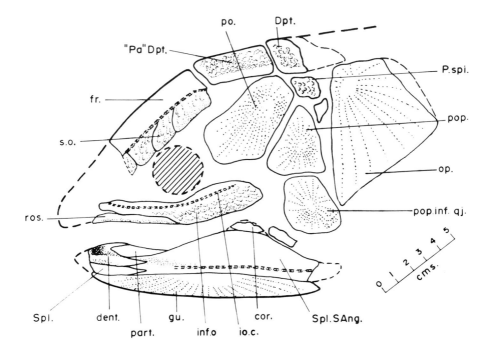

Figure 6.5. Restoration of the coelacanth (*Indocoelacanthus robustus*) from the Kota Formation (from Jain, 1974).

examination of type material along with fresh specimens collected over the years by GSU teams revealed that these were morphologically different from the European *Dapedium,* and so *Paradapedium* has been erected to accommodate the Kota specimens (Jain, 1973). Among other features, *Paradapedium egertoni* (Figure 6.4) has a higher value for the depth of the body below the lateral line canal, a much smaller length and depth of head as a percentage of body length (15–16 percent), less heavily ossified skull bones, and a fewer number of suborbitals, as compared to the species of *Dapedium.* However, as compared to the genera *Hemicalypterus, Tetragonoleps,* and *Dapedium, Paradapedium* is closest to *Dapedium.* None of these genera occurs in horizons lower than Upper Triassic and higher than Liassic. *Hemicalypterus* is from Upper Triassic, *Dapedium* from lower and upper Liassic, *Tetragonolepis* from upper Liassic, and *Paradapedium,* similar to *Dapedium,* is found in the Kota beds along with *Tetragonolepis.* It seems, therefore, that *Paradapedium* need not be geologically younger or older than *Dapedium* but may be an ecological substitute for *Dapedium* in Asia.

Tetragonolepis, another semionotid found in the Kota Formation, is confined to upper Liassic in Europe. Re-examination of the holotype, along with new well-preserved material, has enabled a proper definition of *T. oldhami* (Figure 6.4) from the Kota Formation (Jain, 1973). The other species of the last century, namely *T. rugosus* and *T. analis,* have been rejected because they are based on fragmentary specimens. The restricted geological distribution of *Tetragonolepis* and its presence in the Kota Formation is valuable evidence of upper Liassic age. However, European species of *Tetragonolepis* are known from marine sediments and the Kota species from freshwater deposits. It is, therefore, likely that they were euryhaline forms.

In view of the rarity of coelacanth fish remains from Asia, except a single poorly known record of *Sinocoelacanthus* from China, the finding of coelacanth remains in the field season of 1962–63 from the Kota Formation was exciting. However, this material was poor. In subsequent years, as a result of concerted prospecting efforts, considerable material was obtained. On the basis of these fossils it has been possible to recognize the new monotypic form *Indocoelacanthus robustus* (Jain, 1974a). It is estimated to be slightly larger than *Holophagus.* The skull bones are heavily tuberculated, except the frontals which are slightly ornamented (Figure 6.5). Rostral, parasphenoid, and pterygoid are devoid of any denticles. The lower jaw is robust but also without teeth. Gulars are approximately 3.5 times as long as broad. The orbit is moderate in size, and scale ornamentation is distinctive.

Devonian coelacanths were both marine and freshwater; Carboniferous and Permian genera indicate freshwater and swampy conditions. About 40 percent of the Triassic and nearly all Jurassic coelacanths have been found in marine deposits. *Macropoma* and several other Cretaceous coelacanths were all marine. Beyond Cretaceous there is no fossil record of coelacanths till the living coelacanth *Latimeria.* In view of the composition of the fauna and probable Lower Jurassic age of the Kota Formation, it is of interest to note that another coelacanth genus, *Lualabaea* from Zaire (Stanleyville stage = Middle to Upper Jurassic) is a freshwater form similar to *Indocoelacanthus.*

The shape and disposition of various cranial bones and postcranial structures of *Indocoelacanthus* does not provide any clue to elucidate the age of the Kota Formation. This is primarily due to the fact that *Indocoelacanthus* encompasses some features of coelacanth genera of widely separated geological periods. The cranium of

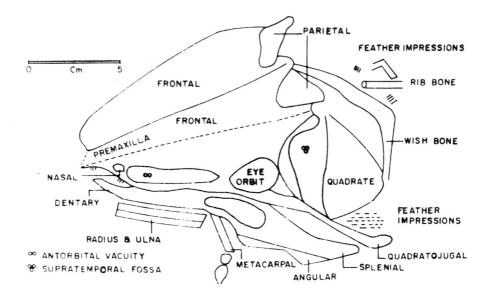

Figure 6.6. The so-called "Kota Bird" from the Kota Formation; for comments please see text (from *India Today,* vol. IV, No. 10, 1979, with permission of the editor).

Lualabaea is more elongate and the scales are quite distinctive, showing no resemblance to *Indocoelacanthus*. Although no complete coelacanth remains have been found in the Kota beds so far, the preserved fossils of *Indocoelacanthus* (most of the head, body squamation, fins) indicate a robust fish about 70 cm long.

Although an account in a scientific journal is so far not available, announcement of the discovery of a "fossil bird" from the Kota Formation has been made through the popular press (*India Today,* 16–31 May 1979). The discovery is attributed to P. Yadagiri (Geological Survey of India, Hyderabad) and it has been reported that "the find is the first record of a Mesozoic fossil bird in India and is being claimed as the oldest fossil bird." In comparison to *Archaeopteryx* it is stated that "while the *Archaeopteryx* measures 15 cm from skull to tail, the Kota fossil bird's skull alone is about 16 cm."

The availability of a reasonably good photograph of the specimen, along with a labelled line drawing (Figure 6.6), makes it possible to assess the report scientifically. It has also been possible to re-examine the specimen independently. There apparently has been gross error in the interpretation of the specimen, to the extent that it has been placed front-side backwards. In the new position, bones labelled as "frontal," "quadrate" and "wish bone" show familiar features of gular, opercular, and cleithrum of a coelacanthid fish. This skull is quite comparable to *I. robustus* and matches in the morphology of gulars, opercular, and cleithrum. Moreover, the size of the alleged "bird" skull is very close to that of *I. robustus.*

The "Kota bird" does not show any resemblance in morphological characters to any fossil bird. It does not exhibit a sclerotic ring in the orbit. Nor does it show any temporal fossa in comparable position. The finely ribbed ornamentation of the coelacanth squamation appears to have been interpreted as the evidence of feathers. The lower jaw however, has two rows of well marked but feeble denticulations. In comparison, *I. robustus* does not have denticulations in the lower jaw. In the absence of the skull roof, it is not possible to make a detailed comparison between *I. robustus* and the "Kota bird" specimen. However, on the basis of available evidence it appears to be a coelacanthid which may prove to be another species of *Indocoelacanthus,* or even a new genus in the Coelacanthidae.

Satsangi and Shah (1973, abstract only) reported the finding of a *Pholidophorus* fish from the Kota Formation which was similar to the species of *P. bechei*. No detailed account of this specimen has been published so far. However, having personally examined the specimen briefly (around 1966) along with a photograph kindly made available by Mr. P. P. Satsangi (Geological Survey of India, Calcutta), it appears to be an excellently preserved fish including an almost perfect head and body. A detailed account of this specimen would be most welcome. The affinity of the Kota pholidophorid to *P. bechei* is interesting because the latter occurs in the lower Liassic (Lyme Regis, Dorset) in Europe. Morphologically this is one of the best known species (Nybelin, 1966).

Yadagiri and Prasad (1977) have reported two additional specimens (GSI type number 19225, plus counterpart numbers 19226 and 19227) which have been assigned to two new species, namely *P. kingii* and *P. indicus*. In addition, three other specimens (SRV 2, 4, 7) have been referred to *P. indicus*. Details of the latter specimens have not been given. The GSI type specimens are almost complete fishes. *P. kingii* is about 110 mm in length, and *P. indicus* is about 160 mm. Of the two species, *P. kingii* is somewhat better preserved and has a length of the head which is almost equal to the depth of the body, and about 25 percent of the body length. In *P. indicus* these proportions are somewhat different; length of head/depth of trunk is 30/50 and length of head/length of body varies from 19 percent to 22 percent as stated by Yadagiri and Prasad (1977).

The specimens described above do certainly have pholidophorid affinities and probably also belong to the genus *Pholidophorus,* but the erection of two species is questionable. In the absence of actual figures of various indices such as length of head, length of body, and depth of body of the three specimens used for the study of *P. indicus,* it is difficult to make useful comparison with other species. In addition, there are obvious errors in the interpretations of the lower jaw morphology and the skull roof. Whereas Figure 2 in Yadagiri and Prasad (1977) shows *P. kingii* without parietals, Figure 3 (obviously the same specimen) shows parietals demarcated from the dermopterotic by vertical division. The opposite is shown for *P. indicus.* Figure 4 in Yadagiri and Prasad (1977) shows parietals as part of the frontals, with large dermopterotics, and Figure 5 (obviously the same specimen) shows no parietals at all, although it is a detailed figure. Judging from the morphology of *Pholidophorus* as interpreted by Nybelin, it seems that extrascapulars have been wrongly interpreted as dermopterotics in all figures. The suture between parietal and dermopterotic which may be transverse has probably been missed.

A description of the specimen of Satsangi and Shah, along with re-examination of the material of Yadagiri and Prasad, plus new undescribed material (three specimens are available in the ISI collection) will go a long way to elucidate the detailed morphology of this important fish group.

Reptiles

Reptiles are well represented in the Kota Formation. These include a rhamphorhynchid pterosaur, sauropod dinosaurs, and fragmentary crocodiles. The fauna is entirely archosaurian.

The pterosaur from Kota Formation is the only example of a flying reptile from the Indian subcontinent, although excellent material has been described from the Lower Cretaceous of China (Young, 1964). The Kota pterosaur, *Campylognathoides indicus* (Jain, 1974b), appears to be a medium-sized animal as evidenced by a portion of the anterior part of the skull and upper jaw and the delicate nature of the dissociated postcranial bones. The two specimens in the collection of Indian Statistical Institute and one with the Geological Survey of India, Calcutta, (Rao and Shah, 1967) have been found in localities close by in the Kota Formation (Figure 6.2). *Campylognathoides* is known from the upper Liassic of Holzmaden in West Germany. The discovery of material assignable to this genus in the Liassic rocks in India extends its geographical distribution. Although there is an excellent record of flying reptiles from Laurasia, these are almost unknown from the Gondwanas, except for some excellent material from Argentina.

The discovery of a dinosaur fauna from the Kota Formation (Jain, Robinson, and Roychowdhury, 1962) has been a major breakthrough in the knowledge of Mesozoic terrestrial vertebrates from India. A major gap in the understanding of the evolution of sauropod dinosaurs has been due to lack of good fossil material from Lower Jurassic rocks, except fragmentary material (*Ohmdenosaurus liasicus*) described recently (Wild, 1978). This gap has been partly filled with the knowledge gained from the study of the Kota dinosaur which is believed to be of Lower

Jurassic age. The first excavation revealed a rich layer of bones at the junction of sandstone with a clay lens, lying about 6 m below a fish-bearing limestone band in the Kota Formation. No complete associated skeleton has been found, although partial associations were obtained. About 300 bones were excavated from an area of approximately 28 x 11 m during the 1960–61 field season. These include sauropod bones having femora almost 15 m long, though smaller individuals are also represented. Presence of a carnosaur was suggested earlier, but further study of the material does not confirm its presence. The bone bed is reminiscent of a log jam in a river as evidenced by the occurrence of logs of wood up to 3 m long along with the bones. Subsequently, further sites have been explored (Kutty, 1969; Rudra, 1972; and Jain, Kutty, Roychowdhury, and Chatterjee, 1975) and are shown in Figure 6.2, except Krishnapur. The Kota dinosaur *Barapasaurus tagorei* is a large sauropod with rather slender limbs. Teeth are spoon-shaped with anterior and posterior keels bearing coarse denticles. The neural spines are not bifurcate and the posterior dorsals are fairly high. The sacrum has four co-ossified vertebrae, with "waisted" amphiplatyan centra, and the sacricostal yokes are set close together (Figure 6.7). Although *B. tagorei* is the earliest known sauropod, it attained a size comparable to later sauropods, being only slightly smaller than *Diplodocus carnegii* (Jain, Kutty, Roychowdhury, and Chatterjee, 1979). However, some of the skeletal elements show that many characters have not reached the stage of development seen in later sauropods.

Among the typical sauropod features of *B. tagorei* are the girdles, especially the ilium (Figure 6.7), the sacrum consisting of four coalesced vertebrae with a well-developed sacricostal yoke, the cervicals which are elongate and opisthocoelous, the anterior dorsals which are opisthocoelous, and evidence of bone reduction by way of excavations on the presacrals. The characters in which it shows prosauropod features, in addition to several features of the vertebral column, are the slenderness of the femur, the acuminate and slightly declined tip of the fourth trochanter, and the smaller size of the pelvic basin.

Yadagiri, Prasad, and Satsangi (1979) have reported excavation of the about 800 specimens representing sauropod dinosaur bones from the Kota Formation. Part of the collection is articulated. This is likely to provide exact relative proportions of fore and hind limbs. Among the collections, ilia of two different dinosaurs are mentioned. One "appeared to be related to the sauropod dinosaur, *Barapasaurus tagorei*" and the other "indicates that it is closer to prosauropods than sauropods." Although detailed description is awaited, Dr. Colbert examined some of this material while visiting Hyderabad during January 1977. He took keen interest in one associated pelvis shown to him by the officers of the Geological Survey of India. The ilium was quite unlike a sauropod ilium. It was tentatively suggested by Dr. Colbert that it shows affinity with an ornithischian ilium. I had an opportunity to examine a photograph taken by Mrs. Colbert. The resemblance is unmistakable. This may give a clue to the presence of a possible ornithischian dinosaur, in addition to sauropods, in the Kota. However, this should be regarded as a very tentative suggestion until laboratory preparation and description of the material.

Fragmentary crocodilian remains have been known from the Kota beds since the last century (Owen, 1852). Although no further material has been reported, considerable new material has been collected during exploration of the Kotas during the sixties. It is awaiting description and will enlarge the understanding of the

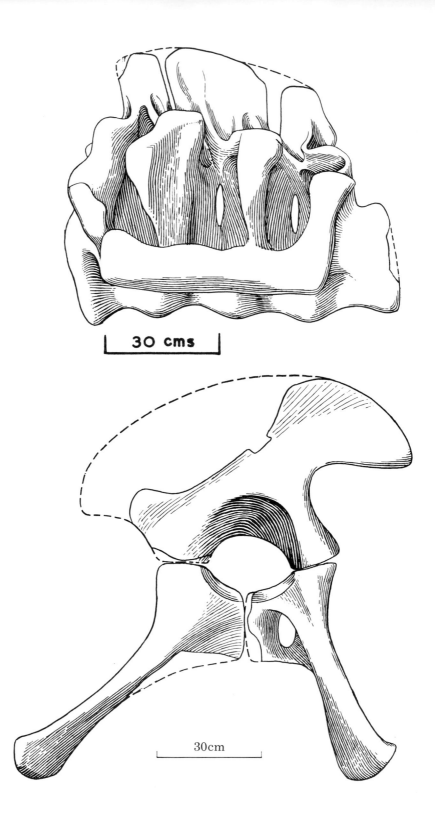

30 cms

30cm

Kota crocodiles. An early Middle Jurassic crocodile has recently been reported from central Oregon, U.S.A. (Buffetaut, 1979). They also occur in the Early to Late Jurassic of Europe, Africa, Madagascar, and South America. They have been referred to the Mesosuchian family Teleosauridae. The Kota crocodile and those from Chile may be the earliest representatives according to Buffetaut.

Owen's material consists of a mass of dermal scutes, with a femur and some fragments of other bones firmly cemented together by the matrix. The scutes are either squarish or oblong, having numerous well-defined and rather small hemispherical pits upon their outer surface which is flat and without any carinal elevation. Owen suggested that these differ from the dermal scutes of the relatively common extant *Gavialis*. In the latter the pits are relatively larger, more frequently confluent, and the middle of the pitted surface in most of the scutes is raised into a keel. The characters of the scutes and the length and slenderness of the femur shows closer resemblance to the marine *Teleosaurus* and amphicoelian crocodiles than to *Gavialis*.

Mammals

While recommending areas for evolutionary studies in India, J. B. S. Haldane (1955) remarked on the existence of continental Jurassic beds in India which are almost unknown in Europe. Because of this fact the knowledge about Jurassic mammals has been very scanty. He further asserted that "There are several areas in India where Jurassic strata of continental origin are to be found. So I think it quite likely that an Indian paleontologist will find the first complete skeleton of a Jurassic mammal. . . ." However, the actual discovery of Early Jurassic mammals from the Kota Formation has been made only recently (Datta, Yadagiri, and Rao, 1978). It may be pointed out that higher mammalian taxa became differentiated from one another long before the Bathonian (Middle Jurassic) as illustrated by Freeman (1979). Therefore, Lower Jurassic was a time of rapid mammalian diversification, yet so little is known from this period that our knowledge is unbalanced.

There is no doubt that the material described by Datta, et al. (1978), comes from the Kota Formation. In fact the material was obtained by screening marly clay through a series of sieves. The marly clay overlies cross-bedded sandstone and was removed from a pit opened for the collection of dinosaur bones near Yamanpalli village (Figure 6.2). No attempt has been made to give the specimens any taxonomic status. The material is microscopic and consists of mandibles with attached teeth, an isolated tooth, limb bones, pelvic and pectoral girdles, vertebrae, and broken parts of a skull. In addition, microscopic eggs (0.35X0.23 mm) have been assigned to the micromammals. I have serious doubts on the reference of the so-called eggs to the mammals primarily because of extreme microscopic dimensions. After all, eggs of oviparous mammals, even of the smallest mammal that ever existed on earth could not possibly be of such minute size. Kielan-Jaworowska (1979) estimated the circular cross-section of the egg of *Kryptobaatar*, a Late Cretaceous multituberculate, to be 3.4 mm in diameter which would mean a smaller egg than any known cleidoic egg. The remaining Kota skeletal material is, however, entirely mammalian.

Figure 6.7. *Barapasaurus tagorei,* the sauropod dinosaur from Kota Formation: *above,* lateral view of sacrum from left side; *below,* pelvic girdle, lateral view of right side (from Jain, Kutty, Roychowdhury, and Chatterjee, 1975).

The isolated tooth, obviously a molar, is well figured and has both roots well preserved. There are four cusps of which the third is longest. The tooth is 0.58 mm and the roots are 0.45 mm long. A spatulate incisor (3.00 mm) is also figured. A portion of the mandible, 2.25 mm long, has four teeth in position. The height of the crown of the teeth varies from 0.10 to 0.25 mm. Among the postcranial bones is a femur, 2.30 mm long, minimum width of the shaft being 0.40 mm. The maximum width of the condyles is 0.66 mm and the head is 0.45 mm. Other bones which have been recorded are tibia, humerus, scapula, pelvic girdle, and a vertebra. The Mesozoic African mammals were no more than 10–17 cm in total length and weighed 20–35 gm. In comparison, it appears that either the Kota mammal material is of juvenile forms or indeed they were extremely small in size. It has been suggested by the authors that detailed work is in progress and that present indications are that these may belong to two or three different orders.

Crompton and Jenkins (1978) while reviewing the Mesozoic mammals have drawn attention to the fact that except for three mammals from the Mesozoic rocks of Africa, *Megazostrodon rudnerae* and *Erythrotherium parringtoni* from Late Triassic, and *Branchatherulum tendagurense* from Late Jurassic, no Mesozoic mammals are known from the southern continents. In view of this, the discovery of an Early Jurassic mammalian fauna from the Indian peninsula is extremely important and the results of detailed investigation are eagerly awaited. Many fundamental questions about mammalian evolution, such as the origin and distribution of monotremes or the differentiation of marsupials and placentals from mammals of "therian grade" can be expected to be answered from the taxonomic evaluation of Kota mammalian fauna.

Invertebrates

Tasch, Sastry, Shah, Rao, Rao, and Ghosh (1973) have identified seven new estheriid-bearing units from the argillaceous limestone of the Kota Formation. Species of *Cyzicus, Paleolimnadia,* and estheriniids occur through various sections. Associates of estherids are fish, plants, and abundant insects. Govindan (1975) discovered a small, well-preserved ostracod faunule occurring in the limestone of the Kota Formation. *Darwinula* is well represented in most of the samples and has been found to be very close to *D. sarytirmenensis* from the Middle Jurassic of the Russian Mangishlala Peninsula, except being slightly smaller in size. The Russian specimens attain a size up to 1.14 X 0.83 mm whereas Indian specimens are up to 0.80 X 0.90 mm. Three forms have been assigned to ? *Limnocythere* pending availability of better material. A third form is assigned to a new species as *Timiriasevia digitalis*. This form differs from all the Russian species of the genus.

Rao and Shah (1959) reported the occurrence of fossil insects in a richly fossiliferous band of limestone in the Kota Formation. Blattids are predominant in the collection, being represented by a single impression of head, pronotum, and forewings, as well as numerous individual specimens of forewings, hind wings, pronota, and body casts. Next in abundance are the Coleoptera. These are represented by a few forms wherein head, prothorax, and elytra are intact. There are also several specimens of elytra of different types. The Hemiptera include mostly wings. The specimens are well illustrated but taxonomic determination is pending. Tasch, et al. (1973), have further affirmed the presence of abundant insects, often excellently preserved. It is believed that the Coleoptera alone may represent as many as twenty species belonging to four to six genera. Blattids from the Kota beds appear to be similar to those found at several Gondwana localities outside India. Tasch, et al. (1973), have recognized eight distinct insect bearing horizons in a vertical sequence, but there could be several more such horizons.

Flora

The term "Kota flora" has been in general use to describe the plant fossils reported from the Kota "group" which was described by King (1881) comprising the limestone and plant-bearing beds of Nowgaon, Balanpur, and Chirakunta. Kutty (1969) has, on the basis of floral evidence, recognized the Gangapur Formation, as separate from the Kota Formation. Shah, Singh, and Gururaja (1973) and Rao (1977) have, in part, cleared up certain controversies with regard to the precise stratigraphic relations of Kota flora. The former have recognized three floral zones in the Kota group: Zone A containing *Glossopteris, Dicroidium,* and *Noeggerathiopsis* as a lower unit of Rhaetic age; Zone B, the limestone horizon, having meager plant fossils which are not diagnostic; and Zone C, generally referred to as Gangapur beds which have yielded a majority of plant fossils such as *Equisetites, Gleichenites gleichenoides, G. rewahensis, Coniopteris hymnophylloides, Cladophlebis indicus, Sphenopteris sp., Hausmannia sp., Pagiophyllum sp., Araucarites sp., Podozamites sp., Ptillophyllum acutifolium, Otozamites sp.,* and others.

Mapping has clearly shown that the Gangapur beds lie unconformably on beds which have yielded the typical Kota fauna, the latter including a limestone band. The Kota beds are equivalent to Zone B of Shah, Singh, and Gururaja (1973) which has meager undiagnostic plant fossils. The only other report of plant remains from a limestone band of the Kota Formation is that of Rao and Shah (1967). The plants reported by them are *Equisetites, Cladophlebis, Otozamites, Sphenopteris, Hausmania,* and *Pagiophyllum.* They are considered to be of Lower Jurassic age. However, in the absence of locality data it is possible that there has been some confusion of locality. All the six genera are also included in Zone C (Gangapur Formation) of Shah, Singh, and Gururaja (1973).

Association of dinosaur bones in the Kota Formation with large logs of wood (up to 3 m) has been mentioned by Jain, Robinson, and Roychowdhury (1962). No leaves or fructifications have been mentioned and the plants have not been identified. Tasch, et al. (1973), mention the occurrence of fossil plants (also fossil wood, leaf, seed coat and carbonized plant) but no identifications are given. The occurrence of "Charophyta" has been mentioned in passing by Datta, et al. 1978). As such, except for the occurrence of fossil wood and "Charophyta," the flora of the Kota Formation is questionable.

Kota Environment and Ecology

An analysis of evidence from fossils, sedimentary rocks, and oxygen isotopes led Hallam (1975) to conclude that conditions were warmer and more uniform in the Jurassic than the present day. The Southern Hemisphere floras differ markedly from those of the Northern Hemisphere during Rhaetic-Liassic. It suggests separate floristic provinces with warm equable climates between 50 degrees and 60 degrees on either side of the equator. The Indian Jurassic climate was evidently of the tropical humid type as suggested by Hallam (1975).

The Kota Formation, like all other members of the Gondwana Group in the northwestern part of the Pranhita-Godavari valley, is a terrestrial deposit. It consists of fluviatile and lacustrine sands and clays, and thin limestones representing deposits in inland lakes. Robinson (1970) on the basis of the geology of the Kota Formation has concluded that "there can be little doubt that the limestones are freshwater in origin." This is confirmed by Tasch, et al. (1973), on the basis of conchostracans, and by Govindan (1975) on the basis of Kota ostracods. In addition, Robinson (1970) has suggested that the Kota represents two different depositional environments. The sandstones might be fluviatile and the limestones and succeeding shales might be lacustrine. There is a suggestion of different climatic conditions during Kota times, from "moderately high relief and rainfall . . . to a low relief and rainfall." The intermediate facies (between arkosic sandstone and limestone), may have been laid down around the margins of "an expanding and very shallow lake. . . ." Tasch, et al. (1973), on the basis of the estherid fauna, which in one locality (K-7) extended over a distance of some 35 m, suggests the presence of "minibasin (ponds)" in the Kota Formation.

The presence of freshwater limestone in the Kota succession implies that at the time of deposition the surrounding terrain had a mature relief so that little clastic material was being brought into the basin of deposition. A moderately hot climate with periods of low rainfall would be necessary for the precipitation of lime in the waters of the lake. Periods of drought are implied by the limestone layers with desiccation cracks. Although dissociated fish remains are not infrequent in the Kota limestone, good associated specimens are often found. The latter imply unusual conditions summarized by Rayner (1958) as due either to abnormally good conditions of preservation such as (1) bottom-waters being foul, (2) absence of natural scavengers, and (3) rapid burial at death. Alternatively, burial can be effected by a natural catastrophe of short duration relative to the rate of sedimentation. Associated fish specimens have been found in mudcracked limestone layers. This suggests the possibility that fishes died either due to lack of water on the drying lake or due to oxygen deficiency in the evaporating water of the Kota lake.

The semionotid fishes dominate in numbers over all the fish remains known from the Kota. Among the semionotids also there is a definite order of abundance (Jain, 1974a), *Lepidotes* abundant, *Paradapedium* uncommon, and *Tetragonolepis* rare. The coelacanthid (*Indocoelacanthus robustus*) description was based on a single partly associated skull, squamation, and other postcranial material. In the meantime one more skull and portions of a body with fins have been found. The pholidophorid is represented by three almost complete specimens and three fragmentary bits in addition to unpublished material. Juvenile specimens of actinopterygian fishes from the Kota are extremely rare in comparison to medium and large fishes (Jain, 1974). The estimated sizes for various members are: *Lepidotes*, 15–35 cm, *Paradapedium* and *Tetragonolepis*, 12–32 cm, *Pholidophorus*, 11–16 cm, and *Indocoelacanthus* up to 70 cm.

Lepidotes has the longest geological history, and one which gives evidence of it colonizing marine or fresh waters. Species occur in the brackish to freshwater Rhaetic, marine upper Liassic, Oxfordian, and Kimmeridgian, brackish to freshwater Purbeckian, freshwater Wealden and in both marine (British) and freshwater (Brazilian) Cenomanian. The genus as a whole may be regarded as euryhaline, and *Lepidotes deccanensis* from the Kota was probably osmotically adaptable and could withstand at least moderate concentrations of salt in the Kota lakes. It had slender, cylindrical, acutely conical teeth in the dentary and premaxilla and more robust teeth in the vomers. The biting apparatus was suited for preying upon fast moving invertebrates including estherids and ostracods, in addition to soft bodied annelid (?) worms (Jain, 1974a). The presence of strong paired and unpaired fins suggest that these were active predators.

Paradapedium and *Tetragonolepis* had relatively poor development of paired and unpaired fins. *Paradapedium* had 15–16 acutely pointed conical teeth in the dentary and slender premaxillary teeth. Similarly *Tetragonolepis* had fine pointed teeth on the premaxilla, maxilla, and dentary. As such it can be presumed that they were predators of fast-moving prey. Although these fishes occurred in a freshwater environment in the Kota Formation, *Tetragonolepis* in Europe is known from marine sediments only and *Dapedium*, which has close similarity to Indian *Paradapedium*, is also marine. It is, therefore, likely that these fishes were euryhaline forms, as mentioned in an earlier section. In addition, Schaeffer (1967) has suggested that these fishes probably assumed ecological roles similar to certain species of *Cyprinodon*, the centrarchids, and the characins.

The Kota coelacanth (*I. robustus*) was a large fish, toothless and probably bottom-living (Jain, 1974a) in a shallow lake. The coelacanths in general appear to have been rather sedentary animals lying in wait for their prey. All fossil coelacanths were near-shore or shallow-water fishes (Thomson, 1969). Only *Latimeria* has been found at a depth of about 73 meters. This fish exhibits a poor dentition, and among its gut contents have been found small intact fishes in addition to other foodstuff. *I. robustus* is assumed to have had similar food habits and, in addition to aquatic plants, may have devoured small fishes. As indicated earlier, juvenile specimens of fishes are indeed rare in the collection.

The Kota pholidophorids have well-developed strong paired and unpaired fins, but poor dentition. They may have been active fishes like *Lepidotes* but may have depended less on fast moving prey for food, rather eating the soft annelids, and nibbling aquatic plants.

It is believed that pterodactyls and pterosaurs lived near the sea coast because all the material comes from marine sediments, except the Kota pterosaur (*Campylognathoides indicus*). In view of the earlier discussion about Kota environment, *C. indicus* can be visualized as inhabiting the trees close by the Kota lake and occasionally swooping down to prey upon small actinopterygian fishes. The Solnhofen pterodactyls were fish-eaters.

The material of Kota crocodiles is extremely fragmentary. Jurassic crocodiles show great dissimilarity in the length of the legs. Buffetaut and Thierry (1977) have suggested that teleosaurid crocodiles preferred shallow or temporarily emerged places where they could be on land, and could be equally at home in the water. The Kota crocodile was probably a "short-nosed" one and, like the extant *Crocodylus*, was probably a fish eater. It is not uncommon to find an occasional live crocodile in the Godavari river today and it seems a somewhat similar ecological role was played by them in the Jurassic Kota lake.

The Kota dinosaur *Barapasaurus tagorei* was herbivorous and was probably similar to better known saurischian Jurassic dinosaurs like *Diplodocus* and *Camarasaurus* in its habits and habitat. The debate on sauropod habitat is as old as the discovery of the earliest dinosaur and this has been aptly summed up in an excellent review by Coombs (1975). The data on sauropod narial morphology indicates either

Figure 6.8. An attempted restoration of the main Kota fossil assemblage in silhouette to indicate suggested ecological role (not to scale). The outlines, except semionotid fishes, are only suggestive and tentative. (1) *Lepidotes deccanensis;* (2) *Paradapedium egertoni;* (3) *Tetragonolepis oldhami;* (4) *Pholidophorus kingii* and *P. indicus;* (5) *Indocoelacanthus robustus;* (6) *Campylognathoides indicus;* (7) *Barapasaurus tagorei;* (8) teleosaurid crocodile; (9) Kota mammal; (10) Charophyte; and (11) fossil wood.

aquatic or terrestrial habits. The axial and appendicular modifications point primarily to terrestrial behavior. The deep thorax is an adaptation to problems of terrestrial weight bearing. Some sauropods probably could rear up on their hind limbs. They definitely entered streams at least sometimes, but sedimentologic evidence does not support immersion in deep lakes as sauropods are frequently pictured. Charig (1979) has suggested that "the sauropods were proper land-dwellers." All in all, it can be assumed that *B. tagorei* was inhabiting land close to Kota lake and largely feeding upon nearby aquatic vegetation.

Knowledge about Kota micromammals is still meager and speculation about their habits and habitat can be only tentative. Jerison (1973) has recently contributed to the knowledge of Mesozoic mammals, and this has been summarized by Crompton and Jenkins (1978). Crompton, Taylor, and Jagger (1978) have suggested that early mammals were probably similar to modern day hedgehogs and Madagascan tenrecs in maintaining a low body temperature (25 degrees-30 degrees C) and possessing a reptilian metabolic rate three to five times lower than most living mammals. In all likelihood these were nocturnal animals, like hedgehogs and tenrecs, secreting themselves in burrows during the day, in which reptilian metabolic rate is economical and sufficient to sustain a constant low body temperature. Like the African forms, the Kota mammals appear to be small insectivores with dentition designed for puncturing and shearing. This is justifiable in view of the abundance of insect remains from the Kota. Whereas blattids are primarily nocturnal, the Coleoptera and Hemiptera are diurnal. In view of the miniature size of Kota mammals it is feasible to suggest that nymphal blattids (nonflying), larval Coleoptera, and nymphal Hemiptera may have formed their main diet.

The faunal elements from the Kota filled diverse ecological niches. An attempt has been made to place the main faunal elements in their suggested ecological habitats in silhouette in Figure 6.8. The outlines are only suggestive of possible body form and *not to scale* with each other.

Significance of Kota Fauna for Continental Fit

The significance of Kota fauna as evidence of lower Jurassic geographic arrangement of southern continents can be discussed on three different counts: (a) the fish evidence, (b) the tetrapod evidence, and (c) the invertebrate evidence.

The Kota semionotid fishes *Lepidotes, Paradapedium,* and *Tetragonolepis* are helpful in suggesting a Lower Jurassic age for the Kota Formation, but their absence in general in the Gondwanaland continents provides little information relevant to land relationships during the Early Jurassic. The coelacanths, however, have been recorded from Gondwanaland continents: *Lualabaea* from the Middle-Upper Jurassic of Zaire, *Whitea* from the Lower Triassic of Madagascar, and *Indocoelacanthus* from the Lower Jurassic of India. *Lualabaea* and *Indocoelacanthus* were freshwater fishes whereas *Whitea* was marine. The extant coelacanth *Latimeria* inhabits deep sea waters adjacent to Madagascar near the Comoro Islands. Although the coelacanths from the mainland of Africa, Madagascar, and India became extinct in the early Mesozoic, they continued to survive for over 200 million years undisturbed in the deep sea. There must have been good land connections between Africa and India for the origin and dispersal of the coelacanth stock.

Another group of fishes from the Kota beds are the pholidophorids *Pholidopho-*

rus kingii and *P. indicus.* Australian Lower Jurassic continental deposits have also yielded pholidophorid fishes. On the basis of scale features, Griffith and Patterson (1963) have divided these fishes into two groups: Group I containing the genera *Aetheolepis* and *Aphnelepis,* and Group II containing *Archaeomanaene* and *Madariscus.* Antarctica has yielded more than 60 specimens of pholidophorid fishes. These have been retrieved from thin sedimentary interbeds in the Kirkpatrick Basalt of the Ferrar Group. An Early Jurassic age has been assigned to these beds on the basis of potassium-argon dating (Schaeffer, 1971, 1972). The Antarctic pholidophorid, *Oreochima ellioti,* is a freshwater form. There is no other record of pholidophorid fishes from these continents in any other geological horizon. Seaways are great barriers to free migrations of freshwater fishes. It is difficult to explain the occurrence of freshwater pholidophorids unless the continents of India, Australia, and Antarctica were in close proximity for migration during the Early Jurassic. Evidence for Early Jurassic land connections of Gondwanaland continents on the basis of tetrapods, especially dinosaurs, has been strongly emphasized by Colbert (1973, 1977, 1978).

The presence of gigantic dinosaurs in the Kota beds is interesting not only because it demonstrates the rapid attainment of gigantic size within this line of reptilian evolution, but also because it shows that India still maintained connections with other parts of the world during Jurassic times. This means that the Indian peninsula had not yet wrenched away from its ligation with Africa to begin its drift towards the Asiatic mainland. For the most part, knowledge about early sauropods from other Gondwanaland continents is fragmentary, except a putative sauropod predecessor, *Vulcanodon,* described by Raath (1972) from Rhodesia, *Rhoetosaurus* from Middle Jurassic (Wallon Series of Queensland) Australia, and *Amygdalodon* from possible Middle Jurassic beds of Patagonia. On this basis Colbert has concluded that during the Early Jurassic peninsular India was a landlocked part of Gondwanaland.

The fact that the pterosaur from Kota, *Campylognathoides indicus,* is closely related to European forms of Early Jurassic age might be regarded as equivocal evidence concerning the continental relationships of peninsular India. Pterosaurs could fly across water barriers, but probably did not cross extensive oceans. They were common during the Jurassic in Laurasia, but left a poor fossil record in Gondwana sediments.

Teleosaurid crocodile remains from the Kota are extremely fragmentary. Their occurrence in the Early Jurassic of Chile and Early to Late Jurassic of Africa, Madagascar, and South America supports the presence of land bridges during the Jurassic among these continents.

Evidence of Kota mammals cannot be considered until their systematic evaluation is available. As mentioned earlier, three mammals from Africa and the Kota specimens are all that is known of Mesozoic mammals from the Gondwanaland continents.

Scattered estheriid occurrences have been reported since the mid-nineteenth century from the Indian Gondwanas, including the Kota beds. Only recently (Tasch, et al., 1973) a systematic biostratigraphic evaluation of the conchostracan-bearing beds and their relevance to the problem of the relations of continental blocks in the Gondwana configuration has become available. Distribution of estheriids in the Jurassic Kota Formation of India, the Lower Cretaceous of Brazil (Bahia), and in the Australian Triassic indicates that estheriids were widely dispersed through-out Gondwana continents during Mesozoic times. Species of *Palaeolimnadia* in the Kota are comparable to palaeolimnadids from Antarctica (Transantarctic Mountains and Mauger Nunatak), western and eastern Australia, and Russia and South America (Tasch, et al., 1973).

The ostracod fauna identified by Govindan (1975) from the Kota has abundant *Darwinula* which incidentally has been also reported from the Upper Triassic Maleri Formation by Sohn and Chatterjee (1979). In addition, species of *Timiriasevia* and ? *Limnocythere* have been identified in the Kota. All are typically freshwater forms. In the absence of data from similar horizons of the Gondwanaland continents, nothing much can be adduced about their relevance to continental connections. Moreover, *Darwinula* ranges from Carboniferous to Recent and commonly occurs in sedimentary rocks both north and south of the Tethys, except Australia and Antarctica.

Fossil beetles (Coleoptera), on the other hand, are distributed in the Triassic of Australia and Africa, Jurassic of Antarctica, India, and South America (Tasch, et al., 1973). The presence of Jurassic palaeolimnadids and beetles in Antarctica and India, and estheriniids in India and Australia strongly suggests that there existed non-marine dispersal routes between these continents during Triassic-Jurassic periods. Smith and Hallam (1970) place India close to Madagascar and Antarctica, while Veevers, et al. (1971), place India close to southwest Australia and not in contact with Antarctica. Tasch, et al. (1973), on the basis of new data, suggest that the Indian subcontinent was close to both Antarctica and Australia.

Sohn and Chatterjee (1979) have suggested that the presence of *Darwinula* in the Indian Triassic and its absence in Australia and Antarctica does not support the models in which the eastern margin of India is connected with Australia or Antarctica. This view is contrary to the conclusions of Tasch, et al. (1973), though the latter have argued partly on the basis of Lower Jurassic conchostracans and beetles and partly on the basis of Triassic beetles from Australia. Yet it would be paradoxical to assume that the position of these continents was drastically different between Late Triassic and Early Jurassic, though it may have been so after the mid-Cretaceous. The Kota vertebrate fossil evidence (especially the presence of pholidophorid fishes), however, supports the placing of the Indian peninsula close to both Antarctica and Australia, as suggested by Tasch, et al. (1973).

The exploration of Antarctica for vertebrate fossils is relatively recent. The results so far have been overwhelmingly encouraging. These include the discovery of *Lystrosaurus*, an important fossil in the Gondwanaland puzzle, in addition to possible labyrinthodonts. Australian Mesozoic vertebrate faunas, especially tetrapods, are poorly known. The faunas include three Triassic and one Jurassic labyrinthodonts, a Triassic lepidosaur, one possible Triassic archosaur, and fragments and footprints of Jurassic and Cretaceous dinosaurs. Active exploratory work is presently going on in India, Australia, and Antarctica. This is likely to bring to light a wealth of new Mesozoic vertebrate faunas during the next decade. This would provide further evidence to review and assess the case fully. It is likely that the issue may be decided with a lesser degree of uncertainty in the future.

Literature Cited

Arkell, W. J. 1956. Jurassic geology of the world. Oliver & Boyd, Edinburgh, 806 pp.

Buffetaut, E. 1979. Jurassic marine crocodilians (Mesosuchia: Teleosauridae) from central Oregon: first record in North America. J. Paleontol., 53(1):210–15.

Buffetaut, E. and Thierry, J. 1977. Les crocodiliens fossils du Jurassique moyen et supérieur de Bourgogne. Geobios, 10(2):151–94.

Charig, A. J. 1979. A new look at the dinosaurs. Mayflower Books, New York, 160 pp.

Colbert, E. H. 1958. Relationships of the Triassic Maleri fauna. J. Palaeontol. Soc. India, 3:68–81.

————. 1973. Wandering lands and animals. E. P. Dutton & Co., Inc., New York, 323 pp.

————. 1977. Gondwana vertebrates. Key paper: Section III, Fourth International Gondwana Symposium, Calcutta: 1–18. Also Fourth International Gondwana Symposium, Calcutta, Proc. (1979), 1:135–43.

————. 1978. Mesozoic tetrapods and the northward migration of India. J. Palaeontol. Soc. India, (J. A. Orlov Mem. Number), 20:138–45.

Coombs, W. P. 1975. Sauropod habits and habitats. Palaeogeog., Palaeoclimat., Palaeoecol., 17:1–33.

Crompton, A. W.; Taylor, C. R.; and Jagger, J. A. 1978. Evolution of homoeothermy in mammals. Nature, 272:333–36.

Crompton, A. W., and Jenkins, F. A. 1978. Mesozoic Mammals. *In* Maglio, V. J. and Cooke, H. B. S. (eds.), Evolution of African mammals. Harvard University Press, Cambridge, pp. 46–55.

Datta, P. M.; Yadagiri, P.; and Rao, B. R. J. 1978. Discovery of Early Jurassic micromammals from Upper Gondwana sequence of Pranhita-Godavari Valley, India. Geol. Soc. India, J., 19(2):64–68.

Freeman, E. F. 1979. A Middle Jurassic mammal bed from Oxfordshire. Palaeontology, 22(1):135–66.

Govindan, A. 1975. Jurassic freshwater ostracods from the Kota limestone of India. Palaeontology, 18(1):207–16.

Griffith, J. and Patterson, C. 1963. The structure and relationships of the Jurassic fish *Ichthyokentema purbeckensis.* Brit. Mus. (Natur. Hist.), Bull., Geol., 6(3):1–43.

Haldane, J. B. S. 1955. Suggestions for evolutionary studies in India. India Nat. Inst. Sci., Bull., 7:25–28.

Hallam, A., 1975. Jurassic environments. Cambridge University Press, Cambridge, 169 pp.

Jain, S. L. 1959. Fossil fishes from the Kota Formation of India. Geol. Soc. London, Proc., 1565:26–27.

————. 1973. New specimens of Lower Jurassic holostean fishes from India. Palaeontology, 16(1):149–77.

————. 1974a. *Indocoelacanthus robustus* n. gen., n. sp., (Coelacanthidae, Lower Jurassic) the first fossil coelacanth from India. J. Paleontol., 48(1):49–62.

————. 1974b. Jurassic pterosaur from India. Geol. Soc. India, J., 15(3):330–35.

Jain, S. L.; and Kutty, T. S.; Roychowdhury, T.; and Chatterjee, S. 1975. The sauropod dinosaur from the Lower Jurassic Kota Formation of India. Roy. Soc. London, Proc., A, 188:221–28.

————. 1979. Some characteristics of *Barapasaurus tagorei,* a sauropod dinosaur from the Lower Jurassic of Deccan, India. Fourth International Gondwana Symposium, Calcutta (1977), Proc., 1:204–16.

Jain, S. L., and Robinson, P. L. 1963. Some new specimens of the fossil fish *Lepidotes* from the English Upper Jurassic. Zool. Soc. London, Proc., 141(1):119–35.

Jain, S. L.; Robinson, P. L.; and Roychowdhury, T. K. 1962. A new vertebrate fauna from the Early Jurassic of the Deccan, India. Nature 194(4830):755–57.

Jerison, H. J. 1973. Evolution of the brain and intelligence. Academic Press, New York.

Kielan-Jaworowska, Z. 1979. Pelvic structure and nature of reproduction in Multituberculata. Nature, 277(5695):402–03.

King, W. 1881. The geology of the Pranhita-Godavari Valley. Geol. Surv. India, Mem., 18:151–311.

Kutty, T. S. 1969. Some contributions to the stratigraphy of the upper Gondwana formations of the Pranhita-Godavari Valley, Central India. Geol. Soc. India, J., 10(1):33–48.

Nybelin, O. 1966. On certain Triassic and Liassic representatives of the family Pholidorphoridae. Brit. Mus. (Natur. Hist.), Bull., Geol., 11(8):353–432.

Owen, R. 1852. Note on the crocodilian remains accompanying Dr. T. L. Bell's paper on Kotah. Geol. Soc. London, Proc., 7:233.

Raath, M. A. 1972. A new dinosaur (Reptilia: Saurischia) from near the Triassic-Jurassic boundary. Arnoldia, 5(30):1–37.

Rao, B. R. J. 1977. Stratigraphic appraisal of "Kota Flora." Geol. Soc. India, J., 18(8):456–58.

Rao, C. N. and Shah, S. C. 1959. Fossil insects from the Gondwanas of India. Indian Minerals, 13:3–5.

———. 1967. On the occurrence of pterosaur from the Kota-Maleri beds of Chanda district. Geol. Surv. India, Rec., 92:315–18. (Dated 1963.)

Rayner, D. H. 1958. The geological environment fossil fishes. *In* Westoll, T. S. (ed.), Studies on fossil vertebrates, pp. 129–56.

Robinson, P. L. 1970. The Indian Gondwana formations—a review. Int. Union Geol. Sci., First Symposium on Gondwana stratigraphy, pp. 201–68.

Rudra, D. K. 1972. A discussion on the Kota Formation of the Pranhita-Godavari valley, Deccan, India. Geol. Min. Metall. Soc. India, Quart. J., 44:213–16.

Satsangi, P. P. and Shah, S. C. 1973. A new fish from the Kota Formation, Pranhita-Godavari basin. Proc. 60th Indian Sci. Congr., Abs., Pt. II:188.

Schaeffer, B. 1967. Late Triassic fishes from the Western United States. Amer. Mus. Natur. Hist., Bull., 135(6):287–342.

———. 1971. Jurassic fishes from Antarctica. Atlantic J., 6(5):190–91.

———. 1972. A Jurassic fish from Antarctica. Amer. Mus. Nov., 2495:1–17.

Shah, S. C.; Singh; and Gururaja, M. N. 1973. Observations on the Post-Triassic Gondwana sequences of India. Palaeobotanist, 20(2):221–37.

Smith, A. G. and Hallam, A., 1970. The fit of southern continents. Nature, 225:139–44.

Sohn, I. G., and Chatterjee, S. 1979. Freshwater ostracodes from the Late Triassic coprolites in central India. J. Paleontol., 53(3):678–86.

Tasch, P.; Sastry, M. V. A.; Shah, S. C.; Rao, B. R. J.; Rao, C. N.; and Ghosh, S. C. 1973. Estheriids of the Indian Gondwanas: Significance for continental fit. Third International Gondwana Symposium: Advances in Stratigraphy and Paleontology, Australia: 445–52.

Thomson, K. S. 1969. The biology of the lobe-finned fishes. Biol. Rev., 44:91–154.

Veevers, J. J.; Jones, J. G., and Talent, J. A. 1971. Indo-Australian stratigraphy and the configuration and dispersal of Gondwanaland. Nature 229:383–88.

Wild, R. 1978. Ein Sauropoden-rest (Reptilia, Saurischia) aus dem Posidonienschiefer (Lias, Toarcium) von Holzmaden. Stuttgarter Beitr. Naturk., Ser. B, 41:1–15.

Yadagiri, P. and Prasad, K. N. 1977. On the discovery of new *Pholidophorus* fishes from the Kota Formation, Adilabad District, Andhra Pradesh. Geol. Soc. India, J., 18(8):436–44.

Yadagiri, P.; Prasad, K. N.; and Satsangi, P. 1979. The sauropod dinosaur from the Kota Formation of the Pranhita-Godavari valley, India. Fourth International Gondwana Symposium, Calcutta (1977). Proc., 1:199–203.

Young, C. C. 1964. On a new pterosaurian from Sinkiang, China. Vert. PalAsiat., 8(3): 221–25.

THE OTIC NOTCH OF METOPOSAURID LABYRINTHODONTS

by Joseph T. Gregory

Introduction

The Metoposauridae are an extinct family of labyrinthodont amphibians characterized by broad flat heads and bodies, small and feeble limbs, stereospondylous vertebrae, and moderately elongate skulls in which the orbits lie in front of the midlength, and small otic notches incise the posterior border. The palate and basicranium display large interpterygoid vacuities, extensive sutures between parasphenoid and pterygoid, double exoccipital condyles, and feeble ossification of the primary braincase which are typical of all Triassic labyrinthodonts (Watson, 1951, pp. 31–78). Metoposaurs are the principal family of labyrinthodonts surviving into the Late Triassic faunas of North America; in Europe they are confined to the earlier faunas of the Keuper; their distribution extends also to the Maleri beds of India (Chowdhury, 1965) and into northern Africa (Dutuit, 1976).

Colbert and Imbric (1956) reviewed earlier studies of the family and concluded that all material, except the indeterminate type of *Dictyocephalus elegans* Leidy, could be assigned to two genera, *Metoposaurus* Lydekker (for all old world forms) and *Eupelor* Cope (for American species). After extensive comparisons of the North American material they recognized three species of *Eupelor*: *E. durus* Cope from the Newark Group in Pennsylvania and New Jersey; *E. fraasi* (Lucas) from the Chinle Formation of Arizona and Dockum Group of Texas (with two geographic subspecies); and *E. browni* (Branson) from the Popo Agie beds of Wyoming. In their analysis, data on morphological variation is related to geographic occurrence, but not to stratigraphic level, within the Upper Triassic deposits.

The comments which follow are based upon material discovered since Colbert and Imbrie published their important analysis of these fossils. It is a pleasure to offer them to Dr. Colbert on his seventy-fifth birthday. My ideas on this subject have benefited from discussions with Donald Baird, Leigh Van Valen, David Wake, and Michael Morales.

Colbert and Imbrie had pointed out that only minor differences in the extent of reticulate sculpture on the clavicles and interclavicles, and in the relationships of the lacrimal with the orbit seemed to separate the American *Eupelor* from the European *Metoposaurus*. Chowdhury (1965) pointed out that the inclusion or exclusion of the lacrimal from the orbital margin varied among American species of metoposaurs and did not constitute a consistent difference between American and Eurasian members of this family. He proposed that all species of the family should be included in the single genus *Metoposaurus*. Dutuit (1976) agreed and noted the practical difficulty of determining whether the lacrimal was included in the orbital margin in specimens with closed or indistinctly preserved sutures.

In 1958 and 1961 vertebrate fossils were collected from the Redonda Formation, the uppermost deposits of the Dockum Group in the region of Tucumcari, New Mexico. Metoposaur skulls in this fauna differ from those found at lower levels in the Dockum and in the Chinle Formation in the absence of an otic notch or tabular horns. The Capitosauridae, another family of Triassic labyrinthodonts, show striking evolutionary changes in the form of their otic notches with time, and it seemed possible that this character might also undergo systematic changes within the metoposaurs. Accordingly the form and size of the notch in various metoposaurs has been reviewed. No trend has been detected, but two contrasting conditions, termed for brevity skulls with or without otic notches, can be recognized. It is concluded that two genera of metoposaurs should be recognized for which the names *Anaschisma* Branson and *Metoposaurus* Lydekker are valid. In the discussion which follows, various specimens or collections of local populations will be referred to by the names under which they were originally described, rather than by their current identification.

Stratigraphic positions of various collections of metoposaur skulls have been determined by the associated phytosaurs (Gregory, 1957; Colbert and Gregory, 1957) and in part by physical evidence of superposition of the deposits. Thus the locality in Scurry County, Texas, from which Case's collection of *Buettneria bakeri* came, lies low in the Dockum and is probably nearly the same age as the Howard County, Texas, deposits which produced *Buettneria howardensis* Sawin and somewhat older than the Crosby County, Texas, *Buettneria perfecta* site. Both phytosaurs and labyrinthodonts indicate that the Sierrita de la Cruz site in Potter and Oldham County, Texas, is close to the faunal level of the Crosby County locality. The Redonda Formation in eastern New Mexico clearly overlies the other portions of the Dockum group. Correlation of the Petrified Forest member of the Chinle Formation in Arizona and northwestern New Mexico with the upper part of the Dockum is established in general but various sites cannot be placed in sequence. Unfortunately the rich deposit of metoposaurs south of Lamy, New Mexico, lies in a complexly faulted area in which stratigraphic levels are unlikely to be established. Its position therefore must be assigned solely on the basis of the morphology of the fossils themselves, which suggest those of the lower levels of the Dockum.

The primitive phytosaur genera of the Popo Agie, *Paleorhinus* and *Angistorhinus*, also occur in the Howard County, Texas, localities and form the basis for placing these at the same stratigraphic level. Other phytosaur genera occur at other localities in the Dockum and Chinle.

Features of Metoposaur Skulls

Many features of metoposaur skulls which are mentioned in published diagnoses are likely to be affected by postmortem distortion of the skull, or to change in a regular fashion with growth. Skull length, and lengths from front or rear of the roof to the orbital borders, pineal foramen, and transverse widths between nares, orbits, or otic notches are little liable to distortion because the central part of the skull roof is nearly plane. Maximum skull width, on the contrary, is easily modified by crushing. Orbits are frequently distorted in many fossil vertebrates, and although the general shape of the metoposaur skull does not suggest that this should be particularly sensitive to deformation, I do not consider differences in orbital diameters as apt to be a useful systematic character for this family. Thickness of skull roofing bones, and depth of pits of the surficial sculpture are strongly correlated with size and are reasonably regarded as growth features.

Branson and Mehl (1929) stressed variations in arrangement of the inner palatal teeth in their diagnoses of various American metoposaurs. Some features which they cited, such as the absence of vomerine teeth in *Anaschisma browni,* may be the result of imperfect preservation or preparation of the specimens. No other comprehensive analysis of these characters has been made and they should not be neglected in the search for significant morphological variation.

Both Case (1932) and Branson and Mehl (1929) stressed the difference in the pattern of sensory canals on the skull roof betwen *Anaschisma,* in which the supraorbital canal does not meet the supratemporal or postorbital canals, and *Buettneria,* in which these canals meet. *Koskinonodon* has an *Anaschisma*-like pattern; in *Borborophagus* the relationships are slightly uncertain from Branson and Mehl's illustrations, but a connection is suggested on the right side. Popo Agie specimens thus are consistent among themselves in the pattern of sensory canals, and differ from the Arizona and Texas populations in this feature.

Colbert and Imbrie (1956) based their conclusions about North American species of metoposaurs upon comparisons of four pairs of measurements:

(1) greatest width of the skull : skull length;
(2) postorbital length of skull : antorbital length;
(3) greatest width of interclavicle : extreme length of interclavicle;
(4) clavicle length : clavicle width.

They also showed that certain regular changes in proportion occur with growth, the skull width increasing faster than length so that the largest skulls are relatively the broadest (1956, p. 441). As their analysis clearly shows, these proportions are as variable within geographic populations as between various samples. They did not, however, address the question of whether these rather general proportions, which characterize metoposaurs in contrast to other families of labyrinthodonts, are the most appropriate criteria for distinguishing generic or specific groups within the family.

So far as the writer is aware, variations in the otic notch of metoposaurs have not previously been suggested as taxonomically significant, although Branson (1905) noted the extremely shallow notch in skulls of *Anaschisma* from Wyoming.

Form of the Otic Notch in Metoposaurs

Metoposaurus diagnosticus (Meyer) and *M. heimi* Kuhn from Keuper deposits of Europe, and *M. ouazzoui* Dutuit from Morocco have well-developed otic notches whose medial borders almost parallel the skull midline. The tabulars extend further back than the postparietals to form gently rounded, weakly developed "horns." The notch itself opens directly backward and is about as wide as deep.

In *Buettneria howardensis* Sawin from Howard County, Texas, and to a lesser extent in the small *B. bakeri* skulls from Scurry County, Texas, and judging from illustrations in the Wyoming skull named *Koskinonodon princeps* Branson and Mehl (1929, pl. 4), the otic notch tends to be somewhat wider, more circular in outline anteriorly than that of *Metoposaurus diagnosticus.* The medial borders converge forward slightly so that the tabular bone projects slightly behind the notch forming a "horn."

The typical *Buettneria perfecta* Case from Crosby County, Texas, differs from these

TABLE 7.1.   Proportions of otic notches of some metoposaurs.

Population Sample	N	R/S	R/C	S/C	S/POL	S/PPL	C/POL	C/PPL	SL	POL	PPL
Anaschisma browni	1	4.27	0.573	0.134	0.077	0.273	0.577	2.034	413	[217]	[62]
Anaschisma n. sp. N.M.	3										
UCMP 63486		3.4	0.466	0.137	0.091	0.245	0.663	1.785	—	113	42
UCMP 63845		3.8	0.408	0.108	0.075	0.184	0.700	1.703	160e	90	37
YPM 4201		5.2	0.400	0.080	0.055	0.148	0.688	1.867	—	122	45
Koskinonodon princeps	1	0.828	0.230	0.278	0.190	0.594	0.684	2.136	540	[285]	[91]
Buettneria howardensis	3										
BEGUT 31100–122		1.064	0.207	0.195	0.134	0.399	0.690	2.054	432	242	81
UCMP 66991		0.950	0.218	0.229	0.164	0.734	0.695	3.207	495	259	76
BEGUT 31099–12 b		1.075	0.210	0.195	0.129	0.385	0.667	2.000	153	81	27
Buettneria bakeri	4										
UMMP 13055		1.000	0.153	0.153	0.104	0.328	0.682	2.145	284	151	48
UMMP 13820		0.805	0.205	0.255	0.146	0.459	0.571	1.729	305	166	55
UMMP 13822		1.183	0.288	0.193	0.162	0.432	0.842	2.260	—	133	50
MCZ 1054		0.679	0.133	0.197	0.127	0.369	0.644	1.874	280	143	49
Buettneria perfecta											
Crosby County, Texas	3										
UCMP 113554		1.262	0.194	0.154	0.099	0.314	0.642	2.039	275	162	51
YPM 1957/14		1.809	0.247	0.136	0.091	0.269	0.669	1.974	—	230	78
YPM 1957/7		—	—	0.238	0.138	0.450	0.581	1.887	440	260	80
Potter County, Texas											
UCMP 13167 (cast)	1	0.928	0.120	0.129	0.086	0.583	0.667	2.571	293	162	42
Metoposaurus ouazzoui	7										
Means		0.882	0.176	0.20	0.130	0.412	0.654	2.088	—	—	—
O.R. smallest		0.682	0.143	0.16	0.096	0.273	0.571	1.727	285	145	48
largest		1.200	0.214	0.25	0.174	0.521	0.759	2.444	510	260	80

Population samples: *Anaschisma browni* Branson, Popo Agie Formation, Willow Creek, south of Lander, Wyoming. Data from illustration of type of *A. brachygnatha* Branson 1905, p. 588, Figure 9. *Anaschisma* n. sp. N.M. original measurements from specimens from Redonda Formation, Apache Canyon southeast of Tucumcari, New Mexico. *Koskinonodon princeps* Branson and Mehl, Popo Agie Formation, Bull Lake Creek, Wyoming, from Branson and Mehl, 1929, Plate 4. *Buettneria howardensis* Sawin, from lower Dockum Formation, near Otis Chalk, Howard County, Texas (original measurements). *Buettneria bakeri* Case, from lower Dockum Formation, Scurry County, Texas, measurements from Case (1932; and from Plates I and II therein). *Buettneria perfecta* Case, original measurements, Dockum Formation, Crosby County, Texas, and cast from Cerrita de la Cruz Creek, Potter County, Texas. *Metoposaurus ouazzoui* Dutuit, base of 15 "grès et argiles silteuses rouges," Azarifen village, Argana, Morrocco (Dutuit, 1976, p. 44, Table 3).

N, size of sample considered; R, width otic notch from squamosal to tabular; S, depth of otic notch from tabular horn parallel midline; C, transverse distance between otic notches; POL, midline distance from rear borders orbits to rear edge skull roof; PPL, midline distance from pineal fenestra to rear edge skull roof; SL, length skull roof; e, estimated from incomplete skull. Measurements of SL, POL, PPL are in millimeters, other figures are ratios. Figures in brackets computed from illustrations and published measurements of these specimens.

principally in having a shallower notch in comparison to overall skull size. This is shown by the lower values of the ratios S/POL and S/PPL in Table 7.1. A subjective impression that the notch in specimens referred to *B. perfecta* is narrower than that of *B. howardensis* is not supported by measurements (ratio R/S). The difference in appearance might better be described as more angular. The radius of curvature near the apex of the notch is shorter.

The large collection of metoposaurs from the Sierrita de la Cruz Creek, Potter and Oldham Counties, Texas, northwest of Amarillo, in the Panhandle Plains Historical Museum, West Texas State University, Canyon, Texas (Colbert and Imbrie, 1956, p. 431, sample 2), shows considerable individual variation in size and form of the otic notch. Of twenty-one skulls examined in 1963 (without making measurements which seemed superfluous at the time, after Colbert's work), all showed tabular horns or lateral projections behind at least the medial part of the notch. On eight of these skulls the horn was characterized as "strong" or "large"; six were termed "small" or "slender." In thirteen the otic notch was called "narrow" or "small" and in eight it was "open" or "large." The latter terms would apply to the entire Howard County sample. None of the Potter County skulls showed the wide, shallow embayment of *Anaschisma*.

Many of the *Buettneria* skulls from south of Lamy, New Mexico (Colbert and Imbrie, 1956, p. 430, sample 1), have large, rounded otic notches and projecting tabular horns. Other specimens from this locality (particularly the U. S. National Museum slab, Colbert and Imbrie, 1956, pl. 28, fig. 2) seem to have broader and less well-defined otic notches and to lack tabular horns. In these respects they approach the Wyoming specimens which were named *Anaschisma* by Branson (1905). Donald Baird has suggested that these differences between the Harvard and USNM specimens from Lamy are due to differences in preparation, and that tabular horns have been added to the Harvard skulls. An alternative possibility is that the "horns" have been broken from some of the USNM specimens. My own observations, made through the glass covers of the exhibits, indicate that a fairly deep and broadly rounded otic notch is characteristic of both samples regardless of the appearance of the tabular horns or their apparent absence. The posterior margin of the Lamy skulls which lack horns is still distinctly incised and unlike either the Wyoming specimens of *Anaschisma* or the small skulls from the Redonda Formation near Tucumcari, New Mexico, which I have referred to that genus.

Anaschisma browni Branson and *A. brachygnatha* Branson from the Popo Agie Formation of Wyoming have broad, shallow sinuses in the position of the otic notches and an essentially straight posterior border to the tabulars and postparietals so the tabulars do not project or form horns of any sort. The otic region thus differs markedly from that of *Metoposaurus* (*Buettneria, Koskinonodon,* etc.). Branson (1905, p. 571) states that there are no otic notches in *Anaschisma*. His Figure 1 (the type of *A. browni*) clearly shows the shallowly concave posterior border of the squamosal. It should be noted that the skulls which form the types of *A. browni* and *A. brachygnatha* were found together in 1902 at the same locality near Lander, Wyoming. Branson (1905, p. 570) says, "The skulls were found closely associated, one slightly overlapping the other." In 1929, Branson and Mehl reinterpreted these specimens and concluded that the *A. browni* and *A. brachygnatha* skulls "both give evidence of having lost an appreciable part of the posterior border about the region of the otic notch. A reasonable restoration of the tabular and squamosal gives ample space for a well-developed otic notch, the lack of which cannot, therefore, be considered a characteristic of the genus or species" (1929, p. 48). Photographs of these skulls (Branson, 1905, figs. 3a and 9) do not indicate such extensive damage. I accept Branson's original interpretation.

The remaining skulls from the Popo Agie beds, to which the names *Borborophagus wyomingensis* Branson and Mehl and *Koskinonodon princeps* Branson and Mehl have been applied, have deep and well-defined notches, narrow in comparison to the two *Anaschisma* skulls. The tabular horns are prominent and similar to those of various species of *Metoposaurus* and *Buettneria*. They were obtained from localities many miles distant from the *Anaschisma* site in the Popo Agie (see Branson and Mehl, 1929, pp. 79–80) and are possibly from a different stratigraphic level.

Before discussing the relationships of these forms to those of eastern North America, Eurasia, and Africa, the possibility of expressing these features of the western North American fossils quantitatively will be examined.

Measurements of the Otic Notch

It is possible to express the relative size and proportions of the otic notch quantitatively. Dutuit (1976, p. 45) has designated the width of the otic notch "R," the depth (measured parallel to the midline) "S," and the minimum distance between notches of opposite sides "C." Skull roof length is obviously the most basic linear measurement with which to compare these dimensions, but is not determinable on many specimens which preserve the otic features adequately. I have tested the correlation of both postorbital length and postpineal length with skull length in several quarry samples of metoposaurs and find the correlation consistently high. Otic notch measurements are compared with postorbital length where size relationships are being investigated.

Points for measurement of the otic notch are difficult to define reproducibly (Figure 7.1). Depth can be most reliably measured from the rear surface of the tabular behind the "horn" (if present) to the most anterior point of the notch. Generally this is close to parallel to the median line; it may exaggerate the depth somewhat, but no other point at the rear of the notch is readily reproducible. Breadth of the notch is more difficult to define, as the outer limit of the otic notch commonly merges smoothly into the posterior curve of the squamosal. A perpendicular from the lateral side of the notch to the closest point on the tabular is reasonably reproducible. In a skull with broad, shallow otic sinuses (*Anaschisma*, Figure 7.2), notch depth is measured from a line tangent to the rear edge of the squamosal and inflection of curvature from the otic sinus to the occipital border of the tabular to the deepest point of the sinus. Notch width is the length of this line between points of contact or tangency with the skull.

That these measurements of the otic notches are not strictly comparable is admitted. The impossibility of obtaining congruent measurements of the otic notches in different specimens of metoposaurs emphasizes the distinctness of the morphology of this region between the different genera.

Most species (or populations) of metoposaurs have well-defined otic notches and are thus readily separable from the species which show only a broad, shallow, ill-defined sinus at the rear of the skull, a condition which for brevity will be termed "without an otic notch." It is not known whether the latter group lost their tympanic membranes or merely moved their auditory structures away from

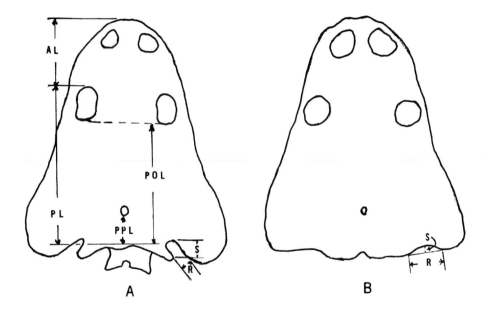

Figure 7.1 A. Outline of skull roof of *Metoposaurus fraasi* (Case) showing positions of measurements used in Table 7.1. Note that postorbital length, *POL*, differs from the posterior skull length, *PL*, of Colbert and Imbrie (1956).

B. Outline of skull roof of *Anaschisma browni* Branson, based on type specimen of *A. brachygnatha*, showing how measurements were taken of otic notch. *AL*, antorbital length; *PL*, posterior skull length; *POL*, postorbital length; *PPL*, postpineal length; *R*, width of otic notch; *S*, depth of otic notch.

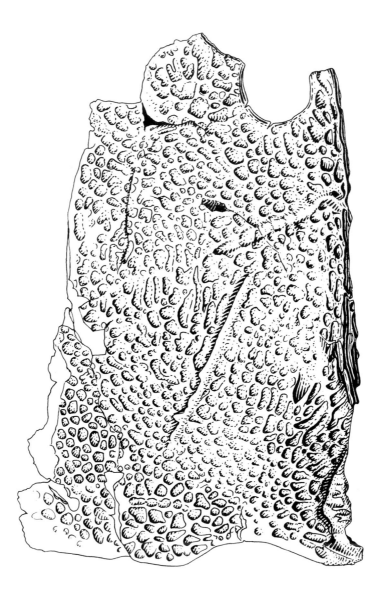

Figure 7.2. *Anaschisma*, undescribed species, Redonda Formation, Apache Canyon, south-east of Tucumcari, New Mexico, Yale Peabody Museum 4,201; x 1.

the skull roof, supporting the eardrum with a frog-like cartilaginous ring. Systematic recognition of these two groups as genera is comparable to recognition of several genera of Capitosauridae upon the basis of a rather different type of modification of the otic notch in that family.

Proper nomenclature for these genera depends upon whether these distinctive conditions of the otic region can be recognized in fossil material from the eastern United States described in the nineteenth century by Leidy and Cope.

Colbert and Imbrie (1956) considered the type specimen of *Dictyocephalus elegans* Leidy 1856 to be indeterminate in its present condition. Early illustrations of this fossil by Emmons (1857, pp. 60–61, figs. 31, 32) show a complete posterior margin of the skull roof with tabulars and postparietals extending appreciably farther posteriorly than the squamosals so that an open otic notch is evident; there is no indication of a tabular horn, but this may be related to the small size of this probably juvenile specimen. An unpublished woodblock engraving of this fossil prepared for Emmons and kindly furnished to me by Donald Baird shows these notches far less conspicuously than the published figures. *Dictyocephalus* has been regarded as a representative of the "notchless" group (D. Baird, personal communication, 1980), but the definite suggestion of otic notches in these early illustrations incline me to regard it as a juvenile of the more common forms which have well-developed notches. If this could be demonstrated (for example by topotypic specimens), *Dictyocephalus* would be the earliest valid name for the genus of metoposaurs with otic notches, since *Metopias* Meyer 1842 is preoccupied by *Metopias* Gory 1832, a coleopteran.

The propriety of this argument is questionable on two grounds: it involves assumptions that are not demonstrable beyond question concerning the nature of the otic notches of *Dictyocephalus*, and concerning the relationship of the small (immature) type specimen of *D. elegans* to larger metoposaurs. Moreover, replacing the long and widely used *Metoposaurus* Lydekker 1890 by the little known and inadequately understood *Dictyocephalus* Leidy would be unnecessarily disruptive of systematic nomenclature.

The same sort of objections apply to adopting *Eupelor* Cope 1866, the next possible name applied in this family, in place of *Metoposaurus*. *Eupelor durus* Cope was based upon fragmentary remains of large metoposaurs from Pennsylvania which lack any indication of the structure of the otic region. As recognized by Cope in 1892 and Colbert and Imbrie (1956), the more completely known specimens from the western United States that have been described as species of *Buettneria, Borborophagus, Kalamoiketor, Koskinonodon,* and *Metoposaurus* all closely resemble *Eupelor* in comparable features. But in absence of positive knowledge concerning the otic notch of *E. durus*, there is no certainty as to whether that name applies to animals with or without the notch. Donald Baird has obtained a well-preserved skull (Princeton University 21,742) from the Wolfville Formation of Nova Scotia which has deeply incised otic notches and distinct tabular horns that compare with those of *Buettneria bakeri* or small *B. howardensis* from Texas. One might assume that this skull is representative of the metoposaur of the Newark group, but one cannot demonstrate that the type of *Eupelor* was also representative. As with *Dictyocephalus*, stability of nomenclature is better served by restricting the name *Eupelor* to its type material and considering it a *nomen vanum*, not identifiable at the generic level in the present state of knowledge of this family.

Discussion

The systematic conclusion of this study, therefore, is simply that two genera of the Metoposauridae can be recognized: *Anaschisma*, for species lacking an otic notch; and *Metoposaurus*, including the remaining forms with well-developed otic notches and frequently horns on the tabular. To *Anaschisma* are referred the specimens from the Popo Agie, *A. browni* and its synonym *A. brachygnatha*, and an unde- scribed species from the Redonda Formation of eastern New Mexico. All other metoposaurs from North America, Europe, Africa, and India, are included in the genus *Metoposaurus*.

Anaschisma and *Metoposaurus* have not been found together. As pointed out above, the Popo Agie representatives of these genera come from widely separated localities. In New Mexico, *Anaschisma* is found in the Redonda Formation and *Metoposaurus* in the underlying Chinle equivalents of the Dockum. The Popo Agie has been correlated with the lowest part of the Dockum (Gregory, 1957, Colbert and Gregory, 1957) on the basis of the occurrence of the primitive phytosaur genera *Paleorhinus* and *Angistorhinus*. If this correlation is correct, *Anaschisma* is present in the earliest and latest Upper Triassic vertebrate fauna of western North America, *Metoposaurus* alone has been found at stratigraphically indeterminate lev- els (as well as in the earliest faunas). The difference in distribution thus appears to be ecological rather than temporal.

For a time the collection from Lamy, New Mexico, was thought to form an exception to this and contain roughly equal numbers of both skull types. Their supposed co-occurrence here posed a problem for the ecological separation hy- pothesis. The deposit contained remains of over twenty of these large amphibians crowded together in a fashion suggesting mass death of the inhabitants of a single pool (Romer, 1939). Such an assemblage of individuals presumably belonged to a single species, and should be so interpreted, unless additional circumstances pro- vide compelling evidence to the contrary. In support of the idea that the Lamy collection is heterogeneous it may be pointed out that the dispersion of several measurements in this relatively large sample is greater than in any other meto- posaur population studied by Colbert and Imbrie (1956). Correlations between antorbital and postorbital lengths, between skull length and width, and between clavicle width and length are much lower than for any other samples (Colbert and Imbrie, 1956, p. 433, table 3). This suggests heterogeneity. Two distinct popula- tions might be mingled in the deposit.

Reexamination of this matter has raised doubts about this interpretation. The initial attempt to separate the Lamy specimens on the basis of presence or absence of tabular horns ignored the occurrence of moderately deep and broad otic notches quite unlike the shallow sinus of the Popo Agie or Redonda specimens. At present, then, all these specimens are considered a single species of *Metoposaurus*.

Even if the Lamy collection is heterogeneous, it does not offer decisive evidence for deciding between the intermingling of two morphologically distinct species which could reasonably be referred to two genera, or the existence of rather pro- nounced sexual dimorphism in a single breeding population.

If metoposaurs were characterized by sexual dimorphism of the otic notch, one might expect to find roughly equal numbers of each form in various Late Triassic localities where these amphibians occur. This is not at all the case. Skulls with the narrow and deep notch of the *Metoposaurus* form are the only kind known from a large number of localities on four continents; the *Anaschisma* form, with shallow otic sinus, is found only at one locality in Wyoming, and one in New Mexico, with the possible exception of the Lamy site. This distribution does not suggest random sampling of a population in which the two variants were in approximatley equal proportions.

Dr. Leigh Van Valen called my attention to the frequent congregation of male frogs and salamanders at breeding sites earlier than the arrival of females. If the local disaster which led to the death and burial of the animals occurred during this interval, the resulting fossil deposit would sample only one sex, which would be male by analogy with the behavior of Recent amphibians. If this applies to metoposaurs, the rarity of deposits containing the two sexes together would seem to indicate that the events leading to fossilization of these animals tended to occur most commonly during this rather short interval of the year in which only males congregated at pond or watercourses.

An objection to interpreting the Lamy deposit of metoposaurs as a breeding congregation, entrapped when the fossilizing event occurred after arrival of the females, and to this hypothesis in general, may be based upon the indications that the animals had crowded into a shrinking pond under conditions of drought, and then perished as the water dried up completely. Amphibians congregate to breed early in the wet season, not when drought is imminent. This unfavorable condition would be more likely to result in the mingling of species which normally occupy separate habitats.

A further difficulty is provided by the occurrence of two deposits which contain only remains of the notchless "female" morphology. This does not accord with behavioral data on modern batrachians, although it might be the result of chance sampling of species whose sexes were spatially segregated most of the time. Furthermore, several of the fossiliferous deposits containing these remains are stream channel conglomerates containing more broken fragments of skulls and skeletons than complete bones, a situation suggesting considerable transportation rather than entombment of a few individuals at their breeding site. Such transported specimens should form a more nearly random sample of the entire population, including members of both sexes.

These considerations cast sufficient doubt upon the hypothesis of sexual dimorphism to make it an unsafe basis for asserting that two distinct morphologies (of the otic notch) represent but a single species. Until a far stronger case for synonymy can be presented, *Anaschisma* and *Metoposaurus* should be recognized as separate genera. This nomenclature, emphasizing known differences, expresses the more probable relationship of the two groups, provides a more succinct specification of the different populations, and in no way interferes with continued investigation of their biological and stratigraphical relationships. Further investigation is needed of the stratigraphic distribution of various character states among metoposaurs before their species can be satisfactorily differentiated.

Literature Cited

Branson, E. B. 1905. Structure and relationships of American Labyrinthodontidae. J. Geol., 13 (7):568–610.
Branson, E. B., and Mehl, M. G. 1929. Triassic amphibians from the Rocky Mountain region. Missouri, Univ. Studies, 4 (2):155–239.

Case, E. C. 1932. A collection of stegocephalians from Scurry County, Texas. Mich., Univ., Mus. Paleontol., Contrib., 4 (1):1–56.

Chowdhury, T. R. 1965. A new metoposaurid amphibian from the Upper Triassic Maleri Formation of central India. Roy. Soc. London, Phil. Trans., Ser. B, 250:1–52.

Colbert, E. H., and Gregory, J. T. 1957. Correlation of continental Triassic sediments by vertebrate fossils. Geol. Soc. Amer., Bull., 68:1456–67.

Colbert, E. H., and Imbrie, J. 1956. Triassic metoposaurid amphibians. Amer. Mus. Natur. Hist., Bull., 110 (6):399–452.

Dutuit, J.-M. 1976. Introduction à l'étude paléontologique du Trias continental Marocain. Description des premiers stegocephales recueillis dans le couloir d'Argana (Atlas occidental). Mus. Nat. Hist. Natur., Paris, Mém., Sér. C, 36:1–253, 70 plates.

Emmons, E. 1857. American geology, containing a statement of the principles of the science, with full illustrations of the characteristic American fossils, with an atlas and a geological map of the United States. Part VI. Albany, x + 152 pp.

Gregory, J. T. 1957. Significance of fossil vertebrates for correlation of late Triassic continental deposits of North America. Congreso geologico internacional, XX sesión, Ciudad de México, Sección II, El Mesozoico del hemispherio occidental y sus correlaciones mundiales: 7–25.

Romer, A. S. 1939. An amphibian graveyard. Scientific Monthly, 49:337–39.

Watson, D. M. S. 1951. Paleontology and modern biology. Yale Univ. Press, New Haven, xii + 216 pp.

THE NORTH AMERICAN SEYMOURIIDAE

by Everett C. Olson

General Overview

Until quite recently the North American family Seymouriidae was known only from well-preserved specimens of *Seymouria baylorensis* Broili and fragments tentatively assigned to this species. Cope in 1896 described *Conodectes favosus* and over the intervening years it has been suggested that this is the proper designation of the North American genus and species. This determination has some merit, but has not been followed by any of the students of the group in recent years, in particular White (1939) and Romer (1947, 1966). Cope's original description of *Conodectes favosus,* being brief and lacking illustrations, is indeterminate and for this reason most, as Romer, have preferred to consider *Seymouria baylorensis* as the appropriate name. White (1939) on the basis of slender evidence recognized *Conodectes* as a distinct genus. Underlying this whole matter has been the desire to maintain the well-known name of *Seymouria* in favor of the lesser used *Conodectes,* which at best can be questioned as having been properly established. *Seymouria* is retained in this paper and *Conodectes* is dropped as not being properly established. *Desmospondylus* Williston was proposed for materials that have been subsequently recognized as *Seymouria* (Williston, 1910; 1911). Williston's description in 1911 and Watson's (1918) were the most complete until White published his definitive study (White, 1939).

The matter of assignment of fragmentary materials of *Seymouria* to species is inevitably a problem, as it is for all Permian vertebrates. The well-preserved materials from the Arroyo Formation, Clear Fork Group, Leonardian (Figure 8.1), all pertain to the one species, *S. baylorensis.* Most of the contemporary fragmentary specimens have been so classified, but without a solid base in morphology.

By far the greatest amount and best materials have come from the Arroyo Formation and the few specimens from pre-Arroyo beds are for the most part rather poorly preserved and fragmentary. The Belle Plains Formation (Wichita Group, Leonardian) and the Putnam and Admiral formations (Wichita Group, Wolfcampian) have produced a few specimens. One partially complete skull and skeleton is known from the Belle Plains and, although little prepared, is sufficiently similar to *S. baylorensis* in revealed features that it can be assigned to this species. The mean width of the posterior zygapophyses of the mid-dorsal vertebrae is about 3.0 cm, small for *S. baylorensis,* but within the adult range. The other pre-Arroyo materials are mostly too incomplete to allow a definitive species assignment.

Williston (1910) mentioned the presence of *Seymouria* in the *Cacops* bone bed of Wilbarger County, Texas. The stratigraphic position of this pocket is somewhat uncertain, but probably is lower Vale and thus overlies the beds from which

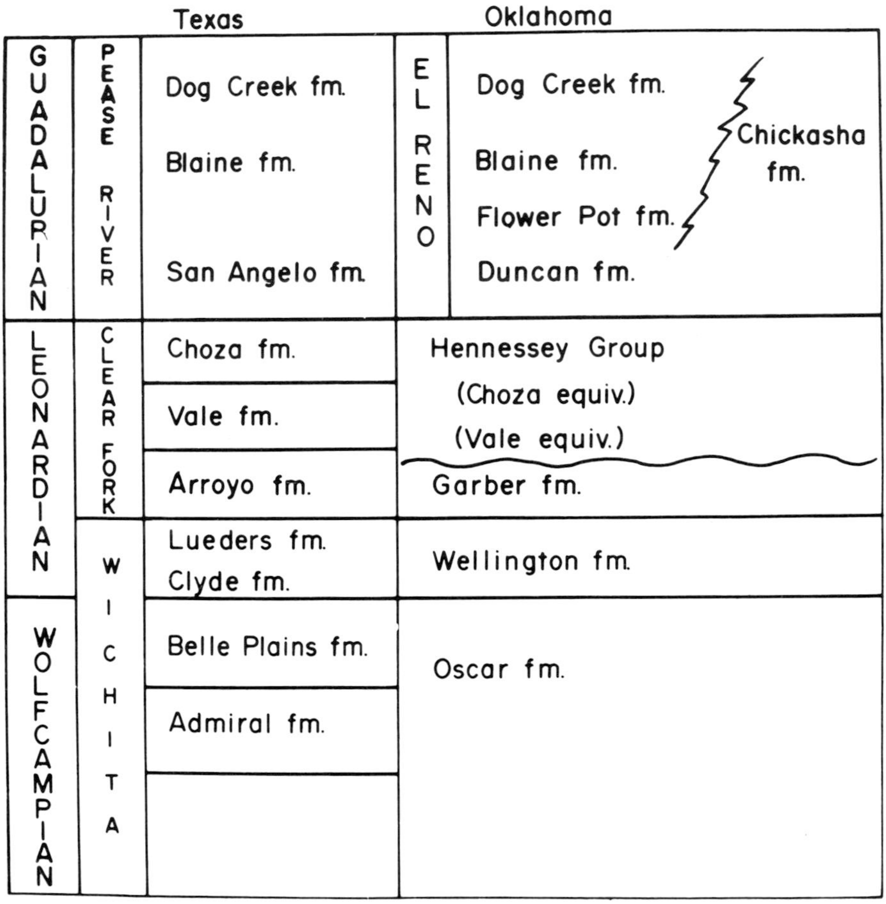

		Texas			Oklahoma	
G U A D A L U R I A N	P E A S E R I V E R	Dog Creek fm. Blaine fm. San Angelo fm.		E L R E N O	Dog Creek fm. Blaine fm. Flower Pot fm. Duncan fm.	Chickasha fm.
L E O N A R D I A N	C L E A R F O R K	Choza fm.			Hennessey Group	
		Vale fm.			(Choza equiv.) (Vale equiv.)	
		Arroyo fm.			Garber fm.	
	W I C H I T A	Lueders fm. Clyde fm.			Wellington fm.	
W O L F C A M P I A N		Belle Plains fm.			Oscar fm.	
		Admiral fm.				

Figure 8.1 Stratigraphic chart of the Lower and lower part of Upper Permian of Texas and Oklahoma, showing beds in which *Seymouria* has been found.

S. baylorensis otherwise is known. The supposed remains of *Seymouria* are associated with *Cacops, Varanops,* and *Casea,* genera that occur as "erratics" in the normal Clear Fork fauna of North Central Texas. A re-examination of the materials from this bone bed, which, except for a few skeletons sent elsewhere, are in the collections of the Field Museum of Natural History, has turned up but one specimen that might pertain to *Seymouria.* It is a poorly preserved and partially prepared femur. Williston did not specify the element or elements upon which his determination was made and it may be that this femur was his primary evidence. The assignment is questionable.

In 1979 a large species of *Seymouria* was described from the Fairmont Shale (Hennessey Group, Leonardian) of Oklahoma and from essentially time-equivalent beds of the middle Vale of Taylor County, Texas. It was named *S. grandis* Olson (1979) in reference to its large size relative to *S. baylorensis.* The braincase is strongly ossified and shows some features similar to those of large, temnospondylous amphibians, in contrast to the braincase of *S. baylorensis.* A tentatively associated partial femur resembles those of some of the European Kotlassiidae, in particular *Kotlassia.* Some modification in locomotor structure may be inferred, but must remain uncertain in view of the tentative nature of the association of the femur with other elements of *S. grandis.* A single large vertebra (Museum of Comparative Zoology 4993) collected in 1913 but undescribed, being considered to pertain to *Labidosaurus,* is a specimen of *Seymouria,* and, on the basis of size, probably *S. grandis.* It comes from Wilbarger County, Texas, with the only locality given being "Sharvar Tank." If it was obtained in the vicinity of this tank, to the south or west where exposures occur, it is either from the lower or middle Vale Formation. The postzygapophysial width of this vertebra is approximately 4.6, within the range of *S. grandis* (mean = 4.73), and well above the recorded range of *S. baylorensis* (mean = 3.67). On the basis of size it is tentatively assigned to *S. grandis* and as such provides a geographic intermediate between the Taylor County, Texas, specimens and the Oklahoma specimens.

A new species of *Seymouria* from the Chickasha Formation, El Reno Group, Blaine County, Oklahoma, is described in this paper. Collected in 1978, it has extended the stratigraphic range of the genus into the lower part of the Upper Permian, early Guadalupian. This species is but slightly older than the earliest Late Permian seymouriamorphs of the Soviet Union, which lie at the base of a moderately extensive adaptive radiation of this group known only from eastern Europe. The only earlier European seymouriamorphs are members of the family Discosauriscidae, partially aquatic animals (Spinar, 1952).

The stratigraphic range of the North American Seymouriidae as now known, including the new species described in this paper, may be summarized as follows:

S. agilis sp. nov.: Chickasha Formation (equivalent to mid–Flower Pot) El Reno Group, Guadalupian, Upper Permian.

S. grandis Olson: Fairmont Shale, Hennessey Group, Oklahoma, and Vale Formation, Clear Fork Group, Texas, Leonardian, Lower Permian.

S. sp.: Admiral and Putnam formations, Wichita Group, Wolfcampian, Lower Permian.

The major taxonomic affiliations of the Seymouriamorpha have long been a matter of debate, with the group being successively placed among the Reptilia and

Amphibia. For many years seymouriamorphs were accepted as constituting a suborder of the Order Cotylosauria, a composite assemblage including as well the Captorhinomorpha and Diadectomorpha. Now the consensus supports assignment to the labyrinthodont amphibians. This assignment has been based on resemblances of morphology, especially of the skull, to various anthracosaur and temnospondylous amphibians and evidence of an aquatic ontogenetic stage in the family Discosauriscidae. The North American seymourians, however, are predominantly terrestrial. Whether or not *Seymouria* had an aquatic stage in its development is an open question. White (1939) identified slight grooves on the skull as lateral lines, in spite of which he made assignment to the Reptilia, swayed by the "cotylosaur" features of the axial column and shoulder girdle and the rather slender evidence of a sexual dimorphism in the anterior haemal arches. The evidence of the presence of lateral lines and dimorphic haemal arches is problematic and, if the morphological interpretation is accepted, the meaning remains obscure. The interpretation that the Seymouriamorpha constitute of group of reptile-like predominantly terrestrial amphibians appears undeniable on the basis of all present information.

A New Species of *Seymouria*

CLASS: AMPHIBIA
  Subclass: Labyrinthodontia
    Order: Batrachosauria
      Suborder: Seymouriamorpha
        Family: Seymouriidae
          Genus: *Seymouria*
            Seymouria agilis sp. nov.

Holotype: University of California, Los Angeles, VP 5329 (Figure 8.2). Much of the skeleton, including vertebral column, girdles, proximal parts of the limbs, abdominal ribs, and a poorly preserved partial skull and lower jaw.

Horizon and locality: Chickasha Formation. Upper Permian, Guadalupian, at level of interfingering with middle part of Flower Pot Formation, El Reno Group. From a red sandy shale at locality BC-8 (Olson, 1965), about 3 miles north of Hitchcock, Blaine County, Oklahoma.

Diagnosis: A small species of *Seymouria*, about three-fifths size of average adult of *S. baylorensis*. Twenty-four presacral vertebrae; three "cervical vertebrae"; atlas with broad wedge-shaped spine, the other two with long, stout spines and narrow zygapophyses. Three well-developed sacral vertebrae each bearing a pair of strong sacral ribs, zygapophyses narrow. Sternal portion of shoulder girdle lightly structured; interclavicle with long stem and narrow wings, grooved anteriorly for reception of clavicle; clavicle with narrow blade and inflected ventral portion; area of attachment to interclavicle restricted. Humerus with well-formed shaft and subdued processes. Radius relatively long and slender. Femur long, slender, with well-formed shaft. Measurements as in Tables 8.1 and 8.2.

Bases for assignments (Figures 8.2–8.6): The vertebrae of the presacral region are characterized by swollen arches with relatively narrow, backswept processes leading to the posterior zygapophyses. This particular shape of the arches is found elsewhere among animals with broad arched vertebrae only in tseajaiids and limnoscelids. Assignment to either the diadectomorphs or captorhinomorphs is thus precluded. The ribs throughout the column are bifid in contrast to those of tsea-

TABLE 8.1

Width measurements of vertebrae based on maximum width of postzygapophyses (1) *Seymouria agilis* sp. nov. UCLA VP 5329; (2) *S. baylorensis* based on average-sized specimen after White (1939). Superscript "a," approximate.

	1	2
Presacral vertebrae		
1	—	12
2	10^a	18
3	12^a	20
4	—	26
5	20^a	31
6	22^a	36
7	25	36
8	25	38
9	25	40
10	25	40
11	25	40
12	25	40
13	—	40
14	—	40
15	—	40
16	—	40
17	24	40
18	25	38
19	27	38
20	27	38
21	26	38
22	26	36
23	24	36
24	24	36
Sacral vertebrae		
1	13	19
2	19	12
3	12	1st caudal

TABLE 8.2

Measurements in mm of limbs and girdle elements. (1) *Seymouria agilis* UCLA VP 5329; (2) *S. baylorensis,* an average individual based on White (1939).

	1	2
Interclavicle		
Total length	43	85
Maximum width	50	56
Depth of anterior blade taken at ½ distance from center to lateral end	8	35
Humerus		
Length	50	66
Proximal width	28	26
Minimum width of shaft	9	14
Radius		
Length	30	40
Minimum width of shaft	3.4	5
Pelvis		
Ventral length	52	80
Anterior width ventral	41	65
Width measured between Iliac blades (maximum)	40	63
Femur		
Length (left for *S. parvus*)	68	63
Proximal width (right for *S. parvus*)	23	30
Shaft width	9	12

jaiids (Moss, 1972) and limnoscelids (Williston, 1911) but as in seymouriamorphs (White, 1939). Assignment to the Seymouriamorpha is based primarily upon these basic features.

Although the presacral vertebrae and ribs are virtually indistinguishable from those of *S. baylorensis* in form, the shoulder girdle, sacral ribs, and limb elements are strikingly different. This raises the question as to whether the new specimen should be assigned to *Seymouria* or made the basis for a new genus. Clearly it is not assignable to any of the European genera. The differences from *S. baylorensis*, as brought out in later parts of this paper, are adaptive, and the strong resemblances of the presacral axial structures to those of *S. baylorensis* suggest that the simplest and clearest assignment is to *Seymouria*. This procedure has been followed here.

The limbs and girdles reveal marked differences between the specimen and representatives of *S. baylorensis* and assignment to a different species is mandatory. The specimen is about one-third the size of *S. grandis* and on this basis can be excluded from the species with reasonable certainty.

Detailed description: Many of the details are shown in Figures 8.2–8.6. The drawings have been somewhat reconstructed as indicated in the figure captions. Although much of the skeleton is present, preservation is only moderate to poor. The clay minerals of the preserving sediments react quickly to surface water, swelling and disrupting the bone, and the enclosing rock is jointed with slippage along the joints. The most poorly preserved part, the skull, lay at and just beneath the surface. Some damage to the limbs occurred during excavation.

Skull (Figures 8.2, 8.4A): Little detail can be deciphered from the partial skull and lower jaw. The right posterior quarter is present and visible in ventral view. The size and general cast of discernible structures are acceptable for *Seymouria*, but not definitive. The basi-occipital is broad and flat, and from it bone of an indeterminate nature passes far laterally. The palate and braincase appear to have been moveably articulated. The adductor fossa is large. Little more can be said.

Vertebral column (Figures 8.2, 8.3, 8.4B): The presacral count of twenty-four vertebrae is typical for *Seymouria*. Williston (1911) and Watson (1918) indicated twenty-three in the specimens they described, but White (1939) with much more material gave the number as twenty-four. The first three anterior vertebrae are quite different from those that follow and can be considered cervicals. They have very narrow zygapophyses and stout relatively long neural spines. The spine of the atlas is broad and more or less wedge-shaped, fairly similar to that in *S. baylorensis* although proportionately broader. The cervical vertebrae are somewhat less "compressed" longitudinally than they are in either *S. baylorensis* or *S. grandis*.

The remaining presacral vertebrae are alike throughout as far as the shape of the arches and neural spines are concerned. No evidence of bifurcated spines, such as occur in the other two species, has been found. This, however, is a feature that varies both along the column and in individuals of *S. baylorensis*.

The transverse processes of the anterior several vertebrae, although present, are not well preserved. More posteriorly, in the mid-presacral region, the processes are short and stout, typically seymouriamorph in form. Presumably the facet for the capitulum of the rib was carried on the intercentrum, but this has not been

---

Figure 8.2 Dorsal view of the preserved remains of *S. agilis* sp. nov., UCLA VP 5329.

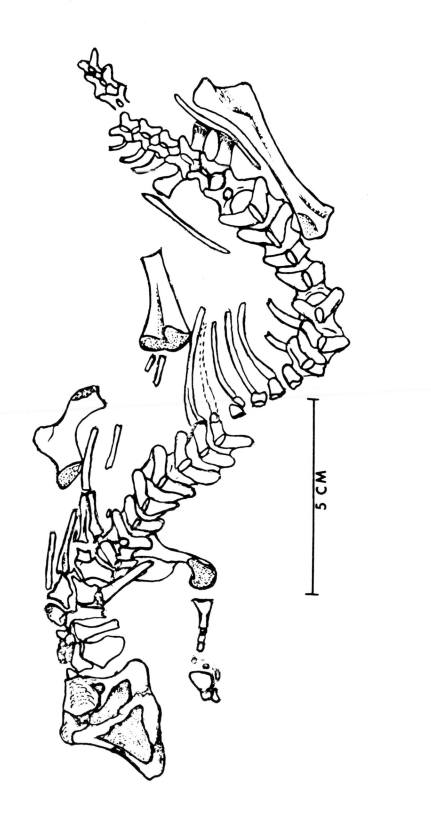

5 CM

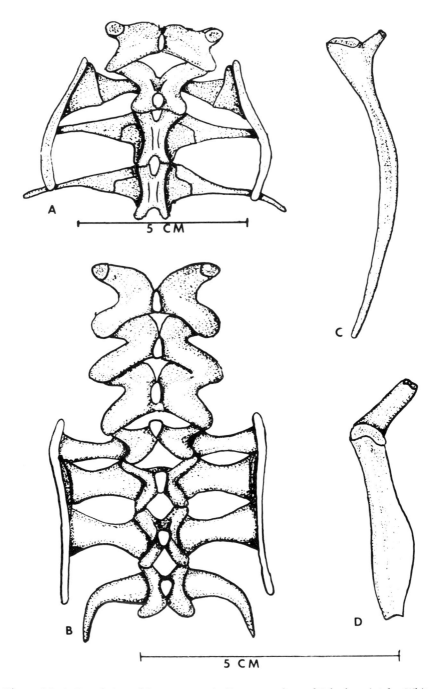

Figure 8.3 A. Dorsal view of the sacrum and adjacent vertebrae of *S. baylorensis* (after White, 1939). B. Dorsal view of sacrum of *S. agilis,* UCLA VP 5329, including first caudal and three presacral vertebrae. Somewhat reconstructed, especially in restoration of the first sacral rib. C. Mid-dorsal rib of *S. agilis,* UCLA VP 5329. D. Subscapular rib of *S. agilis,* UCLA VP 5329.

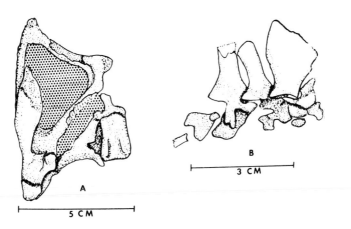

Figure 8.4 A. Ventral view of preserved part of *S. agilis,* UCLA VP 5329. Essentially what is shown by very badly preserved portion of the skull, but with outlines of areas somewhat more precisely defined than in specimen. B. The cervical vertebrae of *S. agilis,* UCLA VP 5329, in lateral aspect, shown as preserved.

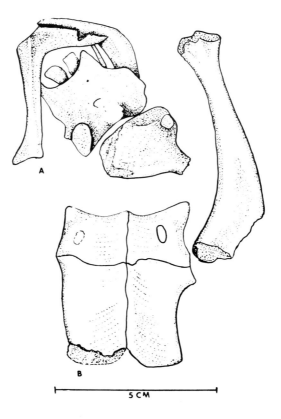

Figure 8.5. *S. agilis,* UCLA VP 5329. A. Shoulder girdle in ventral aspect, with head of humerus, illustrated as preserved. B. Pelvis in ventral view with right femur in place as preserved. In both drawings, anterior to top of page.

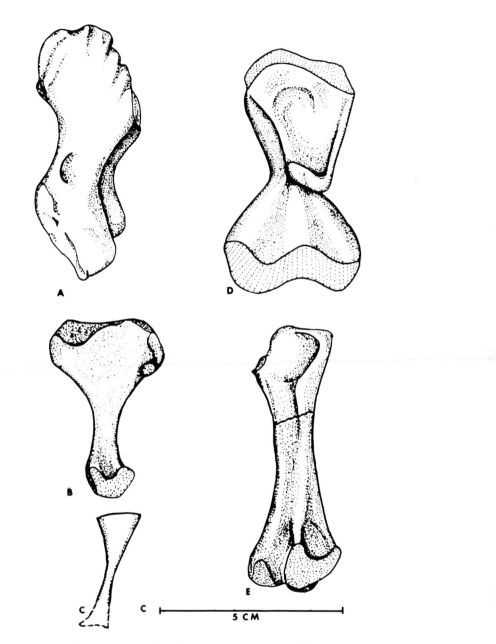

Figure 8.6 A. Left humerus of *S. baylorensis*, in ventral aspect (after White, 1939). B. Left humerus of *S. agilis*, UCLA VP 5329, in ventral view and shown as preserved. C. Left radius of *S. agilis*, UCLA VP 5329, with distal end reconstructed on basis of impression in sediment. D. Left femur of *S. baylorensis* in ventral aspect (after White, 1939). E. Left femur of *S. agilis*, UCLA VP 5329, reconstructed from data from both right and left femora, proximal to top of figure. See Figure 8.2 for two femora.

observed. The centra and intercentra have not been fully exposed. Intercentra were present. The centra, as seen in cross section and basally, are typical amphicoelous structures.

There are three distinctive sacral vertebrae. Their arches are slender and their zygapophyses narrowly spaced. Each carries a pair of strong broad sacral ribs. The ribs of the first sacral were somewhat displaced and broken and have been somewhat reconstructed in the drawing (Figure 8.3B). The other two are fairly well preserved. The pelvis was firmly attached to the vertebral column by these three ribs, which contrasts sharply with the single primary rib and accessory in *S. baylorensis* (Figure 8.3A). Only a few of the caudal vertebrae were preserved. They are normal for *Seymouria*.

Ribs: Ribs throughout the presacral column are bifid. Those of the anteriormost several vertebrae, including the cervicals, are poorly preserved. The ribs in the area of the shoulder girdle are short and broad (Figure 8.3D). The tuberculum is essentially a continuation of the shaft and has a broad articular facet; the capitulum is stout and relatively long. More posteriorly, in the mid-dorsal region, the ribs are long and slender, longer relatively than in *S. baylorensis*. The structure of the last several ribs, if they existed, is unknown, for no certainly identifiable ribs have been found on the posteriormost nine or ten vertebrae. Short ribs, as present in *S. baylorensis*, may have existed. The sacral ribs have been described earlier. The first two caudal ribs are short and strongly recurved.

Shoulder girdle and forelimb (Figures 8.5A, 8.6A,B,C): The shoulder girdle is shown as preserved in Figure 8.5A. The scapula has been badly distorted and partially destroyed. The most distinctive bones of the girdle are the clavicles and interclavicle. The interclavicle has a long stem and narrow, lateral projections which are deeply grooved anteriorly to receive the clavicles. The lateral projections contrast sharply with those at the broad, spatulate, anterior portions of the interclavicle of *S. baylorensis*. Each clavicle is slender throughout and rests in the groove of the wing of the interclavicle. The attachment appears to have been restricted and weak, although probably strengthened by ligaments. This also contrasts with the condition in *S. baylorensis* in which the ventral portion of the broad, spatulate, anterior plate of the interclavicle was firmly attached to the clavicle.

Parts of both the right and left humeri are present, but preservation is poor. The illustration (Figure 8.6B) is based on the left humerus which is shown as preserved. The contrast with the humerus of *S. baylorensis* (Figure 8.6A) is evident in spite of the poor preservation in *S. agilis*. A distinct shaft is present in the latter and the proximal portions resemble those of the humerus of a dissorophid amphibian more than of *S. baylorensis*. The more fragmentary right humerus bears out the general interpretation.

The radius is partly preserved (Figure 8.6C). It is light and relatively long and slender. A few carpal elements are present, showing that this area was ossified but not providing any detailed information.

Pelvis and hind limbs (Figures 8.5B,C, 8.6D,E): The pelvis is similar to that of *S. baylorensis*, differing only in details. The ilium, however, is firmly attached to the three sacral ribs, producing a strong sacrum.

The femur is strikingly different from that of *S. baylorensis* (Figures 8.5C, 8.6D,E) and vaguely similar to that tentatively assigned to *S. grandis* (Olson, 1979). The element in *S. agilis* is long, slender, and has a well-developed shaft. Were it not associated with the pelvis of the specimen it likely would have been taken to pertain to a dissorophid. The ratio of the femur to the length of the base of the pelvis is 1.3 in *S. agilis* as contrasted to 0.79 in *S. baylorensis* (based on White, 1939).

Scalation: The only clear evidence of dermal ossifications is the presence of a well formed series of abdominal or ventral ribs. These formed a midline V-shaped patch with the apex forward and centered under the shoulder girdle. The preserved portion of the V is about 7 cm long and each ramus consists of about twenty rounded elongated rods.

Evolution of the North American Seymouriidae

S. baylorensis persisted from the Belle Plaines Formation through the Arroyo into the very beginning of the Vale (Olson and Mead, in press). During this whole depositional period there was no evident morphological change. The earlier specimens of *Seymouria,* from the Putnam and Admiral formations, show no differences from specimens of *S. baylorensis* and it is probable, but not fully verifiable, that they too belong to the species.

S. baylorensis may be interpreted on the basis of its morphology as a relatively slow-moving, primarily carnivorous, terrestrial animal with short, stocky limbs, a sprawling posture, and a vertebral column capable of considerable flexure but resistant to vertical flexure and torsion (see Olson, 1975). It was a part of the faunal complex that was ultimately dependent upon the plants and animals of the lakes and streams for sustenance. The most abundant, and largest, animal was the top terrestrial predator in the complex, *Dimetrodon.* The large semiaquatic predator *Eryops* persisted throughout as a dominant factor of the ecology. In pre-Arroyo faunas the moderately large reptile *Ophiacodon* also was a prominent semiaquatic predator. *Diadectes* and *Edaphosaurus* were large herbivores. Although certainly competent on land, their occurrences suggest that they lived primarily in the vicinity of standing and running water. Although they were abundant prior to the deposition of the Arroyo beds, they waned as climates became drier during the Arroyo and only *Diadectes* persisted, as a rare constituent of the known Vale fauna. *Captorhinus* was an active, small, presumably omnivorous, terrestrial reptile and various dissorophids were small to moderate sized, seeming terrestrial insectivores and predators. These animals, plus many aquatics, formed a tightly knit community of moderately slow-moving animals. *S. baylorensis* was an integral part of this community.

During the Arroyo climates became increasingly dry and this trend, with some periods of amelioration, continued through the Clear Fork. Aquatic and semiaquatic vertebrates, in particular, became reduced in numbers of species and individuals, but a few such as *Diplocaulus magnicornis* and *Lysorophus tricarigatus* became increasingly abundant. Several genera and species did not survive the end of the Arroyo deposition. In spite of these changes *S. baylorensis* showed no evidence of morphological change. In beds in the upper part of the Arroyo the species is reduced in numbers and the last known specimen has come from the lowermost Vale near the southern limit of the formation (Olson and Mead, in press).

The Vale species, *S. grandis,* probably was derived from *S. baylorensis,* but no transition forms are known. *Dimetrodon* was its most common terrestrial associate, but new genera of captorhinomorphs, *Labidosaurikos* and *Captorhinikos,* apparently herbivores, were contemporaries in the faunal complex. The fauna on the whole was an impoverished remnant of the rich Arroyo fauna, and was dominated on land by *Dimetrodon* and in and near water by *Eryops* and *Diplocaulus*. It may be that the size increase of *S. grandis* over *S. baylorensis*, and the accompanying modifications, were a response to its changed faunal environment.

S. agilis combined the very conservative axial complex of the earlier *S. baylorensis* with a strongly modified locomotor system. It clearly was a more active, fairly rapid-running seymouriamorph. Although the forelimb, based on the length and lightness of the radius, was relatively slender, the epipodial:propodial ratio (based on lenth of radius and length of femur) was 0.67, not greatly different from that of *S. baylorensis* which was 0.60. This ratio has been taken as a crude basis for estimation of stride length in animals with similar axial flexures, humeral-glenoid articulations, and shoulder girdles fixed relative to the axial column (Olson, 1975). The relative stride length of the forelimb in the two species, thus, was about the same.

The elongated femur suggests that the stride length of the hindlimb had increased relative to *S. baylorensis*. In any event, it is clear that the propulsive potential of the hind limbs was enhanced by the strong sacrum which could have provided for much stronger reciprocal thrust between the hind limb and axial column. When held in the axial plane by musles of the vertebrae and ribs, the vertebral column, even with its retained capacity for flexure, could have served as an osseous base for transmission of the thrust. It may be inferred that this was the case during rapid locomotion, whereas the mode of walking, involving considerable axial flexure, may have retained the pattern characteristic of *S. baylorensis*.

The physical and biological environment during deposition of the Chickasha Formation was very different from that under which *S. baylorensis* lived. Deposition of the fossil-bearing beds took place near to the evaporite basins in which the Flower Pot shales and evaporites were formed. The preserved animals appear to have been carried into the sites of deposition by streams but not to have been carried far. The terrestrial members of the vertebrate complex include a proportionately large number of large terrestrial herbivores including *Cotylorhynchus*, *Angelosaurus*, and the large captorhinomorph *Rothianiscus*. Carnivores are known from very few specimens and include the predatory reptiles *Varanodon* and *Watongia* and the amphibian *Fayella*. These were small- to medium-sized active carnivores. In addition a problematic larger carnivore, possibly a sphenacodont, is known. Aquatic elements are few, including rare small amphibians, xenacanth sharks, and palaeoniscoids.

To date, this faunal complex is only poorly known. It is similar in some respects to the better known San Angelo complex from Texas, to the south and across the Wichita mountain range. Some of the elements are found in the Kazanian of the USSR as well, but not in direct association with the seymouriamorphs of this time. In view of the general nature of the Chickasha faunal complex it appears that *S. agilis*, a small, active carnivore, was functionally adapted to this system, which was very different from that in which its predecessors existed.

S. agilis may well have been derived from *S. baylorensis*, but the evidence is far from conclusive. The Chickasha faunal complex represents a stage in what has been called the Caseid chronofauna (Olson, 1971). The first evidence of this chronofauna is the *Cacops* bone bed assemblage, near the Arroyo-Vale boundary. It is known thereafter, until the Guadalupian, only from sporadic remains. *Seymouria* has, as noted, been reported from the *Cacops* bone bed, but the identity of the assigned material is uncertain and the reference to *Seymouria* is problematical. In any event, *S. agilis* would appear to represent an evolutionary line which existed independently of the better known Permo-Carboniferous chronofauna at least since the time of initiation of Vale sedimentation.

Relationships to European Seymouriamorpha

The seymouriamorphs of Europe range from the Early Permian into the base of the Triassic. Much of the material has come from the Kazanian and Tatarian of the USSR. Five families have been recognized: Discosauriscidae, Seymouriidae, Kotlassiidae, Chroniosuchidae, and Lanthanosuchidae. Only the first occurs in the Lower Permian, more or or less contemporary with *S. baylorensis*. It is present in Western Europe and has been reported from the Lower Permian of Tadzhikistan (Efremov and Vjuschkov, 1952; Olson, 1957). The small aquatic to semiaquatic amphibians of this family differ in many respects from *Seymouria* and represent a separate line of development.

Gnorhimosuchus Efremov and *Rhinosaurus* Fisher have been assigned to the family Seymouriidae. Both occur in beds of the Kazanian, Zone II, of Efremov. They may represent a continuation of the family, best known from North America (Romer, 1947). Members of the other families have various specializations not found among the seymouriids, including dermal armor plates along the axis in the Kotlassiidae, elongated skulls in the Chroniosuchidae, and very flat ornamented skulls in the Lanthanosuchidae. An extensive adaptive radiation took place during the Late Permian. Whether there was a comparable radiation in North America, of course, cannot be known at present, in the absence of vertebrates from Permian beds more recent than those of the Chickasha Formation.

Acknowledgments

I want to first express my appreciation to my wife, Lila, for her continuing aid in prospecting and excavating in the Permian and especially in the relatively "barren" beds of much of the post–El Reno beds of Oklahoma. The Field Museum of Natural History, in particular John Bolt, has made available their collections for this study as has the Museum of Comparative Zoology at Harvard University. I am indebted to both for their courtesy and cooperation.

Literature Cited

Cope, E. D. 1896. The reptilian order Cotylosauria. Amer. Phil. Soc., Proc., 34:436–57.

Efremov, I. A. and Vjuschkov, B. P. 1955. Catalogue of localities of Permian and Triassic terrestrial vertebrates in the territories of the U.S.S.R. Acad. Sci. USSR, Paleontol. Inst. Tr., 46:1–185 (in Russian).

Moss, J. L. 1972. The morphology and phylogenetic relationships of the Lower Permian tetrapod *Tseajaia campi* Vaughn (Amphibia: Seymouriamorpha). Calif., Univ. Publ., Geol. Sci., 98:1–63.

Olson, E. C. 1957. Catalogue of localities of the Permian and Triassic terrestrial vertebrates of the territories of the USSR. J. Geol., 65:196–226.

———. 1965. Vertebrates from the Chickasha Formation, Permian of Oklahoma. Okla. Geol. Surv., Circular, 70:70 pp.

———. 1971. Vertebrate paleozoology. Wiley Interscience, New York, 839 pp.

———. 1975. The exploitation of land by early tetrapods. *In* Morphology and biology of reptiles. A. d'A Bellaires and C. B. Cox (eds.), Linn. Soc., Symp. Ser., 3:1–30.

———. 1979. *Seymouria grandis* n. sp. (Batrachosauria: Amphibia) from the middle Clear Fork (Permian) of Oklahoma and Texas. J. Paleontol., 53:720–28.

————, and Mead, J. In press. The Vale Formation, its vertebrates and paleoecology. Univ. Texas.

Spinar, Z. V. 1952. Revuse nekterych moravskych Diskosauriscidu (Revision of some Moravian Discosauriscidae). Roz. Ustred. Ustav. Geol., 15:1–159.

Romer, A. S. 1947. Review of the Labyrinthodontia. Harvard Univ., Mus. Comp. Zool., Bull., 99:3–352.

————. 1966. Vertebrate paleontology. 3rd Edition. Univ. Chicago Press, Chicago, 468 pp.

Watson, D. M. S. 1918. On *Seymouria* the most primitive known reptile. Zool. Soc. London, Proc.:267–301.

White, T. E. 1939. Osteology of *Seymouria baylorensis* Broili. Harvard Univ., Mus. Comp. Zool., Bull., 85:325–409.

Williston, S. W. 1910. *Cacops, Desmospondylus:* new genera of Permian vertebrates. Geol. Soc. Amer., Bull., 21:249–84.

————. 1911. American Permian vertebrates. Univ. Chicago Press, Chicago. 145 pp.

OBSERVATIONS ON TEMPORAL OPENINGS OF REPTILIAN SKULLS AND THE CLASSIFICATION OF REPTILES

by Emil Kuhn-Schnyder

A pure morphology of organic forms is impracticable.

—E. S. Russel, 1916, p. 52

In the Mesozoic Era, reptiles were the dominant terrestrial vertebrates. Today, in contrast, they are a minor constituent of our fauna. The four living orders can all be traced back to the Triassic. Still unquestionable reptiles are known from as early as the base of the Pennsylvanian. According to Carroll (1969) there is no question of the structural identity of even the earliest captorhinomorphs and pelycosaurs as reptiles. Only the arrangement of the temporal bones and the presence of a transverse flange on the pterygoid are in themselves sufficient to identify a reptile at this stage. However, the exact origin of the reptiles among the amphibians remains in doubt.

"It is now impossible to give any definition of the class Reptilia which, whilst including all members of the group, will exclude all other Tetrapods. The essential feature of a reptile is that it can carry out the whole of its life-history on dry land, not producing a gill-breathing larva, and that it is not a mammal or a bird. Reptiles lay a shelled egg except in viviparous forms, in which the egg is hatched before it is laid" (Watson, 1917, p. 171). Nevertheless, the reptiles are distinguished by a series of evolutionary trends, which are usually accompanied by increasing efficiency. Major changes take place in the structure of the skull.

The most primitive reptiles are the Captorhinomorpha. The members of this group are small forms, lacking temporal openings and an otic notch in the skull. From their skull pattern may be derived that of advanced reptiles.

In general the trends in the development of the reptilian skull are as follows (Romer, 1968a): (1) changes in skull proportions; (2) changes in skull openings; (3) palatal modifications; (4) occipital variations; (5) variations in braincase ossification; (6) reduction, loss, or fusion of dermal bones; (7) reduction of cheek covering in relation to action of underlying temporal muscles, including development of the temporal fenestrae, emargination of cheek from below and behind, and loss of arches bordering on temporal fenestrae.

This trend is particularly important in reptilian classification. Fenestrae on the cheek appeared at junctions of three bones (Frazetta, 1968; Halstead, 1969). On the cheek there are two such points (Figure 9.1): (1) the point where jugal, quadratojugal, and squamosal meet; and (2) the point where postorbital, squamosal, and parietal join.

The several explanations for fenestration offered by previous authors include speculations that open spaces in the skull permitted bulging of the jaw-closing muscles, and that fenestrae formed in areas of reduced stress where the presence of bone would be functionally useless. The first of these does not really apply to

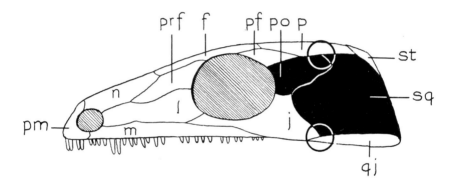

Figure 9.1. *Paleothyris,* a romeriid captorhinomorph, Middle Pennsylvanian. Circles, points of predestined temporal openings. Abbreviations: f, frontal; j, jugal; l, lacrimal; m, maxilla; n, nasal; pf, postfrontal; pm, premaxilla; po, postorbital; prf, prefrontal; q, quadrate; qj, quadratojugal; sq, squamosal; st, supratemporal. (After R. L. Carroll, 1969)

initial evolutionary stages; the second is more satisfactory (Frazetta, 1968). Another factor may also play a part. One of the advances of the reptiles was the development of a distinct neck region with a specialized condition in the first two vertebrae that facilitate movement of the head. This requires a lightly built head. Long-necked animals obtain more information from the surroundings. Since mechanical factors are involved, the appearance of comparable temporal openings does not necessarily mean that the animals concerned must be genetically related. Nevertheless, particular fenestration patterns appear to have remained fairly constant within each of the major radiation lines of the reptiles.

The trend to classify the reptiles by the structure of the temporal region began in 1867. In that year Günther showed that *Sphenodon* (*Hatteria*) differs from the Squamata in the possession of both upper and lower temporal arches, the squamata retaining only the upper arch. Towards the close of the past century, Baur (1889, 1895) and Cope (1892) developed the theory of fenestration. They attributed high value to the absence or presence of temporal openings for phylogenetic relationships. This period was temporarily summed up by the classic work of Osborn, "The Reptilian Subclasses Diapsida and Synapsida and the Early History of the Diaptosauria" (1903). He distinguished two subclasses, Synapsida and Diapsida. The Synapsida have none or a single temporal opening, the Diapsida have two openings. From the Synapsida should arise the mammals, from the Diapsida the birds. With that was introduced a diphyletic classification of the reptiles. Today we know, as Romer (1968a) emphasizes, that an attempt to arrange all reptiles in two groups—one-arched and two-arched—is an oversimplification of a complex problem.

Subsequently classifications of the reptiles based on temporal openings were worked out during the first two decades of our century by Williston (1917) and Versluys (1919). The classification of Williston represents a peak in the utilization of temporal structure as a key character (Table 9.1).

During the last decades Williston's classification was corrected by numerous scholars based on new discoveries in the fossil record. The following modern classification in Table 9.2 presents both a synthesis and a series of compromises of many opinions (Romer, 1970; Starck, 1978). In the classification of Ostrom and Carroll (1979) the order Protorosauria is omitted.

Classifications, as well as phylogenetic speculation, which are based on single characters have always proved to be risky. An example is the exclusive application of temporal openings for the classification of reptiles. It is necessary to ascertain how far the temporal arch character is associated with other distinctive features in different parts of the skeleton, and what supplementary and confirmatory evidence can be adduced for a natural subdivision of the class Reptilia. Moreover, it is necessary to study the whole pattern of the cheek. Now the different subclasses of the Reptilia will be examined according to the structure of their skulls and associated characters of the skeleton.

Subclass Anapsida

Reptiles with non-fenestrated temporal regions of the skull are placed in the Subclass Anapsida, including the two orders Captorhinomorpha and Chelonia.

The order Captorhinomorpha represents the earliest radiation of the reptilian class. Apart from reproductive improvements they are in most regards no more

TABLE 9.1

Williston's classification of reptiles. The doubtful or poorly known groups
are printed in italics.

Subclass Anapsida	Subclass Diapsida
Order Cotylosauria	Order Rhynchocephalia
Chelonia	*Rhynchosauria*
Subclass Synapsida	*Thalattosauria*
Order Theromorpha	*Choristodera*
Therapsida	Phytosauria
Sauropterygia	*Pseudosuchia*
Placodontia	Crocodilia
Subclass Parapsida	Pterosauria
Order Ichthyosauria	Dinosauria
Squamata	*"Eosuchia"*
Protorosauria	
(*Araeoscelis,*	
Acrosauria)	

TABLE 9.2
A synthetic and compromise classification of reptiles.
Class Reptilia

Subclasses	Number and position of temporal openings	Term	Orders
Anapsida	None	anapsid	Captorhinomorpha Chelonia
Synapsida	One vacuity at side of cheek with postorbital and squamosal usually meeting above	synapsid	Pelycosauria Therapsida
Euryapsida (Synaptosauria)	One vacuity, high on cheek, usually with postorbital and squamosal meeting below	euryapsid	Protorosauria Sauropterygia Placodontia
Ichthyopterygia	One vacuity, high on cheek, usually with postfrontal and squamosal meeting below	euryapsid	Ichthyosauria
Lepidosauria	Both vacuities present, usually with postorbital and squamosal meeting between them	diapsid modified diapsid	Eosuchia Rhynchocephalia Squamata
Archosauria	Both vacuities present, usually with postorbital and squamosal meeting between them	diapsid	Thecodontia Crocodilia Pterosauria Saurischia Ornithischia

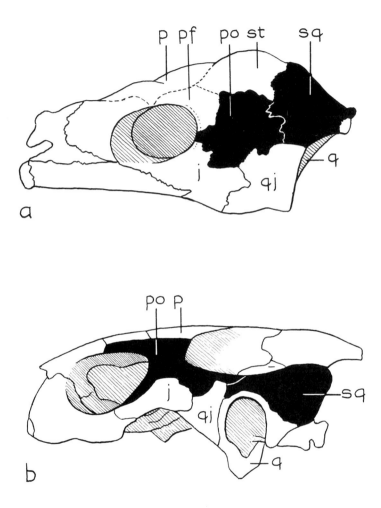

Figure 9.2. Chelonian skulls. a, *Proganochelys* (*Triassochelys*), Upper Triassic; b, *Emys*, Recent. Abbreviations as in Figure 9.1. (After A. S. Romer, 1968a)

advanced than their amphibian ancestors. The oldest known member of that group is *Hylonomus* from the Lower Pennsylvanian of North America. The last survivors are known from Upper Permian. It is likely that many, if not all, later reptiles have been derived from captorhinomorphs.

The order Chelonia has made a conspicuous advance in the development of a protective shell of bone covered by horn guarding both back and belly. This has necessitated remarkable changes in the skeleton and internal organs. The skull is highly modified. The temporal roof is not truly fenestrated but generally emarginated from behind, below, or both, allowing the jaw muscles to expand (Figure 9.2). The specific ancestry of chelonians is not known. A few very primitive forms (e.g., *Proganochelys*) are known from the Triassic of Europe.

To arrange the Chelonia with the Captorhinomorpha as a Subclass Anapsida is open to question. Although they have no temporal openings, their temporal region is very much modified. The skeleton is highly specialized. One cannot consider the basic morphology of the chelonians as primitive reptiles but as an isolated remnant of an early radiation. Creating a separate subclass for the Chelonia is justified.

Subclass Synapsida

The two orders of the Synapsida, Pelycosauria and Therapsida, are characterized as "mammal-like reptiles." Their skull roof of dermal bones behind the eyes had become perforated by a single temporal opening in a lateral position. The pelycosaurs appeared with captorhinomorphs (romeriids) in the Early Pennsylvanian, indicating a prior divergence of the two groups. They were forms distinguished by little more than the presence of a lateral temporal opening (Figure 9.3). In typical synapsids the postorbital and squamosal meet above the opening. In advanced synapsids the temporal opening has extended still farther dorsally, postorbital and squamosal meeting below the opening. The Synapsida seem to be a natural assemblage.

Subclass Euryapsida (Synaptosauria)

A grouping of protorosaurians, sauropterygians, and placodonts in a common major division was customary until the present time. These extinct reptiles should be characterized by a single temporal opening high up on the side of the cheek, usually with postorbital and squamosal meeting below (Figure 9.4). The three orders must be discussed separately.

With the order Protorosauria, Romer (1968a, 1970) entered a puzzling and uncertain territory in reptilian phylogeny. Generally unnoticed, he changed his opinion a few years ago.* In his earlier opinion, today still accepted by many authors, the typical protorosaurs should include a sequence of three well-known forms, which follow one another in time and degree of specialization, *Araeoscelis, Protorosaurus,* and *Tanystropheus.*

Today *Araeoscelis* is considered simply as somewhat specialized captorhinomorph. For a long time, *Araeoscelis* has caused confusion. Williston (1914) believed that the lizards were derived from *Araeoscelis* fairly directly without development of an intermediate diapsid stage. Although the skull of *Protorosaurus* is not entirely known,

---

*In his publication "Unorthodoxies in Reptilian Phylogeny" (1971, p. 110), Romer admitted "with some embarrassment" that he had upheld the category Protorosauria for too long.

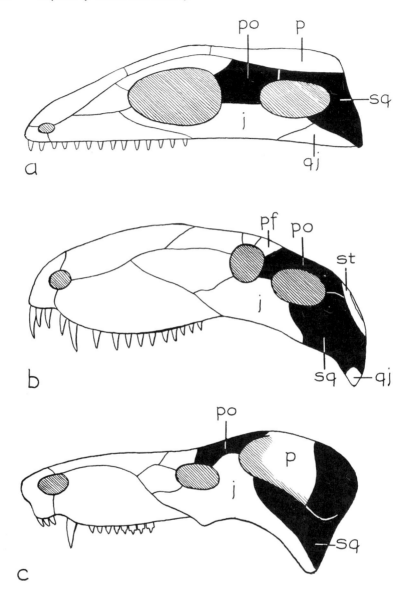

Figure 9.3. a, Diagram of the synapsid skull. Side views of b, pelycosaur; c, therapsid. Abbreviations as in Figure 9.1. (b, c after A. S. Romer, 1970)

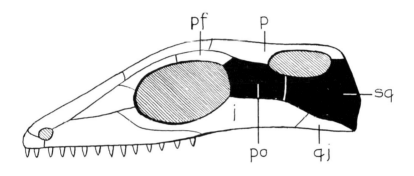

Figure 9.4. Diagram of the euryapsid skull. Abbreviations as in Figure 9.1.

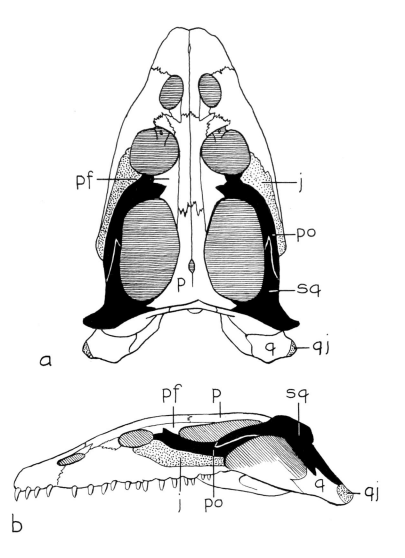

Figure 9.5. *Simosaurus*, Middle Triassic. a, dorsal view; b, side view; Reduced jugal and quadratojugal pointed. Abbreviations as in Figure 9.1. (After E. Kuhn-Schnyder, 1961)

one can be sure that it was diapsid. *Protorosaurus* is an eosuchian (Camp, 1945). *Macrocnemus* (Kuhn-Schnyder, 1962) and *Tanystropheus* (Wild, 1973) belong to the Prolacertiformes. *Trilophosaurus* is a highly aberrant form; its systematic position is unknown. The order Protorosauria should be discarded.

Williston (1925) proposed for the sauropterygians the Subclass Synaptosauria. The name indicates wistful hope that some relationship to mammal-like forms might be proved. Colbert (1955) suggested the term Euryapsida. There is disagreement about the pattern of the temporal region of the sauropterygians. Kuhn-Schnyder (1962, 1963, 1967) attempted to show that the ancestors of the sauropterygians were diapsid, not related to placodonts. Their arched cheek region is due to the loss of a lower temporal arcade. This hypothesis was already advocated by Jaekel (1910). The reduced jugal and rudimentary or missing quadratojugal speak for a former lower arcade (Figure 9.5).

Romer (1971) supposes that from the beginning the sauropterygians had only one temporal opening. The upward curvature of the cheek margin may be due simply to emargination. Who does not remember Williston's idea that the *Araeoscelis* type of skull by the simple emargination of the lower border of the squamosal and the consequent streptostyly gave origin to the Lacertilia? It is now agreed that the situation in lizards is due to the loss of the lower temporal bar with a consequent opening out of the cheek region.

Piveteau (1955) referred to specimens from the Upper Permian of Madagascar as possible plesiosaur antecedents (unfortunately without a figure). Carroll (Ostrom and Carroll, 1979, p. 716) communicates, "Additional excellently preserved material demonstrates clear relationships between these forms and nothosaurs and plesiosaurs on one hand, and eosuchians on the other. Specimens from the Upper Permian and possibly lowest Triassic of Madagascar show an almost complete transition from probably terrestrial eosuchians, through genera showing modest aquatic adaptation, to fragmentary remains of presumably totally aquatic nothosaurs." Sauropterygia are not Euryapsida, but descendants from diapsids.

The Placodontia are marine mollusk-eating reptiles of the Triassic that were for a long time classified as a suborder of sauropterygians. Now they are grouped in a separate order of the Euryapsida. Romer (1971) found many similarities between placodonts and sauropterygians. The resemblances seemed so numerous that at one time he was inclined to go so far as to place them in a single order, and was able to cite more than a full page of common characters (Romer, 1956, pp. 659–60). I was never in agreement with Romer on this placement (Kuhn-Schnyder, 1963, 1967). The skull of placodonts has an upper temporal opening with postorbital and quadratojugal meeting below (Figure 9.6). Therefore, the skull is not euryapsid. Placodonts never possessed a lower temporal opening. They have no relations with sauropterygians. The skeletons of sauropterygians and placodonts demonstrate profound differences too. The vertebrae of the nothosaurs have a zygosphene-zygantrum articulation, the placodonts a hyposphene-hypantrum articulation. Sauropterygians tend to increase the number of vertebrae, the placodonts to reduce it. From the very beginning the placodonts had only one temporal opening. They have no closer relationships with sauropterygians.

Subclass Ichthyopterygia

Williston (1925) has separated ichthyosaurs and sauropterygians into the subclasses Parapsida and Synaptosauria (Euryapsida) respectively on the basis of the temporal opening. In ichthyosaurs the bones forming the lower border of the opening were

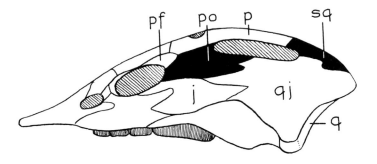

Figure 9.6. *Placochelys,* Upper Triassic. Side view. Abbreviations as in Figure 9.1. (After O. Jaekel, 1907)

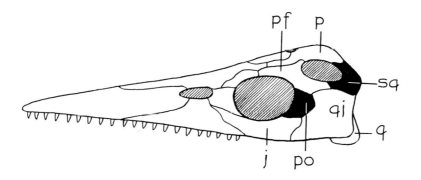

Figure 9.7. *Ichthyosaurus.* Side view. Abbreviations as in Figure 9.1. (After E. S. Goodrich, 1930)

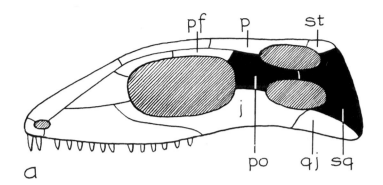

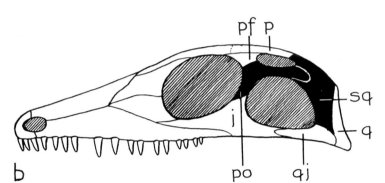

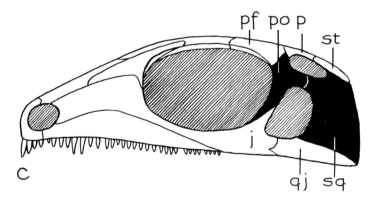

Figure 9.8. a. Diagram of the diapsid skull. b, *Youngina*, Upper Permian. Side view. c, *Petrolacosaurus*, Upper Pennsylvanian. Side view. Abbreviations as in Figure 9.1. (c, after R. R. Reisz, 1977)

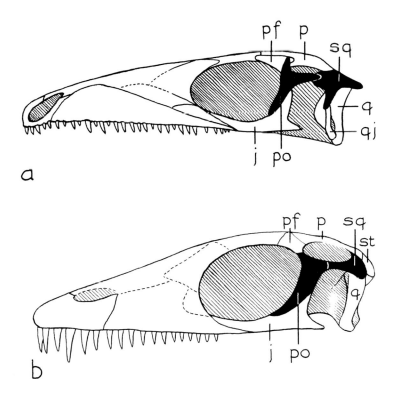

Figure 9.9. a, *Prolacerta,* Lower Triassic. Side view. b, *Tanystropheus,* Middle Triassic. Side view. Abbreviations as in Figure 9.1 (a after C. E. Gow, 1975; b, after R. Wild, 1973)

believed to be the postfrontal and the supratemporal, while in sauropterygians the two bones were clearly the postorbital and the squamosal. Romer (1968b) and McGowan (1973) have shown that the "supratemporal" is the true squamosal (Figure 9.7). Consequently the temporal opening of the ichthyosaurs is bordered by the postfrontal and the squamosal. This is a condition similar to, but not identical with, the Euryapsida. If all characters of ichthyosaurs and placodonts are considered it is clear that they have little in common save that they are marine reptiles.

Subclass Lepidosauria

Lepidosauria are primitive or modified diapsid reptiles, distinct from the archosaurian diapsids. The subclass covers three orders: eosuchians, rhynchocephalians, and squamatans.

To a number of primitive extinct forms from the late Paleozoic and the early Mesozoic, Broom (1924) has applied the term Eosuchia. A typical member of the eosuchians is *Youngina*, a small, primitive, lizard-like diapsid from the Upper Permian of South Aftica (Figure 9.8b). Eosuchians may have evolved from such forms as the Late Pennsylvanian captorhinomorph *Petrolacosaurus* (Figure 9.8c) with a fully developed diapsid skull (Reisz, 1977).

Members of the *Youngina* group have preserved the lower temporal arch, as in rhynchocephalians. In the Lower Triassic, forms are found which are clearly derivable from eosuchians but are in the process of losing the lower temporal arch. The genus *Prolacerta* (Parrington, 1935; Camp, 1945; Gow, 1975) from the Early Triassic *Lystrosaurus* zone of South Africa shows cranial features that suggest an intermediate position between eosuchians and primitive lizards (Figure 9.9a). However, recent descriptions of the postcranial skeleton indicate specializations quite distinct from that seen in primitive lizards. *Prolacerta* is closely related to *Macrocnemus* and *Tanystropheus* (Figure 9.9b) from the Middle Triassic of Europe (Wild, 1973). The Prolacertiformes are neither lacertoid eosuchians nor specialized thecodonts. They represent a separate branch of the eosuchians, mainly characterized by the elongation of the neck and by increased size. Another eosuchian branch is the marine *Thalattosaurus-Askeptosaurus* group of the Triassic.

The Rhynchocephalia are the conservative side branch of the eosuchians. The skull is diapsid with two complete temporal arches. Distinctive characters are the development of acrodonty and of a beak. They have survived with little change in the modern tuatara (*Sphenodon*). According to Carroll (1977) it is advisable to avoid the term Rhynchocephalia in referring to sphenodontids as the name implies association with the unrelated rhynchosaurs.

The Squamata include the suborders Lacertila, Amphisbaena, and Ophidia (Serpentes). They are modified diapsids which have lost the lower or both temporal bars. Another characteristic feature is the mobility of the quadrate. According to Carroll (1975) the Permian-Triassic Paliguanidae (Figure 9.10) may be included with the most primitive known lizards and the possible antecendents of all more advanced members of the Squamata. Remarkable lizards from the Upper Triassic seem to have glided on rib-supported "wings" like the modern *Draco* (*Kuehneosaurus*, Robinson, 1962; *Icarosaurus*, Colbert, 1970). *Daedalosaurus* from the Upper Permian of Madagascar is another small reptile with greatly elongated ribs to support a gliding "membrane." However, the primitive nature of the appendicular skeleton suggest that these genera should be classified among the Eosuchia (Carroll, 1978).

Subclass Archosauria

An array of diapsids such as thecodonts, dinosaurs, pterosaurs, and crocodiles comprises the Subclass Archosauria. They have the diapsid character in common with members of the other great Sublcass Lepidosauria. The archosaurians have a few distinctive features. The skull usually has a vacuity in front of each orbit (Figure 9.11). A skin armor is mostly well developed and osteoderms, probably covered in life by horny scutes, are often present on the back and sometimes on the belly. The Permian pedigree of the Archosauria is obscure. The earliest known members from the lowest Triassic are already far advanced from the primitive reptilian pattern in most aspects of their skeletal anatomy. Carroll (1976) has described an eosuchian reptile *Heleosaurus* from the Upper Permian of South Africa. This species, which is in general similar to *Youngina*, demonstrates that archosaurians could have evolved from eosuchians. Carroll supposes that the eosuchian-archosaurian transition probably took place within a lineage of small, active, terrestrial carnivores.

The most primitive archosaurs are the thecodonts of the Triassic, best characterized by *Euparkeria* (Figure 9.11). In general it is supposed that the thecodonts included the ancestors of all other types of ruling reptiles and perhaps also of the birds. However, according to Wild (1978), the thecodont origin of pterosaurs seems to be doubtful. It is much more probable that the pterosaurs originated in the eosuchians as is shown by the size, morphological features of the skeleton, the geological age (Upper Triassic), and the early radiation of the pterosaurs.

Discussion

The pattern of the reptilian cheek is important in classification. Action of jaw muscles leads to reduction in the cheek covering. This reduction may affect development of temporal fenestrae, loss of arches bordering on temporal openings, and emargination of cheek from below and behind. However, a purely morphological attitude to reptile classification relying on temporal openings alone, such as S. W. Williston's procedure, results in an artificial system. In order to attain a natural system it is necessary to unite detailed studies of form and of function. In addition, affinities must be judged by comparing not only temporal openings but many other features as well. Fenestrae on the cheek appear at the junction of three bones. There are two such points. Therefore, three conditions are to be expected: an upper opening, a lower opening, and two openings—all three conditions in effect exist. In certain groups of reptiles temporal openings may subsequently change position.

Since mechanical factors are involved the appearance of comparable temporal openings does not necessarily mean that the animals concerned must be genetically related. This must be taken into consideration in judging reptiles from the Permian (*Araeoscelis, Bolosaurus, Mesosaurus, Millerosaurus*). Nevertheless, particular fenestration patterns of the Triassic reptiles appear to be constant within the major radiation lines of reptiles. The results of an examination of the reptilian cheek pattern are presented in Table 9.3.

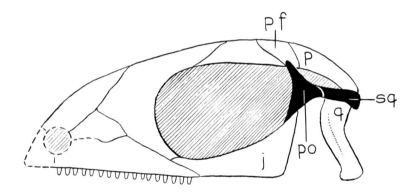

Figure 9.10. *Paliguana,* Lower Triassic. Side view. Abbreviations as in Figure 9.1. (After R. L. Carroll, 1975)

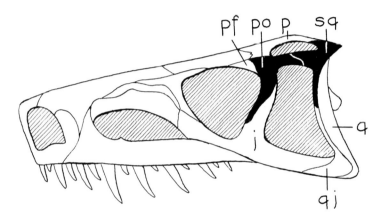

Figure 9.11. *Euparkeria,* Lower Triassic. Side view. Abbreviations as in Figure 9.1. (After B. Krebs, 1976)

TABLE 9.3

Results of examination of the reptilian cheek pattern.

Number and position of temporal openings	Term	Main groups showing the condition
None	anapsid	Captorhinomorpha, etc.
None	modified anapsid	Chelonia
Both vacuities present usually with postorbital and squamosal meeting between them	diapsid	Eosuchia Rhynchocephalia Archosauria
One vacuity high on cheek, postorbital and squamosal meeting below; streptostyl	modified diapsid	Squamata
One vacuity high on cheek, postorbital and squamosal meeting below; monimostyl	modified diapsid	Sauropterygia
One vacuity high on cheek, postorbital and quadratojugal meeting below		Placodontia
One vacuity high on cheek, postfrontal and squamosal meeting below		Ichthyopterygia
One vacuity at side of cheek with postorbital and squamosal usually meeting above	synapsid	Pelycosauria Therapsida

My conclusions are the following:

1. Captorhinomorpha and Chelonia should be separated.
2. The order Protorosauria is to be discarded.
3. The Sauropterygia are descended from diapsids.
4. The Placodontia have a single upper temporal opening, but not of euryapsid type. Placodonts have no closer relationships with sauropterygians.
5. The Ichthyopterygia have an upper temporal opening, neither identical with that of Placodontia nor of synapsid type.
6. The term euryapsid is obsolete.
7. It is very difficult to assign Late Permian and Early Triassic diapsids to either the Lepidosauria or the Archosauria.

The classification of reptiles which seems best to me is given below:

> Subclass: Captorhinomorpha
> Subclass: Chelonia
> Subclass: Sauropsida
> > Lepidosauria
> > Archosauria
> > Sauropterygia
> > Placodontia
> > Ichthyopterygia
> Subclass: Theropsida
> > Pelycosauria
> > Therapsida

Our knowledge of fossil reptiles is far from complete. There will be many interesting forms to be revealed in the future, all over the world. I hope that E. H. Colbert, to whom this paper is dedicated, may still give us many an important contribution to our knowledge of fossil reptiles.

Acknowledgments

Prof. Dr. H. Rieber has been very obliging in providing facilities at the Palaeontological Institute, University of Zurich. Dr. M. Schnitter deserves especial thanks for polishing my English. The credit for the illustration is due to Mr. O. Garraux, Basle.

Literature Cited

Baur, G. 1889. On the morphology of the vertebrate skull. J. Morphol., 3:471–74.
———. 1895. Bemerkugen über die Osteologie der Schläfengegend der höheren Wirbeltiere. Anat. Anz., 10:315–30.
Broom, R. 1924. The classification of the reptiles. Amer. Mus. Natur. Hist., Bull., 51(2): 39–65.
Camp. C. L. 1945. *Prolacerta* and the protorosaurian reptiles. Amer. J. Sci., 243:17–32, 84–101.
Carroll, R. L. 1969. Origin of reptiles. *In* C. Gans (ed.), Biology of the Reptilia 1. Academic Press, London, pp. 1–44.
———. 1975. Permo-Triassic "lizards" from the Karroo. Palaeont. Afr., 18:71–87.
———. 1976. Eosuchians and the origin of archosaurs. *In* C. S. Churcher (ed.), Athlon, Essays on palaeontology in honour of L. S. Russell. Roy. Ontario Mus., pp. 58–79.

————. 1977. The origin of lizards. *In* S. M. Andrews, R. S. Miles, A. D. Walker (eds.), Problems in vertebrate evolution. Linn. Soc. Symposium, 4, pp. 359–96.

————. 1978. Permo-Triassic "lizards" from the Karroo System. II. A gliding reptile from the Upper Permian of Madagascar. Palaeont. Afr., 21:143–59.

Colbert, E. H. 1955. Evolution of the vertebrates. J. Wiley & Sons, New York, 479 pp.

————. 1970. The Triassic gliding reptile *Icarosaurus*. Amer. Mus. Natur. Hist., Bull., 143: 89–142.

Cope, E. D. 1892. On the homologies of the posterior cranial arches in the Reptilia. Amer. Phil. Soc., Trans., 17:11–26.

Frazetta, T. H. 1968. Adaptive problems and possibilities in the temporal fenestration of tetrapod skulls. J. Morphol., 125:145–58.

Goodrich, E. S. 1930. Studies on the structure and development of vertebrates. Macmillan, London, 837 pp.

Gow, C. E. 1975. The morphology and relationships of *Youngina capensis* Broom and *Prolacerta broomi* Parrington. Palaeont. Afr., 18:89–131.

Günther, A. C. 1867. Contribution to the anatomy of *Hatteria* (*Rhynchocephalus* Owen). Roy. Soc. London, Phil. Trans., 157:595–629.

Halstead, L. B. 1969. The pattern of vertebrate evolution. Oliver and Boyd, Edinburgh, 209 pp.

Jaekel, O. 1907. *Placochelys placodonta* aus der Obertrias des Bakony. Resultate d. wiss. Erforschung d. Balatonsees, 1(1) Pal. Anhang., 90 pp.

————. 1910. Ueber das System der Reptilien. Zool. Anz., 35:324–41.

Krebs, B. 1976. Pseudosuchia. *In* Handbuch der Paläoherpetologie, Teil 13:40–98.

Kuhn-Schnyder, E. 1961. Der Schädel von *Simosaurus*. Palaeontol. Z., 35:95–113.

————. 1962. Ein weiterer Schädel von *Macrocnemus bassanii* Nopcsa aus der anisischen Stufe der Trias des Monte San Giorgio (Kt. Tessin, Schweiz). Palaeont. Z., H. Schmidt-Festband:110–33.

————. 1963. Wege der Reptiliensystematik. Palaeont. Z. 37:61–87.

————. 1967. Das Problem der Euryapsida. Problèmes actuels de Paléontologie (Évolution des Vertébrés) CNRS, Paris, 163:336–48.

McGowan, C. 1973. The cranial morphology of the Lower Liassic latipinnate ichthyosaurs of England. Brit. Mus. (Natur. Hist.), Bull., Geol. 24(1):1–109.

Osborn, H. F. 1903. The reptilian subclasses Diapsida and Synapsida and the early history of the Diaptosauria. Amer. Mus. Natur. Hist., Mem., 1:451–507.

Ostrom, J. H., and Carroll, R. L. 1979. Reptilia. Encyclopedia of paleontology. Dowdon, Hutchinson & Ross, Stroudsburg, pp. 705-20.

Parrington, F. R. 1935. On *Prolacerta broomi* gen. et sp. n., and the origin of lizards. Ann. Mag. Natur. Hist., (10) 16:197–205.

Piveteau, J. 1955. L'origine des Plésiosaures. Acad. Sci. C. R. hebd., 241:1486–88.

Reisz, R. R. 1977. *Petrolacosaurus*, the oldest known diapsid reptile. Science, 196:1091–93.

Robinson, P. L. 1962. Gliding lizards from the Upper Keuper of Great Britain. Geol. Soc. London, Proc., 1601:137–46.

Romer, A. S. 1968a. Osteology of the reptiles. Univ. Chicago Press, Chicago, 772 pp.

————. 1968b. An ichthyosaur skull from the Cretaceous of Wyoming. Wyo. Univ. Contrib. Geol., 7(1)27–41.

————. 1970. The vertebrate body. 4th ed. W. B. Saunders, Philadelphia, 601 pp.

————. 1971. Unorthodoxies in reptilian phylogeny. Evolution, 25:103 112.

Russell, E. S. 1916. Form and function. J. Murray, London, 383 pp.

Starck, D. 1978. Vergleichende Anatomie der Wirbeltiere auf evolutionsbiologischer Grundlage, 1, Springer, Berlin, 274 pp.

Versluys, J. 1919. Ueber die Pylogenie der Schläfengruben und Jochbogen bei den Reptilia. Sitzungsber. Heidelb. Akad. Wiss., math.-natur. Kl. (B), 1919, 13:1–29.

Watson, D. M. S. 1917. A sketch classification of the pre-Jurassic tetrapod vertebrates. Zool. Soc. London Proc., 1917:167–86.

Wild, R. 1973. *Tanystropheus longobardicus* (Bassani) (Neue Ergebnisse). Schweiz. Paläont. Abh., 95:1–162.

————. 1978. Die Flugsaurier (Reptilia, Pterosauria) aus der Oberen Trias von Cene bei Bergamo, Italien. Soc. Paleontol. Ital., Boll., 17, 2:176-256.

Williston, S. W. 1914. The osteology of some American Permian vertebrates. J. Geol., 22: 364–419.

————. 1917. The phylogeny and classification of reptiles. J. Geol., 25:411–21.

————. 1925. The osteology of reptiles. Harvard Univ. Press, Cambridge, Mass., 300 pp.

REMARKS ON A NEW
OPHIOMORPH REPTILE
FROM THE LOWER CENOMANIAN
OF EIN JABRUD, ISRAEL

by G. Haas

Introduction

A second snakelike reptile from the lower Cenomanian quarries of Ein Jabrud, near Ramallah, north of Jerusalem, was acquired from a local quarry worker about two years ago. The specimen is preserved in a slab of limestone (Figure 10.1), and provides information on the origin of snakes.

Unfortunately, an anterior portion of unknown extent and length is missing between an isolated series of about 12 nuchal vertebrae embedded at the right side of a lengthy series of about 115 vertebrae. The greater part of the tail is also missing. Remains of the scattered skull are revealed through 6–7 contiguous intercostal intervals, preceded by 5 intervals revealing about 6 nuchal vertebrae. At the posterior end of the fossil the interesting rudimentary right hind limb and pelvic girdle and some traces of the left one can be seen. This area is followed by the first of about 6 caudal vertebrae. The length of the missing tail can only be guessed.

The description can be followed with the accompanying illustrations. Abbreviations used in the figures are as follows:

1—Ilium	Fi—Fibula	Pt—Pterygoid
2—Ischium	In—Intermedium (tarsi)	Q—Quadrate
3—Pubis	J—Jugal	R—Rib
Bt—"Bladed" tooth	Mx—Maxilla	S—Sacral vertebra
De—Dentary	N—Nasal	Sp—Splenial
E—Ectopterygoid	P—Parietal	Sq—Squamosal
Ep—Epipterygoid	Pf—Postfrontal	St—Supratemporal
Fb—Fibulare	Pl—Palatine	Ti—Tibia
Ft—Furrowed tooth	Po—Prootic	V—Nuchal vertebra
Fe—Femur	Pr—Prefrontal	*—Postdental area
		of mandibles

Measurements are given in Table 10.1.

The Axial Skeleton

The preparation of the ribs was rather easy, whereas the vertebrae were less tractable and only limited regions were completely prepared, especially around the scattered skull elements. Four to seven vertebrae in front of the skull region were prepared. The isolated nuchal piece was cleaned thoroughly. Unfortunately, the region in front of the limb rudiments proved to be especially intractable due to the hardness of the matrix. This region is marked on Figure 10.1 by two arrows.

The isolated contiguous fragment of the nuchal vertebrae shows surprisingly small and delicate vertebrae compared with those of the adjacent anterior thoracic

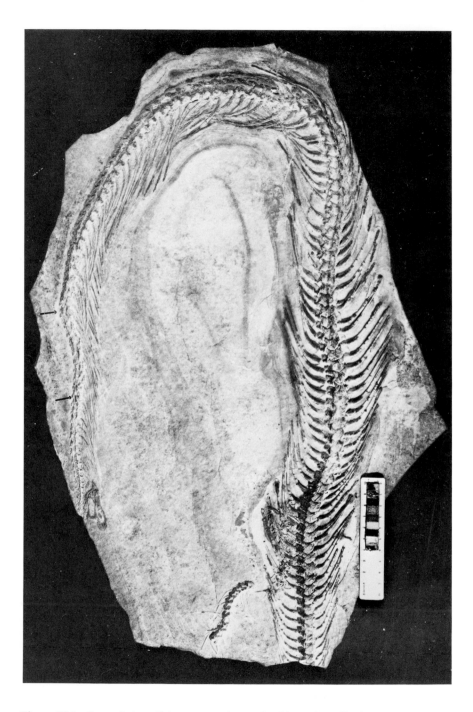

Figure 10.1. General view of the type specimen of *Ophiomorphus colberti* gen. et sp. nov.

TABLE 10.1

Measurements (cm) of *Ophiomorphus colberti.*

	Length	Width
Series of 115 vertebrae	107	
12 neck vertebrae	8	
Missing portion of neck (estimated)	15	
Missing portion of tail (estimated)	20	
Anterior portion of neck (estimated)	5	
Total (estimated)	155	
Right hind limb	3.00	
Right femur	1.85	.48 proximal .60 distal .30 minimum
Right tibia	1.05	.40 proxial .22 distal
Right fibula	1.00	
Right tarsal elements (each measured separately along the proximo-distal axis)	.20	

regions. The cranial-most vertebra has just a quarter of the width of the thoracics nearby. All these vertebrae carry slender and rod-shaped ribs with a single articular facet. Two more ribs can be seen close to the periphery of the right-hand rib basket. Such minute ribs are assumed to continue up to the skull with only a few ribless nuchal vertebrae. The vertebral diameter gradually diminishes posteriorly from the first third of the contiguous piece of the fossil so that the vertebrae facing the limb rudiments are about the same size as the posterior vertebrae of the isolated nuchal fragment. This fact shows that the missing tail was certainly rather short and thin. The complete absence of pachyostosis in the ribs or in the vertebrae is a striking difference from the previously described *Pachyrhachis* (Haas, 1979).

The ribs of the anterior half (approximately) of the body show a characteristic curvature between the first and second quarter of their total length. They are straight in the neck region and then gradually straighten in the posterior half of the body. As in snakes, the ribs are terminally flattened and somewhat widened distally and have abrupt ends. This widened end might indicate a cartilaginous end, perhaps anchored in the skin as in many ophidians (Hardaway and Williams, 1975, 1976). At several places, especially in the posterior half of the fossil, ribs are crowded together in small groups, overlapping the set of the other side of the body (Figure 10.1). Globular diapophyses can be clearly seen in the thoracic area, as in snakes, especially in the region where the skull remains are hidden between the rib intervals. They are in front and above the ventral rib articulation and have a concave elongated proximal joint with the parapophyses.

The dorsally exposed vertebrae have mostly incompletely preserved neural spines. Those preserved are rather low and exhibit a sharp longitudinal ridge. Pre- and postzygapophyses are ophidian in character. Zygosphene-zygantral links cannot clearly be seen, but almost certainly exist. Until further preparation, the occurrence of hypapophyses cannot be documented. There are certainly no lymphapophyses in the pelvic region. Knoblike processes bulging dorsally off the parapophyses in the anterior thoracic region seem to be free diapophyses, not connected with a dorsal branch of the rib. This branch was probably replaced by a costo-diapophysis ligament. Such diapophyses are absent in the pelvic region.

The fossil described here differs drastically from *Simoliophis* in the neural spines, which are laterally compressed and probably shorter. An inner pair of intervertebral articulations was almost certainly present, but the contiguity of the vertebral series does not permit definite determination. Without ventral preparation the hypapophyses cannot be described. The area of possible lymphapophyses, caudal to the limb rudiment, shows no trace of bifurcation.

The inner angle of the ribs is no less than 120°–140° and gradually increases caudally to 170° near the limb rudiment. This angle migrates gradually toward the vertebral articulation of the ribs so that at least three pairs of ribs in front of the ilium are completely straightened.

Skull Remains

The remains of the skull (Figures 10.2, 10.3, 10.4) are covered by the rib basket along the right side of the body and comprise a scattered disarray of isolated bones, visible in a contiguous series of intercostal areas. For convenience, the intercostal spaces under consideration will be designated N, N+1 to N+6 from anterior to posterior (see Figures 10.3 and 10.4). Scattered parts of the first vertebrae and ribs

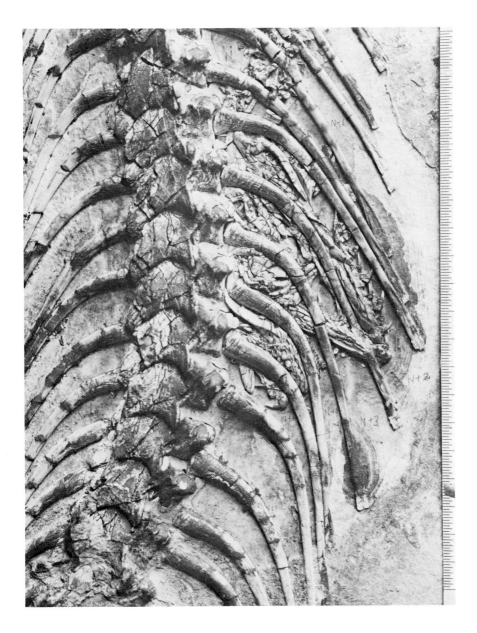

Figure 10.2. View of the scattered skull remains. Scale in mm.

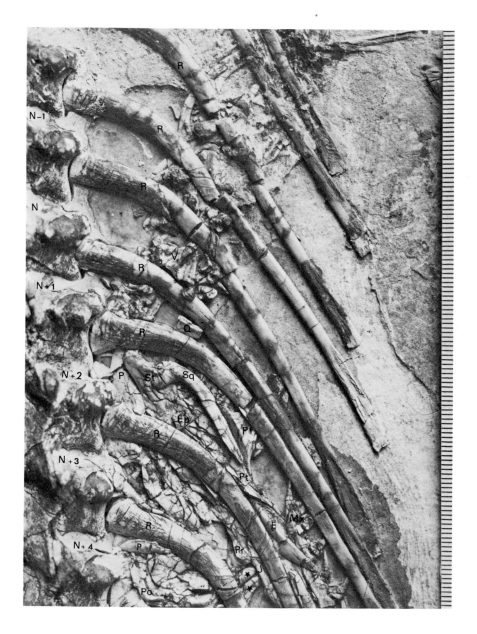

Figure 10.3. Enlargement of the anterior part of the skull. Scale in mm.

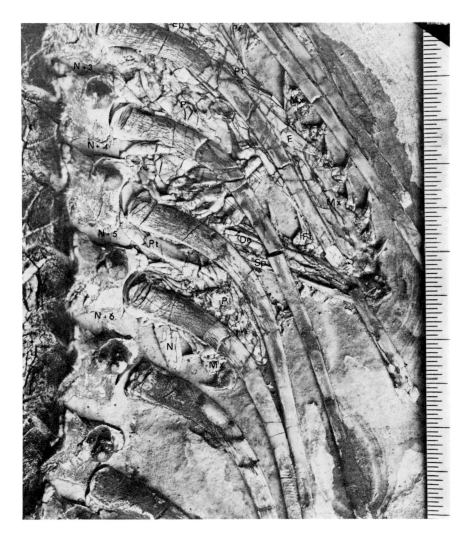

Figure 10.4. Enlargement of the posterior part of the skull. Scale in mm.

are in intercostal intervals N–1 to N–6, whereas skull elements are observed in N+, covering a total length of about 55 mm. The readily identified toothed elements will be described first.

A long bone with teeth is exposed in interval N+2 in lateral view, and is clearly the left maxilla, with the rostral end pointing laterally. It is provided with a long series of teeth, most slightly displaced, but some others still in natural position. All these teeth, except the last three of the series closely resemble those of snakes in their slender laterally compressed shape and their caudally receding sharp tips. They are fixed to the bone in a typically acrodont to a slightly pleurodont way. The six preserved teeth of this type are followed by three posterior densely implanted ones with rather broadened bases (tips are missing). Such broad bases are also seen in the third and fifth of the anterior larger teeth. The maxilla, toothed almost to its tapering posterior end, bends slightly ventral at the area of the smaller posterior three teeth. Two or three trigeminal foramina pierce the central third of the bone at short intervals. The rounded anterior end of the maxilla indicates a free end and precludes a sutural connection with the premaxilla (which is missing); this is certainly an ophidian character. The festooned lower border of the maxilla indicates the position of twelve to fifteen shed or alternating teeth.

Another elongated bone starts in interval N+2 overlapping the anterior part of the maxilla and can be followed across the next two intervals. This bone is certainly the left dentary seen in lingual view. The rounded anterior end in N+2 shows no indication of a rigid symphysis. There is no trace of a roughened facet for contact with the partner. This is certainly a familiar condition in ophidians but also is known in varanoids, especially mosasaurs. The rather flattened inner side of this bone seen in N+3 and N+4 has a thickened bulging lower margin. The dentary gradually deepens posteriorly and is excavated medially for the attachment of the splenial and for providing room for a probably extensive Meckelian cartilage dorsal and medial to the splenial. The dentiferous border shows no festoons such as those in the maxilla, but is straight. The anterior end shows a strongly recurved tooth of ophidian shape. In space N+3 three teeth can be seen.

The anterior end of the right maxillary with two teeth is seen in N+6. Most of this bone is hidden under the preceding rib and adjacent bones.

The splenial is closely applied to the medial face of the dentary. Its gradually tapering anterior part is seen in N+3. Ventral to it the Meckelian fossa forms a longitudinal gap. In N+4 the splenial reaches its maximum height and covers most of the internal face of the dentary. Just in front of the following rib the height is slightly reduced. Unfortunately the posterior border of the bone is hidden under a rib.

The right palato-pterygoid series is preserved in N+5. The sharply pointed caudal end of the pterygoid points medially and is seen near the vertebral centrum. The alveolar border of the palatine has festoons for tooth bases, but no tooth is preserved. Laterally the palatine is pierced by a series of five foramina. The complicated and partly blurred suture with the pterygoid shows a deep notch which separates a lateral process at the caudal end of the palatine from the main body of the bone. This process shows clearly that it is the right palatal series that is preserved. It formed a link with the maxillary as in *Varanus* or *Python*. The pterygoid, with its characteristic sigmoid twist, has a longitudinal keel on its anterior half. Along its medial contour there are five displaced teeth. A rostral group of three pointing medially is succeeded by two parallel teeth which lie closer to the axial skeleton in their dislodged position. Because no alveoli are evident in the pterygoid, the teeth could have been dislodged from the close palatine, where about five festoons can be seen. They could also be displaced from the concealed posterior part of the right maxilla. No furrows are present on these teeth. All tips point toward the axial skeleton.

The posterior hooked end of the squamosal, seen in N+2, points axially. Its long tapering anterior arm points distally. The enlarged and thickened posterior hook has a bend of about 120° relative to the long arm, and is rounded as in *Varanus*. It forms the main part of a free jugal arch, which probably has a suture posteriorly with the supratemporal and anteriorly with a postorbital.

The element that articulated posteriorly with the squamosal, and which juts posteriorly beyond this joint, is probably the supratemporal. It forms a terminal swelling. Its cranial end is hidden under a rib, but indicates a rostral tapering so characteristic for a supratemporal.

The single or combined postfrontal-postorbital element is in close contact with the anterior half of the rostrally tapering jugal arch formed mainly by the squamosal. It might be displaced in a caudal direction, because this bone is usually in front of the squamosal. Even after displacement, it lies a little rostral to this bone. Its elongated contact with the squamosal is clear, but its ventral narrowing wing is hidden under two closely placed ribs. Rostrally a deeper postocular wing is clearly visible as well as a caudal wing which faces the open lower temporal fossa. This part is deeper than the cranially jutting tapering process and ends bluntly. Sharp ridges separate the orbital wing from the main body and a niche in the bone (not pierced) can be seen (N+2).

The ectopterygoid (N+4, distally) encloses, laterally more than medially, the tapering posterior end of the left maxilla. Both laminae are separated by a sharp keel. The outer lamina is partially covered by a rib and hidden at its posterior end. The postero-ventral margin of the ectopterygoid is in close contact with the partly visible left pterygoid. Its posterior limit is not clearly defined (between the squamosal and a limited part of the parietal).

In N+2 parts of an extensive parietal are exposed between the pterygoid and the supratemporal. More of the parietal is seen in N+3 and N+4 (proximally).

A bone stump, visible in N+1 near two distally approaching ribs, is considered as a part of the quadrate. The upper head is hidden and the lower articulation missing. The tentative identification is based on the topography of the protruding stump.

The prefrontal and the jugal are seen in N+3. The first element is seen distal to the big tooth of the dentary at the angle between the latter and the preceding rib. This bone is only partially visible and, on the basis of its position, is certainly a prefrontal.

The jugal is an elongated sliver of bone lying along the preceding rib. Its caudal border is not clear but is probably parallel to the rib terminating in a long suture at the upper posterior margin of the maxillary.

N+3 and N+4 are filled by an array of cracked and flattened bone which certainly contains the posterior part of the skull table. This probably includes an azygous parietal, which certainly was a flattened bone without any trace of a longitudinal crest and probably with little or no descensus, as in *Varanus*. These structures are in sharp contrast with those of *Pachyrhachis* from the same area and of the same age. Only a short stretch of an unaffected free margin in N+3 is clearly shown. Most probably this represents the left anterior lateral margin, and a very short stretch of the left anterior margin. A wide cleft filled by matrix separates these borders from any neighboring bone. An almost identical border at the right side of the parietal can be traced in N+4. Here a laterally adjoined and fractured complex bone could be a dorsal part of a flattened prootic (see Figure 10.4).

In N+6 an elongated flat bone with a gently pointed anterior (?) end protrudes from the preceding rib. It could be a nasal or a frontal. The partner is missing. A medial prominence near the rib contour to the right has a spoonlike shape. I cannot explain this structure. Several other cranial elements will remain difficult to interpret until preparation from the ventral side is complete.

Several bones seen mainly in N+4, but also exposed in the preceding space, are unexplained for the time being. They lie in front of the area filled by the indistinct parietal in the position of paired and dislodged frontals, but they certainly cannot be the frontals. The following interpretation is just a possibility and not too convincing. The distal element in N+4 and N+5 has a protruding bulge postero-proximally, followed distally by a depression along the long axis of the bone. Comparison with *Varanus* points to the possibility that these are parts of the lingual face of the caudal half of a left mandible with its coronoid process. Under it are parts of the surangular and more ventrally the angular. If this explanation is correct, the postdental part of the lower jaw must have been of relatively moderate size, but the caudal part of the mandible was much more extensive and a great part is hidden under adjacent bones. The parallel bony bar along the preceding element in N+4 could also be a fraction of the postdental region. The coronoid elevation seems to be small, but much of it could be hidden.

Another bone is connected to the coronoid proximally in N+4. It has a deep incisure near the rib. In the branch distal of this incisure there is probably a natural cleft which widens toward the succeeding rib. Is this bone a displaced prootic or part of a descensus frontalis? Conditions in *Varanus* are completely different. There is barely any descensus, at least in *V. griseus,* the species at my disposal.

The caudally tapering bony wing in N+2 is considered the right posterior wing of the parietal. It seems to correspond to the deepest bone proximal to the supratemporal in N+2. It is broken at and near the posterior tapering end. In N+4 a small part of the transverse anterior border of the parietal which originally was in sutural contact with the (here invisible) pair of frontals is seen. Only a short stretch of this border is visible at the right anterior corner of the skull table.

Another element whose interpretation is not clear extends in N+2 from the right corner of the jugal arch toward the following rib. A rounded bone with a slightly expanded upper end lying vertical to the arch could be a caudally dislodged left epipterygoid. Compressed as it is, its bulging stem suggests an originally rounded column of bone. The partner of this element is hidden. In the intervals N+1 and N anterior axial elements are seen. They are clearly not parts of the basicranium (basioccipital or basisphenoid), but rather anterior nuchal vertebrae, with a clear neural arch in N+1. The preceding space contains an incomplete larger nuchal vertebra.

Six teeth of the maxilla all have a lateral cutting edge running distally from the enlarged bases of the teeth. The last three preserved tooth bases indicate a smaller size, but their distal parts are missing (N+2). The third anterior tooth shows a very shallow furrow in front.

The dentary (N+2, 3, 4) shows one tooth in N+2 and three in N+3. Of these, the anteriormost and the third show clearly a sharp lateral edge. The five preserved, but dislodged, teeth near the palato-pterygoid series show neither clear furrows nor sharpened edges. There is in the depth of the matrix an almost hidden tooth, the sixth of the group. Of the two teeth belonging to the exposed part of the other maxilla, the first at the rostral end shows a distal keel in profile. The other is more evenly rounded. No clear furrows such as present in *Pachyrhachis* can be observed, with the possible exception of a very shallow one seen in a single tooth in the maxilla. The teeth of *Varanus griseus* have a clear posterior keel at the posterior curvature. The longitudinally enlarged tooth bases in the fossil described here are remarkable, both in the maxilla and the dentary. This type of base seems to be wanting in the displaced group of teeth close to the palato-pterygoid series. In pythonid teeth, as least in the specimens at my disposal, no trace of any keel exists. Drastically enlarged tooth bases are not seen in *Varanus* or *Python*.

The Rudimentary Hind Limb and Pelvis

The relatively well-developed hind limb (Figures 10.5, 10.6) surpasses in complexity all rudiments described so far from Recent primitive snakes, as Leptotyphlopidae, Typhlopidae, Ilysiidae, Anilidae, and Boidae. There are also pelvic elements.

The femur is a nearly straight bone. The narrowest part (40 percent the total length from the proximal end) is crossed lengthwise by a groove which is clearly developed for a distance of 4.5 mm, and becomes less distinct toward both ends. The surface of both apophyses is somewhat damaged (more so proximally).

Between the elements of the zygopodium an elliptical space, pointed at both ends, remains open. At the widest point the distance between tibia and fibula is 2.2 mm. The stouter tibia has a proximal end much wider than the distal. The external (lateral) contour of the tibia is completely straight, the medial gently curved. The narrowest part of the tibia lies approximately at a point one-third the length from the distal end. The contact with the distal end of the femur is in the nature of an amphiarthrotic joint. This contrasts with the rather loose contact of the fibula with the medial slightly shorter half of the distal femoral apophysis. Here the round proximal end of the fibula approaches a similarly rounded medial end of the tibia. In short there is a sharp contrast in the nature of the "knee joint" between the tibia and the adjoined following elements. The proximal end of the tibia overlaps the fibula slightly at its medial edge. Distally, both bones are in smooth contact: the fibular facet, laterally, with the medio-distal end of the tibial shaft. The fibula is narrower and somewhat longer than the tibia, and evenly expanded at both ends. The narrowest point (1.1 mm) lies one-third of the way from the distal end of the strongly bent fibula.

The limb seems to have protruded behind the last thoracic rib. Much shorter ribs are visible in the preserved part of the tail. The distal ends of the tibia and fibula seem to be coossified terminally so that a third of the distal fibular end is more or less fused with the medial side of the tibia for a short stretch proximal to the terminal face of articulation. The fibula, thus, ends in a close connection with the tibia, prohibiting any movement in this area. Two fused tarsal elements are them-

Figure 10.5. The right hind limb; parts of left also visible. Scale in mm.

Figure 10.6. The right hind limb in greater enlargement. Scale in mm.

selves fused to the fibula medially, and a shallower tarsal element, certainly a tibiale, is welded to the whole terminal articular surface of the tibia. It is not quite clear if or how much of the very thin tarsal region is missing; probably little or nothing. Certainly both tarsal elements, a tibiale and intermedium (perhaps with a fused fibulare) are rigidly united to each other and also to the ends of the zygopodium. This whole complex functioned as one entity.

An elongated bone jutting dorso-medially and slightly in front of the proximal end of the femur is considered an ilium. Its originally rostro-ventral (acetabular) end is obliquely rounded and expanded, but no trace of an acetabulum is developed. The upper dorsal end was probably disconnected from its natural position in contact with a sacral vertebra. The abruptly truncated and slightly enlarged dorsal end speaks in favor of a rather close contact with a reduced sacrum. The bone is slightly curved in the caudal direction. The truncated end has two distinct facet surfaces of about equal extent. An elongated caudally tapering bone closely attached to the medial and (hidden) dorsal surface of the proximal femoral apophysis is most probably an ischium. The left-hand partner of the same bone is better exposed and shows the distended acetabular end in its full extent, parallel in position to the right and partly hidden ischium.

Remnants of a pubis (?) can be made out immediately in front of the caput femoris, but its surface is damaged and indistinct. It can be said with certainty that such an element did exist, but no details are clear. It was probably a flat bone in the area between the ilium, the last long thoracic rib, and the caput femoris. The ilium of the other limb is hidden, but a crooked pubis (?) and most of the left ischium are clearly defined. There was probably no clear acetabular depression at the converging three pelvic elements.

Comparing this rudimentary leg with drawings of reptilian, and especially ophidian, rudiments, the first strange fact is the relatively great length of the autopodium which has almost one-and-a-half to two times the length of the zygopodium. Just the reverse of this proportion is seen in *Pygopus*. In Recent snakes, no rudiments comparable to this specimen in the complexity of their parts are known. Pelvic elements seem to be more stable than those of the limb (see, e.g., *Leptotyphlops nigricans, Trachyboa boulengeri*). In many typhlopids pelvic elements only are developed, but nothing remains of the limb. In contrast, this specimen clearly shows a femur, the zygopodium, and at least two fused tarsal elements.

This specimen is a new genus which I dedicate to my dear friend Professor E. H. Colbert and give it the name:

Ophiomorphus colberti gen. et sp. nov.

The type is catalogued H.U.J. Pal. 3775, and kept in the collection of Paleontology of the Department of Zoology at the Hebrew University of Jerusalem. Measurements are given in Table 10.1.

Discussion

The genus *Ophiomorphus* is clearly distinct from the genus *Pachyrhachis* from the same lower Cenomanian level at Ein Jabrud. This form is certainly much closer, as far as the skull is concerned, to a more generalized varanoid type. On the other hand, many characters of *Pachyrhachis* resemble more closely those of ophidians, particularly the descensus and crista of the parietal, the snakelike quadrate, the quadrato-stapedial close contact, which indicates the loss of a tympanic membrane and cavity, the elongated squamosal as the main base for the upper end of the quadrate, the presence of some furrowed, very snakelike teeth, presence of an intramandibular articulation, and the absence of a jugal.

Ophiomorphus most probably had a broad and flat skull table formed by the parietal which had no descensus. It had a normal upper jugal arch, certainly involving the squamosal and supratemporal, and also most probably in front, the postorbital. The two anterior laminae of the ectopterygoid enclosed the posterior end of the maxilla. The teeth are not furrowed (with one possible exception), but some exhibit a peculiar external longitudinal keel and a very broad basal part. The pterygoid was most probably edentulus, but the palatine bore about six or more teeth. Posterolateral wings of the parietal were present. There are no signs of an anterior widening of the parietal at its transverse suture with the (missing) frontals. A jugal was present in broad contact with the dorsal margin of the maxilla.

The dentary and maxilla were toothed. The premaxilla is missing. Many details are unknown, for instance, the shape of the frontals, the prootics, and the occipital and basicranial elements, and the vomer-septomaxillary complex. More details of the skull may be determined after preparation of the ventral side of the slab.

The vertebral column and the ribs, which lack any traces of sternal structures, however, are rather snakelike. The rudimentary skeleton of the hind limb exhibits more primitive features than found even in the most primitive Recent snakes.

What could be the systematic position of *Ophiomorphus?* All details point to a certain relationship to varanoids, especially to the Cretaceous marine groups like aigialosaurids and dolichosaurids. But the first has more normal vertebral numbers, well-developed limbs, and a relatively short and clearly defined neck. Even the purely aquatic/marine dolichosaurids cannot be considered ancestral. They have four limbs, a well-defined neck (fourteen vertebrae), and zygosphenic vertebrae like *Ophiomorphus.* But *Ophiomorphus* has no traces of forelimbs and rudimentary hind limbs. In other words, *Ophiomorphus* is more snakelike than the known dolichosaurids. However, I think *Ophiomorphus* should be considered as loosely related to the dolichosaurids.

Hoffstetter (1959) described three vertebrae of the terrestrial snake *Lapparentophis defreunei* from the Lower Cretaceous of the Sahara. This snake is considered as an ancestor of the Upper Cretaceous genus *Simoliophis.* The vertebrae have well-developed zygosphene-zygantral articulations in addition to the normal zygapophyses and, as in Booidea, elongated rib bases. This snake demonstrates the early development of the characteristic "inner part" of intervertebral articulations by the Lower Cretaceous.

In the early Cenomanian several groups of seemingly snakelike reptiles were marine or at least estuarine. Nopcsa (1923, 1924), in describing vertebrae of *Simoliophis* and of the new genus *Pachyophis,* considered that snakes evolved in the marine realm. This idea, however, appears to conflict with many anatomical features seen in the most primitive Recent snake groups which point to a strictly terrestrial and semi-burrowing ancestry.

Ophiomorphus seems to be one more link between a varanoid and an ophidian stage. It points to the fact that the snakelike body and the loss of limbs did develop in a marine surrounding. But why should we bluntly exclude the possibility of manyfold types of origin? Both marine and terrestrial environments could have been important in the broad origins of snakes, in spite of the fact that most of the Recent primitive snake groups are, in fact, more or less fossorial.

Acknowledgments

I have to thank Mrs. Mary Rosenthal for the difficult preparation of the specimen and for typing the manuscript and lettering the illustrations. Thanks also to Mr. Aby Neev who produced the high quality photographs.

Literature Cited

Haas, G. 1979. On a new snakelike reptile from the lower Cenomanian of Ein Jabrud, near Jerusalem. Mus. Nat. Hist. Natur., Paris, Bull., 1(1):51–64.

Hardaway, T. E., and Williams, K. L. 1975. A procedure for double staining cartilage and bone. Brit. J. Herpetol., 5(4):473–74.

———. 1976. Costal cartilages in snakes and their phylogenetic significance. Herpetologica, 32(4):378–87.

Hoffstetter, R. 1955. Squamates de type moderne. *In* J. Piveteau (ed.), Traité de Paléontologie, Vol. V. Masson et Cie, Paris, pp. 606–62.

Nopcsa, F. 1923. *Eidolosaurus* und *Pachyophis,* zwei neue Neocom Reptilien. Paläontol., 65(4):97–154.

———. 1924. Die *Simoliophis* Rests. Bayer. Akad. Wiss., Math.-Naturwiss. Kl., Abh., 30(4): 1–27.

A NEW GENUS AND SPECIES OF CROCODILIAN FROM THE KAYENTA FORMATION (LATE TRIASSIC?) OF NORTHERN ARIZONA

by A. W. Crompton and Kathleen K. Smith[1]

Introduction

The earliest true crocodiles are members of a group generally recognized as the Protosuchia (Kälin, 1955, Walker, 1970). Animals belonging to this group have been found in the Late Triassic (or Early Jurassic[2]) of South Africa (i.e., *Notochampsa*, Broom, 1904; *Orthosuchus*, Nash, 1968, 1975), North America (i.e., *Protosuchus*, Colbert and Mook, 1951), and South America (i.e., *Hemiprotosuchus*, Bonaparte, 1971). Although the skull of these early crocodiles has recently been discussed in some detail (Walker, 1968; Bonaparte, 1971; Nash, 1975), a number of features have not been precisely or clearly described. These features include the relations of the structures of the otic region, details of the specialized contacts of the quadrate, and information on the eustachian tube system and accompanying cranial pneumatization. This information is essential for, as Walker (1972) points out, these features, to a large extent, characterize the skull of modern crocodiles.

Further, in published accounts on the Protosuchia (Colbert and Mook, 1951; Walker, 1968; Bonaparte, 1971; Nash, 1975) a number of inconsistencies and questions about cranial features exist. These include questions on the position of the tympanic membrane and the internal choanae, the relation of the quadrate to the supratemporal fenestra, the presence or absence of an antorbital fenestra, and the structure of the basisphenoidal region.

The purpose of this paper is to describe new crocodilian material from the Kayenta and Moenave formations of Northern Arizona which provide information on the braincase and quadrate of these Triassic crocodilians. Included is a discussion of new cranial material of *Protosuchus* and a description of a new genus and species of protosuchid.

Description of New *Protosuchus* Material

A new *Protosuchus* specimen (MCZ 6727; MNA Pl. 2461, Figure 11.1) provides details not present in previous specimens described by Brown (1933), and Colbert and Mook (1951). This information is particularly significant in resolving the apparent differences in the morphology of *Protosuchus* and other protosuchids such as *Orthosuchus* (Nash, 1975) and *Hemiprotosuchus* (Bonaparte, 1971), and will aid in

---

1. Listed alphabetically; no seniority of authorship implied.
2. There is current discussion on the age of the formations containing this material (*see* Olsen and Galton, 1977, for a review). The age of the Kayenta in particular is discussed in Colbert and Mook (1951), Welles (1954), Harshbarger, et al. (1957), and Lewis, et al. (1961). As the specific age of the material is not important to our discussion, we will not refer to this issue.

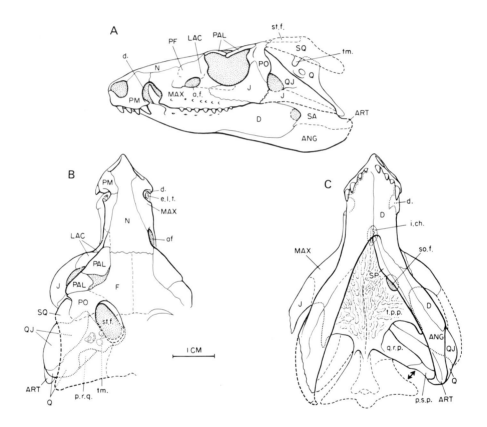

Figure 11.1. *Protosuchus* sp. A: lateral; B: dorsal and C: ventral view of a new specimen of *Protosuchus*. The specimen is slightly distorted and the double arrow indicates the separation of the quadrate away from the braincase.

comparison of *Protosuchus* with the new genus described below.

Locality

The new specimen was found in the Dinosaur Canyon Member of the Moenave Formation by Mr. C. R. Schaff on June 28, 1979. The locality (79/2A) is in Tonahakaad Wash, fifteen miles southeast of Cameron, Coconino County, Arizona. Approximate coordinates are 35°47' latitude; 111°07' longitude. The specimen is preserved in soft, reddish-brown sandstone and consists of a nearly complete skull and portions of the post-cranial skeleton. The posterior region of the skull (parietal and occiput) was either damaged, crushed, or lost. This has exposed the dorso-posterior surface of the quadrate. The remainder of the skull is also slightly distorted and no attempt has been made in Figure 11.1 to correct for distortion.

Morphology of the quadrate

Colbert and Mook (1951) reconstructed the quadrate of *Protosuchus* as being inclined moderately forward to meet the squamosal near the posterior border of the supratemporal fossa. In both *Orthosuchus* and *Hemiprotosuchus* the quadrate and quadratojugal reach the anterior margin of the supratemporal fossa, and Bonaparte (1971) modified Colbert and Mook's reconstruction of *Protosuchus* to conform to *Hemiprotosuchus*. In the new specimen of *Protosuchus* Bonaparte's reconstruction is supported. The heads of both the quadratojugal and the quadrate (Q) meet the postorbital and squamosal at the antero-lateral border of the supratemporal fossa (Figures 11.1A and 11.1B). In the new specimen the quadratojugal (QJ) is a slender, broad plate much wider than is shown by Bonaparte. The dorsomedial quadrate contact with the prootic and laterosphenoid cannot be seen.

Numerous sinuses and canals penetrate the dorso-lateral surface of the quadrate as in *Hemiprotosuchus* and *Orthosuchus*. In the protosuchids the tympanic membrane (tm) was most likely supported by the quadrate anteriorly, ventrally, and posteriorly, and the squamosal dorsally as in modern crocodilians. This is contrary to Nash's reconstruction (1975, p. 241) showing a post-quadrate tympanic membrane. The pterygoid ramus of the quadrate (p.r.q.), as shown in ventral view (Figure 11.1C), is short and abuts against a very robust quadrate ramus of the pterygoid (q.r.p.). Unfortunately, the occipital region of the new specimen is severely damaged.

Presence of an antorbital fenestra

Colbert and Mook figured no antorbital fenestra in *Protosuchus* and propose that the loss of the antorbital fenestra is a feature which distinguishes crocodilians from ancestral thecodonts. An antorbital fenestra (a.f.) is present in some mesosuchians. Antunes (1967) questioned Colbert and Mook's interpretation of this area in protosuchids, stating that it is unlikely that the antorbital fenestra was redeveloped in the Mesosuchia. Bonaparte showed an antorbital fenestra in *Hemiprotosuchus*, as does Nash in *Orthosuchus*. An antorbital fenestra is clearly present in the new specimen. It is thus likely to be a primitive feature of protosuchians and only lost later in some of the mesosuchian and in all eusuchian crocodiles.

Enlarged tooth in the lower jaw

In the new specimen of *Protosuchus* a large excavation is present on the external surface of the snout between the premaxilla and the maxilla (d). This houses an enlarged tooth of the lower jaw (e.l.t.). Unfortunately, the teeth immediately anterior and posterior to this large tooth are not visible but they must have been considerably smaller. Anterior to the large tooth the dentary is shallow and shovel-shaped, and the symphysis is elongated. Nash did not illustrate an enlarged lower tooth in *Orthosuchus*, but she illustrated a pocket between the premaxilla and the maxilla similar to that present in the new protosuchian specimen. This suggests that an enlarged lower tooth was present and that it was either lost or in the process of being replaced. *Hemiprotosuchus* appears to lack an external pocket between the maxilla and premaxilla but Bonaparte illustrated an enlarged lower tooth in this position. The suggestion of an enlarged lower tooth in all three protosuchian genera suggests a close affinity. An interesting feature of the new specimen is the clear differentiation between the four anterior premaxillary teeth and the eight or nine maxillary teeth. The premaxillary teeth are caniniform while the maxillary teeth are spatula-shaped and flattened medio-laterally. This heterodont condition has not been previously reported in other protosuchians.

Participation of the frontal in the supratemporal opening

Colbert and Mook reconstructed the supratemporal opening to be bordered anteriorly by the postorbital and parietal as in modern crocodiles. In many mesosuchians (i.e., telosaurids, metryriorhinchids, Kälin, 1955; Steel, 1973) as well as the protosuchians *Orthosuchus* (Nash, 1975), *Notochampsa* (Walker, 1968), and *Hemiprotosuchus* (Bonaparte, 1971), the anterior margin of the supratemporal fenestra is formed by the frontal bone. In the new *Protosuchus* specimen the skull roof between the temporal fossae is damaged and the suture in this region cannot be confirmed. However, a suture cannot be identified in the position figured by Colbert and Mook for *Protosuchus*. It is, therefore, likely that it is present in the more typical protosuchid position.

Position of the internal choanae

Internal choanae (i.ch.) bordered by the palatines and vomers are present adjacent to the anterior maxillary teeth. Either the vomers or palatines apparently arch downwards and inwards immediately below the choanae to form a rudimentary secondary palate. Posterior to the choanae the pterygoids and palatines form a relatively flat palate with a rugose surface. There is no indication of a median groove that could have been floored by soft tissue to extend the actual openings of the internal choanae further posteriorly, as proposed by Nash for *Orthosuchus*.

The new *Protosuchus* specimen is smaller than the type and differs from published accounts in many features. It should perhaps be placed in a separate species or new genus. This decision will be made when the new specimen and the type material of *Protosuchus* are more fully prepared.

**Description of a New Genus and Species
of a Late Triassic Crocodilian**

Class: Reptilia
  Order: Crocodilia
    Suborder: Protosuchia
      Family: Protosuchidae
        Eopneumatosuchus gen. nov.

Derivation of name: In recognition that in this crocodilian an early stage in the pneumatization of the skull by an extension of the eustachian tube system is present.

Diagnosis: As for *E. colberti*

Eopneumatosuchus colberti sp. nov.

Derivation of name: In honor of Dr. E. H. Colbert whose numerous contributions have resulted in a greater understanding of the terrestrial fauna of the Late Triassic including the crocodiles of this time.

Material

Only the holotype in the Museum of Northern Arizona (MNA Pl. 2460) consisting of a braincase and part of the roof of the skull. The bone was covered by a thick layer of hematite and preserved in a limy sandstone. The specimen was prepared by Mr. W. W. Amaral of the MCZ. It was partially prepared in a 5 percent solution of mercaptoacetic acid ($HSCH_2$ COOH). Exposed bone and breaks in the bone were coated with polystyrene dissolved in ethylacetate. The specimen was repeatedly treated with acid for approximately six hours followed by approximately eighteen hours in running water. Following acid preparation the remaining matrix was removed with dental burrs and carbide needles.

Locality

The specimen was discovered by Mr. T. Rowe on July 5, 1979, in the silty facies at the base of the Kayenta Formation in Coconino County, Arizona. The site is eleven miles NE of Cameron (approximately five miles north of "Dinosaur Canyon" of Colbert and Mook, 1951) between the southern two tributaries of Five Mile Wash. Approximate coordinates are 35°58' latitude; 111°15' longitude (locality 79^A/7).

Diagnosis

The braincase is low and long. The upper temporal opening is considerably longer than the orbits, whereas the opposite is true for all the other known protosuchians. The quadrate does not appear to have reached the anterior border of the temporal opening. The braincase is greatly elongated anterior to the fenestra ovalis and pituitary fossa with the result that the trigeminal foramen is situated far forward of the fenestra ovalis. Coupled with this expansion is a great elongation of the basipterygoid processes which fuse with the basisphenoidal rostrum and with one another in the midline. The basisphenoid forms a long secondary floor to the braincase between the pituitary fossa and the trigeminal foramen. This floor joins the elongated prootic with the result that the pituitary fossa is completely walled off from the exterior. The laterosphenoid is also expanded antero-posteriorly with the result that the IIIrd and IVth cranial nerves exit far ahead of the Vth.

The nature of the contacts of the quadrate and the braincase clearly establishes this animal as a crocodilian.

Description of material (Figures 11.2–6)

The species is represented by a nearly complete braincase. The loss of the quadrates has exposed the lateral surface of the braincase; an area that has not been adequately described or figured for other protosuchians. The squamosal on the left side is nearly complete. With the exception of part of the central region of the pterygoid, the remainder of the palate has been lost. Anterior to the braincase only the dorsal surface of the skull is partially preserved.

Dorsal view (Figure 11.2)

The dorsal surface of the squamosal (SQ), frontal (F), prefrontal (PF), postorbital (PO), and the broad sagittal crest formed by the frontal and the parietal (P) are deeply sculptured (Figure 11.6). This is in contrast to a broad nearly horizontally oriented rim to the temporal fossa which is smooth; probably indicating a broad area for muscle insertion (stippled in Figure 11.2). This insertion area is formed by the postorbital, frontal, parietal, dorsal edge of the laterosphenoid (LS), and a thin strip of squamosal. A median suture separates the frontals while the parietals appear to have been fused. Only a short section of the frontal forms part of the medial rim of the orbital border. A groove and slightly depressed area on the lateral surface of the prefrontal presumably housed the anterior extension of a large palpebral (PAL) bone. In sharp contrast to other Triassic crocodilians and most modern crocodilians (Figure 11.7), the orbits are considerably shorter anteroposteriorly than the temporal openings. Both the frontal and parietal bones contribute to the medial border of the upper temporal opening. Only the medial tip of the postorbital is preserved and the extent to which it forms the lateral border of the supratemporal opening cannot be determined. The supraoccipital (SO) has a pronounced broad medial crest. This bone and the extensive fused exoccipital and opisthotic (EX & OP) are visible in dorsal view. The posterior surface of the parietal, behind the sagittal crest, is slightly damaged, but lambdoidal crests, presumably extending from the midline to the parietal-squamosal contact, were probably present.

A deeply interdigitated (tongue and groove) suture, which joined the ventral surface of the squamosal to the quadrate (Q), is partially visible in dorsal view (Q/SQ). Medial to this, part of the dorsal edge of the extensive prootic contact (PR/Q) for the quadrate is also visible. Consequently, as in *Orthosuchus* and *Hemiprotosuchus*, the quadrate forms the ventral edge of the posterior border of the supratemporal opening. A horizontal groove is present in the external surface of the squamosal.

Occiput (Figure 11.3)

The lateral and ventral regions of the occiput are missing, but a sufficient proportion remains to indicate that the occiput was low and broad. The foramen magnum (f.m.) in this view is transversely ovoid. The condyle (con.) of the craniovertebral joint is formed almost exclusively by the basioccipital (BO), but the articular surface is lost.

It is difficult to distinguish all the sutures of the occiput with confidence. The dorsal edge of the occiput is formed by the parietal and squamosal. The supraoccipital is transversely widened and terminates below the squamosals. The exoccipital and opisthotics are fused. As in most crocodilians, the exoccipitals meet in a short suture above the foramen magnum. A small foramen (XII), ventro-lateral to the foramen magnum, appears to be for the hypoglossal nerve. The basioccipital is extended laterally to surround and form secondary foramina for the internal carotid, the IX and Xth nerves, as well as some vascular structures (vas.). This is in contrast to modern crocodilians (Figure 11.8) where these structures penetrate a ventro-

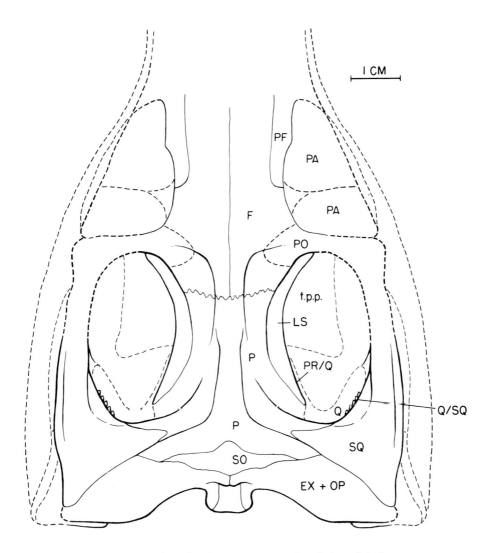

Figure 11.2. *Eopneumatosuchus colberti* gen. et sp. nov.—dorsal view of skull.

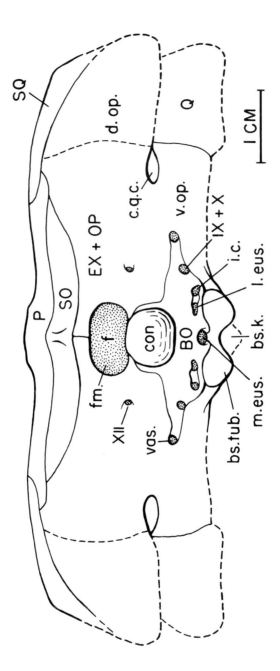

Figure 11.3. *Eopneumatosuchus colberti* gen. et sp. nov.—occipital view of skull.

medial extension of the fused exoccipital and opisthotic. Three eustachian canals are enclosed between the basisphenoid and the basioccipital; one on the midline (m.eus.) adjacent to the foramina for the internal carotid artery in the basioccipital. Two basisphenoid tubera (bs.tub.) were present on either side of the medial eustachian canal, but the ventral extension of these tubera is missing. The fused exoccipital-opisthotic extends ventro-medially below the lateral extension of the basioccipital to meet the basisphenoid. The opisthotic-exoccipital extends laterally as two well-defined regions; a longer dorsal region (d.op.) which meets the squamosal, and a shorter but more robust ventral region (v.op.). The lateral termination of the dorsal part of the opisthotic is a thin lamella of bone, plastered against the posterior surface of the squamosal. The ventral region, as seen in occipital view, terminates in a nearly vertical face which meets the internal surface of the quadrate below the reduced cranio-quadrate canal (c-q.c.). The posterior opening of the cranio-quadrate canal is bordered above, below and medially by the two regions of the opisthotic and laterally by the quadrate.

Braincase (Figures 11.4 and 11.5A)

In modern crocodilians (Figure 11.9) the laterosphenoid meets the basisphenoid (dorsum sellae) above and slightly behind the pituitary fossa. The incisura prootica is closed in front by the laterosphenoid to form the foramen for the trigeminal nerve. This foramen lies postero-dorsally to the dorsum sellae and the prootic behind this foramen is relatively narrow. In lateral view only a narrow bridge of bone separates the fenestra ovalis and the trigeminal foramen.

The characteristic feature of *Eopneumatosuchus* is that the prootic (PR) (Figure 11.5A) is greatly expanded between the fenestra ovalis and trigeminal foramen. The trigeminal foramen thus lies far anterior of the presumptive position of the pituitary fossa above the sella turcica (sel.tur.) and the fenestra ovalis. The exits of the IIIrd and IVth cranial nerves are also carried forward. In *Eopneumatosuchus* this pituitary fossa is not open externally because the basisphenoid extends forwards dorsally to the basisphenoidal rostrum and fuses with the prootic. This forms a solid secondary floor to the braincase (s.f.). A further result of the expansion of the braincase is that the basipterygoid processes (bp.p.) are lengthened and fuse with one another and the basisphenoid rostrum (bs.r.) to form a cylindrical bar running forward to meet the pterygoids (PT).

The basisphenoid and basioccipital bones are expanded ventrally to enclose sinuses associated with the eustachian tube system. The roof of the sinuses is formed by the original ventral surface of the basioccipital. In Figure 11.5A the dotted lines indicate the position of this floor and of the sella turcica. In Figure 11.4 the basisphenoid rostrum, as seen in ventral view, is also shown in dotted lines. A series of slender bony struts (b.s.) joint the basipterygoid processes to the secondary floor of the braincase and pterygoid.

The primary facial foramen (VII) lies anterior to the fenestra ovalis. Shallow grooves extend dorsally and ventrally from this foramen on the external surface of the prootic. The dorsal groove presumably housed the hyomandibular ramus of the facial nerve. The nerve must have passed above the fenestra ovalis (f.o.) and the fenestra pseudorotundum (f.pr.) to exit via the cranio-quadrate canal. The ventral groove housed the palatine ramus of the facial nerve (p.r.). This nerve appears to have traversed a long canal in the lateral wall of the basipterygoid

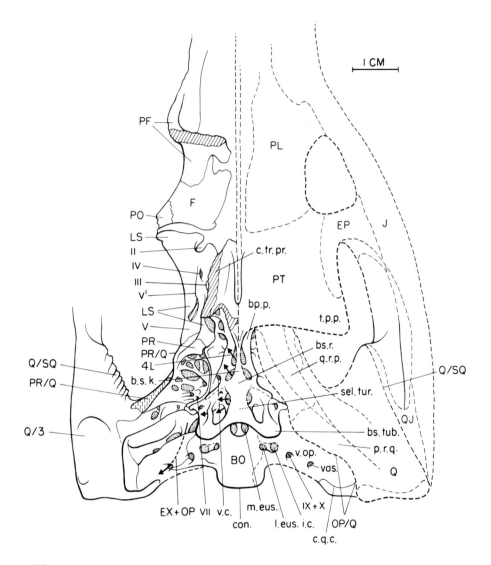

Figure 11.4. *Eopneumatosuchus colberti* gen. et sp. nov.—ventral view of palate: as preserved on the left and reconstructed on the right.

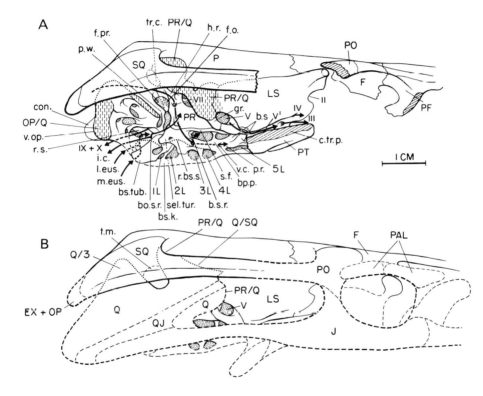

Figure 11.5. *Eopneumatosuchus colberti* gen. et sp. nov. A: lateral view of skull as preserved;
B: reconstruction.

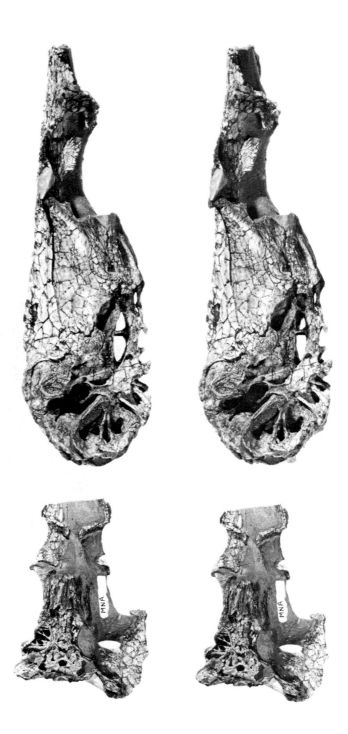

Figure 11.6. Stereo photographs of the braincase of *Eopneumatosuchus colberti* gen. et sp. nov.

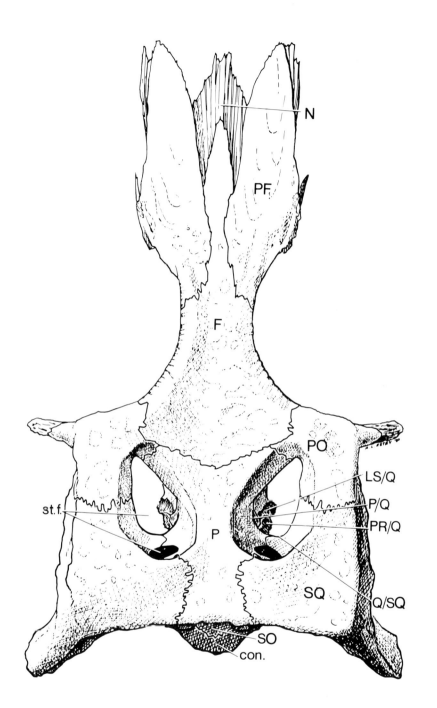

Figure 11.7. *Caiman* sp. Dorsal view of skull after removal of quadrate, quadratojugal and pterygoids.

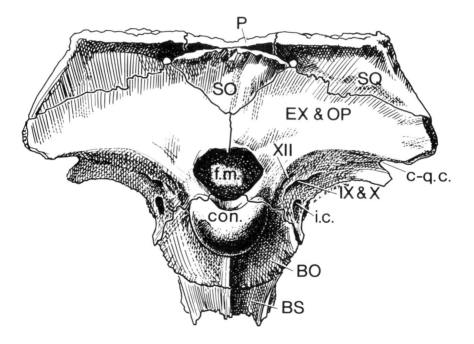

Figure 11.8. *Caiman* sp. Occipital view of skull after removal of quadrate, quadratojugal and pterygoids.

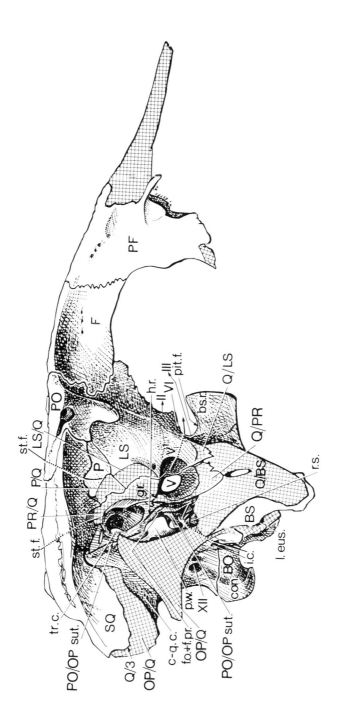

Figure 11.9. *Caiman* sp. Lateral view of braincase after the removal of the pterygoid, quadrate, quadratojugal, palatine and ectopterygoids.

process, which is probably homologous to the vidian canal (v.c.) of lizards. Unlike lizards, in modern crocodilians and *Eopneumatosuchus* the carotid artery does not pass through part of the vidian canal. In modern crocodilians (Figure 11.9) the fenestra ovalis lies on the boundary between the opisthotic and prootic, and the fenestra pseudorotundum lies within the opisthotic immediately behind the fenestra ovalis. A thin flange of bone forms a posterior wall (p.w.) to a deep recess into which the fenestra pseudorotundum opens. The IXth cranial nerve traverses the posterior part of this pocket. In *Eopneumatosuchus* the thin flange of bone separating the f. pseudorotundum from the f. ovalis was damaged during fossilization or during preparation, but a flange of bone (p.w.) formed by the prootic and forming a posterior wall to the recess into which the fenestra pseudorotundum opens is similar in structure and position to that formed in modern crocodiles.

A large kidney-shaped facet for the primary head of the quadrate is present on the antero-dorsal region of the prootic (PR/Q). This facet extends on to the laterosphenoid and parietal, as in modern crocodiles (Figures 11.9 and 11.10).

The trigeminal foramen (V) lies in the border between the prootic and the laterosphenoid with both bones lining a deep trumpet-shaped foramen. A groove or channel (gr.) extending upwards from the trigeminal foramen along the ventral edge of the prootic-quadrate contact is present. A similar groove for V_1 extends forward from the antero-ventral edge of the trigeminal foramen. A ventral process of the laterosphenoid lies laterally to this groove. In life this process (ossification of the pila antotica spuria) probably contacted the pterygoid, converting the groove into a foramen. Anterior to this process two foramina opened into the longer anterior extension of the groove for V_1. These are presumably for cranial nerves III and IV. The laterosphenoid extends forward to contact the frontal and postorbital at the front edge of the supratemporal fossa. These three bones are elongated relative to modern crocodiles (Figure 11.9), resulting in the expanded supratemporal opening. In ventral view it can be seen that the laterosphenoid contact with the frontal and postorbital is expanded transversely. As in modern crocodilians, this contact is rounded and appears to have been covered with cartilage. A marked notch (II) in the anterior border presumably served for the exit of the optic nerve.

In this specimen all but the part of the pterygoid on or near the midline is missing. Broken edges indicate the attachment areas of the transverse process (t.p.p.) and the quadrate rami of the pterygoid (q.r.p.).

In the present specimen, because the quadrate is missing, it is possible to study the pneumatization of the bones forming the braincase. As in modern crocodiles (Figure 11.10), there are both median (m.eus.) and lateral eustachian tubes (l.eus.) transversing foramina lying in the suture between the basisphenoid and the basioccipital (Colbert, 1946). These two canals appear to have led to a complex system of sinuses and canals in the bones surrounding the braincase. The system is more open than the one encountered in modern crocodilians. The sinuses in the basisphenoid, opisthotic, and the middle ear cavity have expanded into the surrounding bones, resulting in a meshwork of slender bony struts and canals. The complexity of these canals and sinuses is well shown in the stereo photographs. Both the median and lateral eustachian canals, or foramina, lead to a fairly deep basioccipital sinus, walled principally by the basioccipital and partially by the basisphenoid. The roof of this sinus (r.bo.s.), as would be seen in lateral view, is shown in dotted lines in Figure 11.5A. A narrow ventral keel (bs.k.) of the basisphenoid is

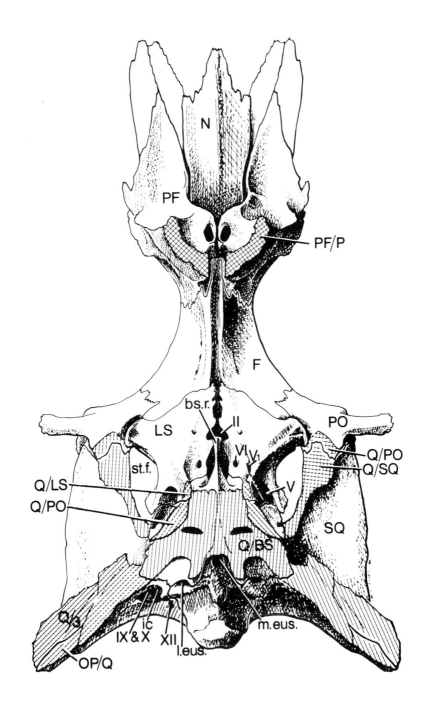

Figure 11.10. *Caiman* sp. Palatal view after removal of the pterygoid, quadrate, quadrato-jugal, jugal and ectopterygoid.

missing, but the struts of bone (b.s.) which connected the ventral keel to the main body of the basisphenoid (BS) and the basipterygoid process (bp.p.) are preserved.

A second sinus, mostly contained within the basisphenoid, is present in the midline in front of the basioccipital sinus. A wide foramen connects the two in the midline. A transverse ridge on the basioccipital, presumably the sella turcica (sel.tur.), forms a partial division between the two sinuses. The basisphenoidal rostrum (bs.r.), which extends forward in the central line from the sella turcica, is fused with both the basipterygoid process and the osseous floor (s.f.) to the braincase formed by the basisphenoid in front of the pituitary. The position of the basisphenoidal rostrum and the sella turcica above the basisphenoidal keel is indicated in the palatal view (Figure 11.4). The basisphenoidal rostrum partially divides the basioccipital sinus into right and left sides. The basisphenoidal sinus extends forward into the ventral part of the basipterygoid process right up to the contact of this bone of the pterygoid. Five large foramina open out of this forward extension of the basisphenoid sinus in a ventro-lateral direction. These foramina are bordered by bony struts extending from the median ventral keel of the basisphenoid to the main body of this bone and the basipterygoid process. (These foramina have been numbered 1*l* through 5*l* in Figures 11.4 and 11.5.) Two large dorso-lateral foramina lying on either side of the basisphenoidal rostrum connect the basisphenoidal sinus with the space between the basipterygoid processes and the ossified floor of the braincase in front of the pituitary fossa. The position of this foramen is indicated by a bold arrow in Figure 11.5A.

The basioccipital sinus opens on either side through a slit-shaped aperture into a large sinus (r.s.) lying below the ventral region of the opisthotic. This sinus appears to be similar to the rhomboidal sinus in modern crocodiles. The posterior wall of this sinus is formed by the lateral expansion of the basioccipital and the ventral extension of the fused opisthotic and exoccipital to meet the basisphenoid. The approximate course of the median and lateral eustachian tubes into the basioccipital sinus and through the slit-shaped foramen into the rhomboidal sinus is indicated in Figure 11.5A. The external wall of the rhomboidal sinus is presumably formed by the pterygoid ramus of the quadrate. The internal carotid artery (i.c.) appears to have entered the sinus via a secondary foramen in the basioccipital and exited anteriorly via a large foramen in the basisphenoid. (This course is also shown in Figure 11.5A.) This artery continued forward through the basisphenoid to reach the pituitary fossa. The IXth and Xth cranial nerves appear to have entered the rhomboidal sinus via one or two foramina in the opisthotic above the foramen for the carotid and exited via a secondary canal lying on the border between the opisthotic and basioccipital. The rhomboidal sinus extends up into the opisthotic and exoccipital in a maze of large and small sinuses and canals. A wide groove antero-dorsal to the rhomboidal sinus leads to the middle ear cavity which is walled medially by the opisthotic, prootic, fenestra ovalis, and fenestra pseudorotundum, and laterally by the main body of the quadrate. As in the case of the rhomboidal sinus, the tympanic cavity expands anteriorly and medially into a series of sinuses and canals in the prootic and the body of the opisthotic. A large canal (tr.c.) opening above the fenestra ovalis and fenestra pseudorotundum extends medially into the opisthotic. This may have extended through the skull bones above the brain to the tympanic cavity on the opposite side. A large canal above the tympanic cavity on both sides of the skull is found in a similar position in modern crocodiles.

In modern crocodiles (Figure 11.9) the median and lateral eustachian canals also open into sinuses in the basioccipital and from there into the rhomboidal sinus. The basioccipital sinus also extends forward to open into a basisphenoidal sinus. In contrast to modern crocodiles, in *Eopneumatosuchus* the basisphenoidal sinus does not appear to establish an independent connection with the rhomboidal sinus. Also, in modern crocodiles the IXth and Xth cranial nerves are separated by bone from the rhomboidal sinus.

Quadrate (Figure 11.5B)

A diagnostic feature of crocodiles is that the primary articulation of the head of the quadrate has shifted forwards in an antero-medial direction onto the prootic and laterosphenoid, anterior to the fenestra ovalis and fenestra pseudorotundum. In most reptiles this contact is with the opisthotic (paroccipital process) and squamosal, posterior to the fenestra ovalis. The primary articulation of crocodiles also extends in an antero-lateral direction to the ventral surface of the squamosal and postorbital. The quadrate also establishes secondary contacts with the opisthotic and squamosal behind the tympanic membrane and above and below the hyomandibular ramus of the facial nerve to greatly restrict the dimension of the cranioquadrate passage to a canal.

In *Eopneumatosuchus* the antero-medial contact of the quadrate head with the prootic, parietal, and laterosphenoid is well preserved. Most of the suture is of the interdigitating type and this contact is the same as that seen in modern crocodilians. Antero-laterally the primary head of the quadrate establishes a deep tongue and groove suture with the ventro-medial surface of the squamosal. There is no indication that this suture extends forwards onto a postorbital bone. As much of this bone is missing, this could not be confirmed; however, because of the abrupt termination of the quadrate contact posterior to the preserved anterior end of the squamosal, an extension of the quadrate to the postorbital is unlikely.

The quadrate ramus of the pterygoid (q.r.p.) is not preserved but based upon conditions in *Protosuchus*, this was probably a broad sheet of bone with a deep ventral keel closely applied to the basisphenoid. It probably overlapped the pterygoid ramus of the quadrate (p.r.q.). The latter must have been closely applied to the antero-lateral surface of the ventral region of the opisthotic to form an antero-lateral wall to the sinuses and canals in this part of the opisthotic. This contact (OP/Q) between the opisthotic and the quadrate formed a ventral border to the cranio-quadrate canal. This is in contrast to the condition in lizards and rhynchocephalians, where the ventral border of the large cranio-quadrate passage is formed by the quadrate ramus of the pterygoid.

A third quadrate contact with the squamosal lies behind the fenestra ovalis and fenestra pseudorotundum and as in modern crocodiles served as a posterior attachment area for the tympanic membrane (t.m.). This third contact forms an additional lateral wall to the cranio-quadrate canal. The evidence for this quadrate squamosal contact is a shallow depression in the ventral surface of the squamosal (Q/3, Figure 11.4). There is no evidence of a deeply interdigitating suture in this area that characterizes this contact in modern crocodiles.

Discussion

A number of characteristic features of the braincase of modern crocodiles distin-

shaped and flattened medio-laterally.

Eopneumatosuchus may be included with the Protosuchia because it is clearly crocodilian, is of a similar age, and resembles *Orthosuchus* in general size and proportions. A major difference between *Eopneumatosuchus* and the Protosuchia is the size of the supratemporal fenestra. However, in this feature the Protosuchia are quite variable. In *Protosuchus* the supratemporal fenestra is circular and much smaller than the orbit, in *Hemiprotosuchus* it is elongated to form a narrow, long, oval opening, and in *Orthosuchus* it is circular but equal to the size of the orbit. Kälin (1955), Langston (1973), and Nash (1975) discussed the significance of the relative size of the supratemporal fenestra in crocodilians. Expanded supratemporal fenestra are generally found in longirostral forms, and relate to the development of musculature in the anterior portion of the adductor fossa, in particular the pseudotemporalis (Iordansky, 1964; Langston, 1973). The variability in the size of the supratemporal fenestra in Triassic crocodiles, coupled with the apparent dental variability suggests that, although at this stage the major diagnostic cranial features of crocodiles were well established, there existed a greater diversity in the feeding apparatus than has previously been supposed. Evaluation of these features may only be made after further preparation and study of the protosuchians.

The diagnostic feature of *Eopneumatosuchus* is its expanded prootic region. Unfortunately, this region is not exposed in other protosuchids. However, Nash's reconstruction of the palate in *Orthosuchus* suggests a lengthening of the basisphenoid region and an anterior position of the trigeminal foramen similar to that in *Eopneumatosuchus*. Nash illustrated a very long pterygoid ramus of quadrate. This is unlike the situation in *Protosuchus* and may be in error.

Walker (1972) suggested that *Sphenosuchus*, a form from the Late Triassic Red Beds of southern Africa, is a crocodile because the primary head of the quadrate meets the prootic rather than the opisthotic, because the basisphenoid and basioccipital contain large sinuses, and because of the presence of a fenestra pseudorotundum. On the basis of the similarity of *Sphenosuchus* to birds he suggested a common origin between birds and crocodiles. It is possible that a form similar to *Sphenosuchus* could have been ancestral to the Protosuchia, but as Walker (1972) himself pointed out, *Sphenosuchus* existed contemporaneously with the Protosuchia. We have discussed a number of diagnostic features of crocodiles which are present in the Protosuchia. As most of these are absent in *Sphenosuchus*, we would not consider this form as a crocodile.

Eopneumatosuchus resembles the marine Mesosuchia (Deslongchamps, 1884; Andrews, 1913; Antunes, 1967; Wenz, 1968) in the extreme expansion of the supratemporal fenestra. In these mesosuchids the expansion of the supratemporal opening involves lengthening the anterior elements, primarily the frontal and laterosphenoid medially and the postorbital laterally. Thus, although the quadrate is restricted to the posterior region of the supratemporal fenestra, it maintains its contacts with the prootic and laterosphenoid medially and the squamosal and postorbital laterally in a characteristic crocodilian manner. In *Eopneumatosuchus* the supratemporal expansion appears to have taken place in a similar manner, although, as stated above, the lateral border of the supratemporal fenestra is missing and the extent of the postorbital cannot be determined.

Eopneumatosuchus differs from the mesosuchids in that the braincase expansion has involved not only the laterosphenoid but also the prootic, which is lengthened so that the distance between the foramen ovale and the trigeminal foramen is large. Wenz (1968) figured the braincase of a mesosuchid (*Metryriorhinchus*), and in this animal the two foramina are quite close as in modern crocodilians. Mesosuchids further differ from *Eopneumatosuchus* in that the prootic has expanded in a dorso-ventral direction and is exposed on the external surface of the skull above the quadrate (Deslongchamps, 1884; Wenz, 1968). In this, the mesosuchids are unique amongst the Crocodilia.

Thus, despite the superficial similarities between *Eopneumatosuchus* and the mesosuchids, in details of its braincase the former resembles the Protosuchia much more closely than the latter. It is likely that supratemporal expansion took place independently in these two groups. Antunes (1967) claimed that it is unlikely that the Mesosuchia evolved out of the highly specialized Protosuchia in the short time span between the Late Triassic or Early Jurassic and the later Jurassic and suggests that the mesosuchids were a polyphyletic group. However, the mesosuchids possess all the diagnostic features of the Crocodilia, which suggests that the Crocodilia are a monophyletic group. *Eopneumatosuchus* demonstrates that the Triassic crocodiles were more diverse in structure than previously supposed, and the ancestor for the highly specialized marine Mesosuchia might yet be found to be a protosuchian.

Acknowledgments

This material was collected by a joint Harvard University and Museum of Northern Arizona expedition supported by National Science Foundation grant number DEB 78-01327. We thank F. A. Jenkins, Jr., for collecting and allowing us to work on his material, W. W. Amaral for his excellent preparation, L. Meszoly and M. Estey for illustrations, A. Coleman and P. Chandoha for photography, and M. Reynolds for typing several drafts of the manuscript.

Literature Cited

Andrews, C. W. 1913. A descriptive catalogue of the marine reptiles of the Oxford Clay. Part II, British Museum (Nat. Hist.), London, 206 pp.

Antunes, M. T. 1967. Sur quelques caractères archaiques des crocodilians a propos d'un mésosuchien du Lias supérieur de Tomer (Portugal) Colloq. Int. du Cent. Nat. Rech. Scient., 163:109 14.

Bonaparte, J. 1971. Los tetrapodos del sector superior de al formacion Los Colorados, La Rioja, Argentina. 1 parte Op. lilloana, 22:1–183.

Broom, R. 1904. On a new crocodilian genus (*Notochampsa*) from the upper Stormberg beds of South Africa. Geol. Mag., (5)1:582–84.

Brown, B. 1933. An ancestral crocodile. Amer. Mus. Nov., 683:1–4.

Colbert, E. H. 1946. The eustachian tubes in the crocodilia. Copeia, 1946:12–14.

Colbert, E. H., and Mook, C. C. 1951. The ancestral crocodilian *Protosuchus*. Amer. Mus. Nat. Hist., Bull. 97:147–82.

Deslongchamps, J-A Eudes. 1884. Mémoires sur les Téléosauriens de l'epoque Jurassique. Soc. Linn. Normandie Mem. 13(3):1–138.

Harshbarger, J. W.; Repenning, C. A.; and Irwin, J. H. 1957. Stratigraphy of the uppermost Triassic and the Jurassic rocks of the Navajo Country. U.S. Geol. Surv., Prof. Paper, 291:1–74.

Iordansky, N. N. 1964. The jaw muscles of the crocodiles and some relating structures of the crocodilian skull. Anat. Anz., 115:256–80.

———. 1973. The skull of the crocodilia. *In*, C. Gans and T. S. Parsons (eds.) Biology of the Reptilia. Vol. 4. Academic Press, New York, pp. 201–61.

Kälin, J. 1955. Crocodilia. *In*, J. Piveteau (ed.) Traite de Paleontologie. Vol. 5. Masson, Paris, pp. 695–784.

Langston, W. 1973. The crocodilian skull in historical perspective. *In,* C. Gans and T. S. Parsons (eds.) Biology of the reptilia. Vol. 4. Adacemic Press, New York, pp. 263–84.

Lewis, G. E.; Irwin, J. H.; and Wilson, R. F. 1961. Age of the Glen Canyon Group on the Colorado Plateau. Geol. Soc. Am. Bull., 72:1437–40.

Nash, D. 1968. A crocodile from the upper Triassic of Lesotho. J. Zool., 156:163–79.

———. 1975. The morphology and relationships of a crocodilian, *Orthosuchus stormbergi,* from the upper Triassic of Lesotho. S. Afr. Mus. Ann., 67:227–329.

Olsen, P. E., and Galton, P. M. 1977. Triassic and Jurassic tetrapod extinctions: are they real? Science, 197:983–86.

Steel, R. 1973. Crocodilia. *In,* O. Kuhn (ed.), Encyclopedia of paleoherpetology. Gustav-Fischer Verlag, New York, 16:1–116.

Walker, A. D. 1968. *Protosuchus, Proterochampsa* and the origin of phytosaurs and crocodiles. Geol. Mag., 105:1–14.

———. 1970. A revision of the Jurassic reptile *Hallopus victor* with remarks on the classification of crocodiles. Roy. Soc. London, Phil. Trans., Ser. B., 257:323–72.

———. 1972. New light on the origin of birds and crocodiles. Nature, 237:257–63.

Welles, S. P. 1954. New Jurassic dinosaur from the Kayenta Formation of Arizona. Geol. Soc. Amer., Bull., 65:591–98.

Wenz, S. 1968. Contribution à l'etude due genere Metriorhynchus—crâne et moulage endocrânien de *Metriorhynchus superciliosus.* Ann. de Paleontologie (verts), 54:149–83.

Whetstone, K. N., and Martin, L. D. 1979. New look at the origin of birds and crocodiles. Nature 279:234–36.

List of Abbreviations Used in the Figures

a.f.—antorbital fenestra
ANG—angular
ART—articular

BO—basioccipital
bp.p.—basipterygoid process
BS—basisphenoid
bs.k.—ventral keel of the basisphenoid
bs.r.—approximate position of basisphenoidal rostrum above the basisphenoid keel
b.s.—bony struts
bs.tub.—basisphenoidal tubera

con.—basioccipital condyle
c-q.c.—cranio-quadrate canal
c.tr.p.—contact of the transverse process of the pterygoid with main body of the bone

D—dentary
d.—depression on premaxilla and maxilla suture for the enlarged lower tooth
d.op.—dorsal portion of the fused exoccipital and opisthotic

e.l.t.—enlarged lower tooth
EP—ectopterygoid
EX & OP—fused exoccipital and opisthotic

F—frontal
f.m.—foramen magnum
f.o.—fenestra ovalis
f.pr.—fenestra pseudorotundum

gr.—groove leading dorso-posteriorly from the trigeminal foramen

h.r.—hyomandibular ramus of the facial nerve

i.c.—secondary foramen for the internal carotid artery
i.ch.—position of internal choanae

J—jugal

LAC—lacrimal
l.eus.—foramen for the lateral eustachian tube
LS—laterosphenoid
LS/Q—contact of the primary head of the quadrate and laterosphenoid

MAX—maxilla
m.eus.—foramen for the median eustachian tube

N—nasal

OP/Q—secondary contact between the opisthotic and quadrate below the cranio-quadrate canal.

P—parietal
P/Q—contact of the primary head of the quadrate with parietal
PAL—palpebral bone
PF—prefrontal
PF/P—prefrontal contact with the palatine
pit.f.—external opening of the

pituitary fossa
PL—palatine
PM—premaxilla
PO—postorbital
PO/OP sut.—suture between the prootic and opisthotic
p.r.—palatine ramus of the facial nerve
PR—prootic
p.r.q.—pterygoid ramus of quadrate
PR/Q—contact between the primary head of the quadrate and prootic
PT—pterygoid
p.w.—posterior wall of the recess into which the fenestra pseudorotundum opens

Q—quadrate
Q/3—secondary contact between the quadrate and squamosal behind the tympanic membrane
Q/BS—secondary contact between the quadrate and basisphenoid
QJ—quadratojugal
Q/LS—secondary contact between the quadrate and laterosphenoid
Q/PO—secondary contact between the quadrate and prootic
q.r.p.—quadrate ramus of the pterygoid
Q/SQ—contact between primary head of the quadrate and the squamosal

r.s.—rhomboidal sinus
r.bo.s.—roof of the basioccipital sinus

r.bs.s.—roof of the basisphenoid sinus

SA—surangular
sel.tur.—approximate position of the sella turcica above the basisphenoidal keel
s.f.—solid secondary floor to the braincase
SO—supraoccipital
so.f.—suborbital fenestra
SQ—squamosal
st.f.—supratemporal fossa

t.m.—position of the tympanic membrane
t.p.p.—transverse processes of pterygoid
tr.c.—transverse canal

vas.—possibly a foramen for a vascular vessel
v.c.—vidian canal
v.op.—ventral portion of the fused exoccipital and opisthotic

1*l*–5*l*—five lateral foramina opening out laterally from the long basisphenoidal sinus
II—foramen for the optic nerve
III—foramen for the oculomotor nerve
IV—foramen for the trochlear nerve
V—trigeminal foramen
V$_1$—groove for profundus branch of the trigeminal nerve
VII—foramen for the facial nerve
IX & X—secondary foramen for the glossopharyngeal and vagus nerves
XII—foramen for the hypoglossal nerve

A PROSAUROPOD DINOSAUR TRACKWAY FROM THE NAVAJO SANDSTONE (LOWER JURASSIC) OF ARIZONA

by Donald Baird

Introduction

The presence of prosauropod dinosaurs in the Navajo Sandstone of Arizona was first reported by the late Mr. Lionel F. Brady (1935, 1936), who referred a partial skeleton to *Ammosaurus*, a genus otherwise known only from the Portland Formation of the Connecticut Valley. A detailed restudy of this and subsequently collected material by Dr. Peter M. Galton (1971, 1976) has confirmed Brady's identification and established beyond question the prosauropod nature of *Ammosaurus*.

From an analysis of the manus structure in this and other genera of prosauropods Galton has concluded that, contrary to previous reconstructions, the large falciform pollex claw was normally carried with its sagittal plane horizontal and its tip pointed medially. I have independently verified Dr. Galton's functional reconstruction by manipulating the actual bones in various articulations. While facultatively bipedal, *Ammosaurus* must normally have walked on all fours. Its large hind feet were tetradactyl and its smaller front feet were functionally tridactyl with the second and third digits carrying most of the weight except on soft ground, where the side of the great pollex claw might be expected to impress.

This new insight into the locomotor pattern of *Ammosaurus* and its relatives prompted me, as an ichnologist, to search the footprint literature and my extensive collection of unpublished material for trackways which might be attributed to prosauropod dinosaurs. The only pertinent specimen, by a happy coincidence, proved to be another discovery of Major Brady's: an undescribed trackway which he had collected in the Navajo Sandstone, Arizona, (Figure 12.1), and of which he had sent me a photograph and partial mold in 1960. His label for it states, "Part of track of dinosaur—impressed on 25° slope of dune—less than 10 percent of the weight seems to have been on front feet—may be a sort of protosauropod?" Analysis of this trackway and comparison of the result with the foot structure of *Ammosaurus* convinces me that Brady was correct, and that his find is the first authentic prosauropod trackway to be recognized. This conclusion has been reported in two papers by Galton (1971, p. 783; 1976, p. 6).

Age of the Navajo Sandstone

Until recently most authors have assigned the Navajo Sandstone, wholly or in part, to the Upper Triassic. Galton (1971, p. 791) has usefully tabulated the succession of interpretations. His assignment of a Late Triassic age to the Navajo was based on its reptile fauna. The Navajo prosauropod dinosaur, correctly identified by Brady as *Ammosaurus*, is congeneric with *Ammosaurus major* from the Portland Formation of the Connecticut Valley. The crocodilian *Protosuchus* has been shown

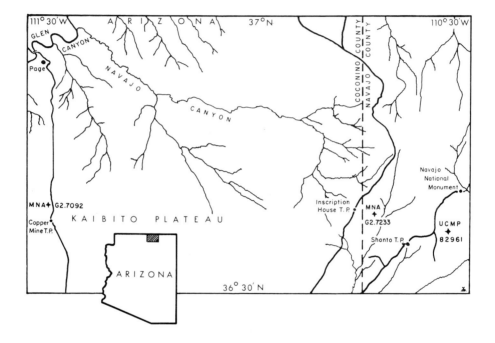

Figure 12.1. Prosauropod dinosaur localities in the Navajo Sandstone of northern Arizona. MNA G2.7092, type trackway of *Navahopus falcipollex*. MNA G2.7233, Flagstaff skeleton of *Ammosaurus* cf. *A. major* found by L. F. Brady. UCMP 82961, Berkeley skeleton of *Ammosaurus* cf. *A. major* found by M. Wetherill. Base map after Cooley, et al. (1969). Abbreviations: MNA, Museum of Northern Arizona; UCMP, University of California Museum of Paleontology; T. P., Trading Post.

by Walker (1968) to be closely related to *Stegomosuchus longipes* from the Long-meadow Sandstone of the Portland. (The coelurosaurian dinosaur *Segisaurus*, being unique as well as a member of a long-lived group, is not useful for correlation). As the Portland had been universally assigned to the Upper Triassic, Galton reasoned that the Navajo must be the same age.

In the last few years much work has been done on the faunas, palynology, and potassium argon dating of the Triassic-Jurassic transition as recorded in the Newark Supergroup of eastern North America. While Galton's correlation of the Navajo with the Portland remains valid, the latter is no longer Triassic but has been redated as lower Liassic ("not older than Sinemurian") on the basis of its palynoflora and fish fauna (Cornet, Traverse, and McDonald, 1973). This new age-determination has been extended to the Southwest and elsewhere by Olsen and Galton (1977), who conclude that "all of the Glen Canyon Group of the southwestern United States . . . [is] Early Jurassic." In their correlation chart the Glen Canyon (from which they exclude the basal phytosaur-bearing Rock Point Member of the Wingate Formation) is shown as beginning in Hettangian time and continuing through the Sinemurian until well along in the Pliensbachian.

Because of the intercontinental resolution possible with palynological analysis, considerable weight is given to a microflora from the Whitmore Point Member of the Moenave, which Bruce Cornet (quoted in Olsen and Galton, 1977, pp. 18, 34) correlates with the lower Liassic on the basis of a preponderance of *Corollina* (more than 90 percent of palynoflora). This evidence suggests a Sinemurian to Pliensbachian age for the Navajo Sandstone.

Systematic Paleontology
Genus *Navahopus* Baird, gen. nov.

Diagnosis. Quadrupedal trackways referable to the saurischian family Plateosauridae. Pes tetradactyl, clawed, with digits in order of increasing length I–II–IV–III; manus functionally tridactyl with short, clawed digits II and III directed forward, and horizontally recumbent, falciform pollex claw directed medially.

Type-species. *Navahopus falcipollex* Baird, sp. nov.
(Figures 12.2, 12.3)

Diagnosis. As for the genus, which is monotypic at present.

Etymology. *Navaho* (Spanish *Navajo*) for the source formation and the Navajo Indian Reservation, plus *pus*, a foot, the standard termination of generic names in vertebrate ichnology; *falx*, a sickle, plus *pollex*, thumb, the specific name being a noun in apposition.

Type specimen. P3.339 (MNA G2.7092) in the Museum of Northern Arizona at Flagstaff, a trackway of six pedes and six manus collected by Lionel F. Brady in 1958.

Source. Cross-stratified dune sand of the upper Navajo Sandstone, Glen Canyon Group, Lower Jurassic (Sinemurian-Pliensbachian). Two miles north of Copper Mine Trading Post on the Kaibito Plateau, Navajo Indian Reservation, Coconino County, Arizona; about 36°10' North, 111°12' West. MNA Locality 226.

My method of analysis was to chalk the outlines of the tracks on the latex mold, then project these by means of a camera obscura onto tracing paper. Each of the four feet was analyzed separately, the outlines of its successive imprints were super-

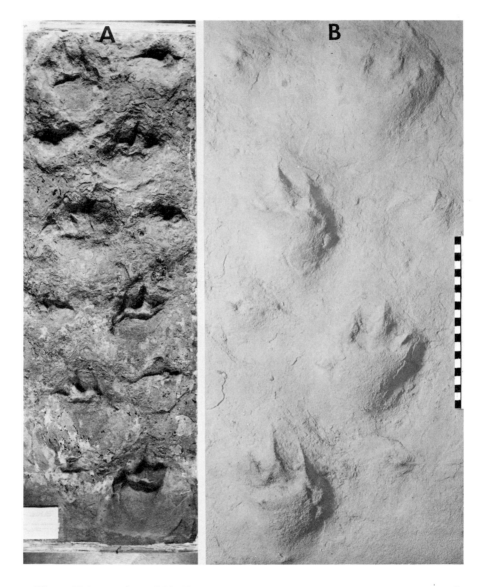

Figure 12.2. *Navahopus falcipollex*, type trackway, MNA P3.339 [G2.7092]. A, entire track-way as preserved. B, enlargement of last four manus-pes sets photographed from a latex mold (and thus the reverse of A). Scale in centimeters. Museum of Northern Arizona photo-graphs.

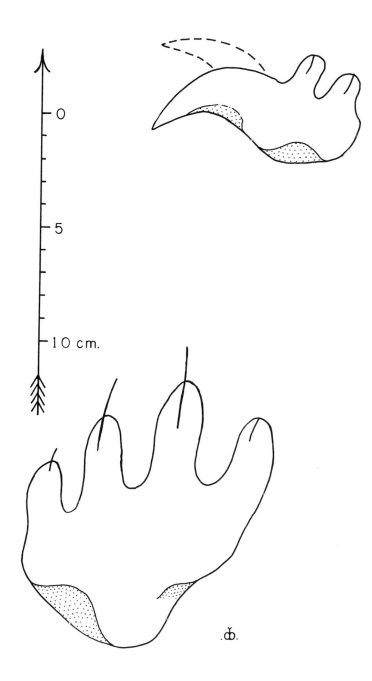

Figure 12.3. *Navahopus falcipollex,* composite outline of right manus-pes set, MNA P3.339. Arrow represents midline of trackway. Dashed line shows occasional, extended position of pollex claw.

imposed and a composite outline was drawn and checked against the specimen. This method helps to distinguish the morphological constant of the foot from the vagaries of impression. Outlines of the left feet were then collated with those of the right to produce the composite shown in Figure 12.3. This reconstruction, of course, represents the footprints rather than the feet that made them. We must never forget that a footprint is not the natural mold of a morphological structure but is, instead, the record of that structure in dynamic contact with a plastic substrate.

Like most trackways made in dune sand, this one is obscured by the way in which sand was displaced and then slumped back as each foot was implanted and withdrawn. Marks left by the soles and the ventral surfaces of the digits show little morphological detail, and the claws are represented mainly by scratch-marks. Despite these obscurities, however, the basic structure of manus and pes can be made out.

The pes is clearly tetradactyl, although the imprint of its first digit is generally distorted by slumping. Digits I to III evidently bore strong claws which left conspicuous scratches, but the claw of digit IV appears to have been small. In most pes prints the sole is rather shapeless, but at least two show a postero-lateral heel which may represent padding associated with a vestigial fifth digit. The pes is outturned slightly so that digit II points directly forward.

The manus prints of *Navahopus* are unique among vertebrate ichnites in being small and rather hoof-like, with two short, clawed digits pointing antero-laterally and a great hook—the pollex ungual—arcing toward the midline in the horizontal plane. An expansion of the sole lateral to digit III suggests the presence of additional but rudimentary digits which were carried clear of the ground. In the third imprint of the left manus the pollex claw has been drawn forward by the *M. extensor pollicis* into the position indicated by a dashed line in Figure 12.3. While the hind limbs were held well under the body, the forelimbs had a wider straddle.

Trackway parameters are recorded here with the warning that they are probably not typical for the genus. Reference points for measurement were the bases of pes and manus digits III:

Pace, angular: 19.5 to 20.5 cm; mean of five, 19.8 cm.
Stride: 31.3 to 33.5 cm; mean of four, 32 cm.
Pace angulation of pedes: 108° to 112°; mean of four, 109°.
Pace angulation of manus: 60° to 69°; mean of four, 64°.

The gleno-acetabular trunk length (or "wheelbase") of the trackmaker can be estimated on the assumption that, whenever the four feet were simultaneously in contact with the ground, the center of the shoulder axis occupied a position midway between the carpi while that of the hip axis lay midway between the tarsi. This assumption is valid for crocodilian trackways (cf. Shaeffer, 1941, fig. 17B) and cannot be far wrong for the trackway of a quadrupedal dinosaur. (In practice the gleno-acetabular length can be determined equally well by measuring along the trackway midline from a point opposite one tarsus imprint to a point opposite the next-succeeding carpus of the other side.) The track sequence in the type specimen of *Navahopus falcipollex* is long enough to permit three independent measurements of the gleno-acetabular length to be derived from three sets of simultaneously occupied footprints: 38.5, 38.0 and 37.5 cm, which gives 38 cm as a working figure. This parameter provides a useful index for comparisons between trackways and skeletons.

Let me emphasize that the type specimen cannot be considered a "normal" specimen of *Navahopus*. If the same individual reptile had been walking on a level mud-flat instead of a sandy dune-slope, its footprints and trackway proportions would surely appear very different from those on the slab at hand. Its stride would be longer; its pes imprints would lack deep heel-marks and would end posteriorly with the pads of the metatarso-phalangeal articulations; and the digits would show more clearly-defined articular pads and claws. Its manus imprints would probably lie closer to the midline of the trackway, and might be faint or missing altogether, as the trackmaker appears to have been facultatively bipedal. Nevertheless, despite differences in gait and in the nature of the recording substrate, the fundamental characteristics of the *Navahopus* trackway should make it readily identifiable.

Zoological Comparisons

Comparison of Figure 12.3 with Galton's reconstruction of the manus and pes of *Ammosaurus* cf. *A. major* from the Navajo Sandstone (Figure 12.4) should convince even the skeptical that *Navahopus falcipollex* is indeed the trackway of a plateosaurid prosauropod (Galton, 1971, 1976; Galton and Cluver, 1976). The correspondence is too close to require argument. In its pedal proportions the trackmaker is less similar to the relatively slender-footed *Anchisaurus* than to the broad-footed *Ammosaurus*. As two skeletons of *Ammosaurus* have been found in the same formation, 40 and 48 miles east of the track locality, we have reason to assume that the trackway was made by an individual of that genus. The trackmaker was about four-fifths the linear size of Brady's *Ammosaurus* specimen (MNA G2.7233).

The nomenclaturally orthodox reader may inquire why the ichnologist, if he correlates bones and footprints with such apparent confidence, does not apply the same Linnaean name to both under the provisions of the International Code of Zoological Nomenclature, Article 24(b)(iii). My answer is that such correlations are almost never certain; and even if certainty existed—e.g., if a skeleton were found at the end of its trackway—a separate nomenclature should be retained for the tracks. Ichnogenera are in many cases equivalent to families or even high categories of reptiles, and, conversely, the members of a reptilian family may differ so much in foot structure that their tracks would be assigned to different ichnogenera. Under these circumstances the only zoologically realistic solution is to maintain a separate system of form- or organ-taxa for ichnites, whether or not such a system is sanctioned by the Code.

To help visualize the geo-biological event which the *Navahopus* trackway represents, Figure 12.5 attempts to reconstruct the scene. The prosauropod dinosaur is restored on the basis of the known skeletal parts of *Ammosaurus*, supplemented where necessary from its relative *Anchisaurus*. Brady's original interpretation—that the trackway had been made up the slope of a sand dune—is supported by McKee's (1979) conclusion that the Navajo Sandstone is eolian, an example of an ancient "sand sea."

Ichnological Comparisons

The subject of prosauropod trackways has been confused by two mistaken correlations which were made long ago by Richard S. Lull and which have been accepted in much of the uncritical secondary literature. In 1904, in proposing the name *Anchisauripus* for certain tracks from the Upper Triassic of the Connecticut Valley,

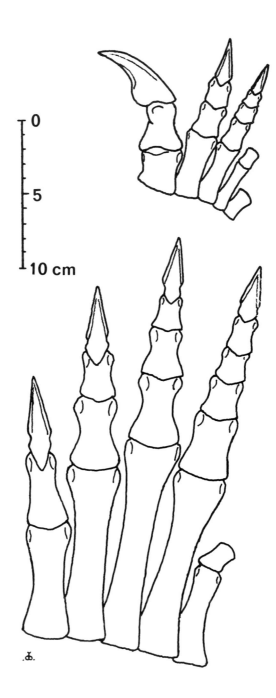

Figure 12.4. Manus and pes of *Ammosaurus* cf. *A. major* from the Navajo Sandstone. Restored and redrawn after Galton (1976).

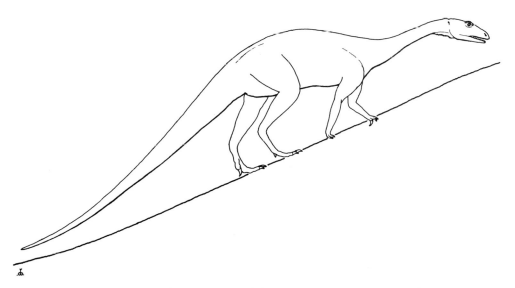

Figure 12.5. Restoration of the prosauropod dinosaur *Ammosaurus* in the act of making a trackway similar or identical to *Navahopus*.

Lull expressed his conviction that the dinosaurs *Anchisaurus* and *Ammosaurus* made bipedal trackways of footprints in which digits II–III–IV form a forward-pointing group, digit III being much the longest, while the hallux is rotated to a postero-medial position. This correlation was based on a double error. In restoring the foot structure of the *Anchisauripus* trackmaker Lull drew articulations to correspond with the crease-constrictions in the footprint instead of with the knuckle-pads, a misinterpretation which Heilmann (1927, pp. 179–83) was the first to correct. The trackmaker's foot was thus made to look a little more prosauropod-like, although when correctly restored it is obviously that of a coelurosaur or carnosaur. Second, in order to adapt the foot of *Anchisaurus* or *Ammosaurus* to the *Anchisauripus* foot-print, Lull (1915, pp. 141, 154) had to dislocate and rotate the hallux (metatarsal and all) and do procrustean violence to the lengths of the other three digits. His correlation is thus doubly untenable, and footprints of the *Anchisauripus* type have nothing to do with Prosauropoda.

A second source of confusion arose from Lull's (1953, pp. 187–89) identification of the gigantic Connecticut Valley trackmaker *Otozoum* as a prosauropod of the *Plateosaurus* type. There is indeed a resemblance between the pes of *Otozoum* and those of *Navahopus, Ammosaurus,* and *Plateosaurus. Otozoum,* however, is in fact functionally pentadactyl rather than tetradactyl, as Nopcsa (1923, p. 144) first pointed out, "Erhielt sich der Ballen der fünfte Zehe von *Chirotherium* bei *Otozoum* recht gut." In other words, the postero-lateral heel pad of *Otozoum* represents the rudiment of a chirotherioid fifth digit, the familiar "thumb" of the "hand animal," which becomes reduced to a pad in the brachychirotherian group of chirotheres. The short-fingered manus is also chirotherioid in form as well as crocodilian-like in its outward-turned position. For this reason I interpret *Otozoum* and its diminu-tive relatives *Batrachopus, Cheirotheroides,* and *Comptichnus* as members of the pseu-dosuchian/proto-crocodilian branch of the Archosauria (Baird, 1946, p. 485). Their manus structure is quite incompatible with that of prosauropods.

Having disposed of these false leads we may proceed to search the ichnological literature for *Navahopus*-like footprints. Among previously described tracks from the Navajo Sandstone only those designated as B4.43 by Faul and Roberts (1951) are even superficially similar, and the manus structure is quite different. No other ichnite from the North American Jura-Trias is at all similar to *Navahopus*. The only comparable form known to me is a trackway from the Lower Stormberg (Zone A 3, upper Molteno) of Seaka-Falatsa (Ouest), Lesotho (Basutoland), South Africa, that was named *Tetrasauropus unguiferus* by Ellenberger (1973, pp. 22, 68–69, fig. 36, pl. 7 bas, dépliant 1). This ichnogenus resembles *Navahopus* in being quad-rupedal with a narrow, short-striding trackway in which the manus imprint lies anterolateral to the pes. The long-soled pes is tetradactyl (its supposed fifth digit is improbable) while the much smaller manus is functionally tetradactyl with claws on digits I–III only. The medially directed pollex is large and looks somewhat fal-ciform, though its appearance may be exaggerated in the imprints, as the pes claws appear similar. Ellenberger has not assigned this trackway to any reptilian group but he mentions the prosauropods among other possible correlatives. Although better material would be required for any definitive comparison, I suspect that *Navahopus falcipollex* and *Tetrasauropus unguiferus* may be the trackways of similar types of reptiles.

Sauropod trackways such as those known from the Lower Cretaceous should afford useful comparisons with the prosauropod *Navahopus,* but unfortunately the published accounts of them (e.g., Bird, 1944, 1954; Ginsburg, et al., 1966) are descriptive rather than analytical.

That a family of reptiles which is known from a number of skeletons should be represented by only a single trackway may seem paradoxical, but not to an experienced ichnologist. A trackway and a skeleton are different kinds of phenomena, and a sedimentary environment that will preserve one is unlikely to preserve the other. To take a reverse example, the great majority of footprints known from the formations of the Newark Supergroup in the Connecticut Valley and New Jersey —footprints numbering in the tens of thousands—belong to numerous species of *Anchisauripus, Grallator,* and *Eubrontes,* ichnogenera which are reliably correlated with the coelurosaur-carnosaur group of dinosaurs. Yet these same formations have yielded only two partial skeletons of coelurosaurs, both assigned to the genus *Coelophysis* (Colbert and Baird, 1958; Colbert, 1964). Thus the osteological and ichnological records supplement each other far more often than they overlap, and the presence of both skeletons and footprints of *Ammosaurus* in the Navajo Sandstone of Arizona is a happy exception to the usual order of things.

Summary

A unique trackway of a quadrupedal dinosaur shows a tetradactyl pes and a functionally tridactyl manus with a sickle-shaped, laterally recumbent claw on the thumb. The trackmaker's morphology correlates closely with that of the prosauropod dinosaur *Ammosaurus* (family Plateosauridae) from the same formation and area, and confirms Galton's functional interpretation of the prosauropod manus. As the first authentic record of a prosauropod trackway, the specimen is assigned to a new ichnogenus and species, *Navahopus falcipollex.*

Acknowledgments

It gives me special pleasure to render posthumous honor to my friend Major Brady by publishing this latest of his contributions to paleontology—a task that he delegated to me in 1960 with the observation, "I'm over eighty and am getting lazy." Drs. Edwin H. Colbert and Eugene S. Gaffney kindly made and sent me a latex mold of the entire trackway so that a full analysis could be made. The slab was photographed by Mr. Parker Hamilton of Flagstaff, Arizona. Messrs. William J. Breed and Milton Wetherill provided precise information on collecting localities. Dr. Hartmut Haubold generously supplied valuable comparative data from his unpublished ichnological work. For stimulating discussion at all stages of the study I am pleasantly indebted to Dr. Peter M. Galton. This research was supported by the William Berryman Scott Fund of Princeton University.

Literature Cited

Baird, Donald. 1957. Triassic reptile footprint faunules from Milford, New Jersey. Harvard Univ., Mus. Comp. Zool., Bull., 117:447–520, 4 pls.

Bird, R. T. 1944. Did *Brontosaurus* ever walk on land? Natur. Hist., 53:61–67.

———. 1954. We captured a "live" brontosaur. Nat. Geogr. Mag., 105(5):707–22.

Brady, L. F. 1935. Preliminary note on the occurrence of a primitive theropod in the Navajo. Amer. J. Sci., (5)30:210–15.

———. 1936. A note concerning the fragmentary remains of a small theropod recovered from the Navajo Sandstone in northern Arizona. Amer. J. Sci., (5)31:150.

Colbert, E. H. 1964. The Triassic dinosaur genera *Podokesaurus* and *Coelophysis.* Amer. Nus. Nov., 2168:1–12.

Colbert, E. H., and Baird, Donald. 1958. Coelurosaur bone casts from the Connecticut Valley

Triassic. Amer. Mus. Nov., 1901:1–11.

Cooley, M. E.; Harshbarger, J. W.; Akers, J. P.; and Hardt, W. F. 1969. Regional hydrogeology of the Navajo and Hopi Indian Reservations, Arizona, New Mexico and Utah. U. S. Geol. Surv., Prof. Paper, 521–A:A1–A61, pls. 1–5.

Cornet, Bruce; Traverse, A.; and McDonald, N. G. 1973. Fossil spores, pollen, and fishes from Connecticut indicate Early Jurassic age for part of the Newark Group. Science, 182:1243–47.

Ellenberger, P. 1973. Contribution à la classification des pistes de vertébrés du Trias: les types du Stormberg d'Afrique du Sud (I). Palaeovertebrata, Mém. Extraord., 1972:1–152.

Faul, Henry, and Roberts, W. A. 1951. New fossil footprints from the Navajo(?) Sandstone of Colorado. J. Paleontol., 25:266–74.

Galton, P. M. 1971. The prosauropod dinosaur *Ammosaurus,* the crocodile *Protosuchus,* and their bearing on the age of the Navajo Sandstone of northeastern Arizona. J. Paleontol., 45(5):781–95.

———. 1976. Prosauropod dinosaurs (Reptilia: Saurischia) of North America. Peabody Mus. Postilla, 169:1–98.

Galton, P. M., and Cluver, M. A. 1976. *Anchisaurus capensis* (Broom) and a revision of the Anchisauridae (Reptilia, Saurischia). S. Afr. Mus., Ann., 69(6):121–59.

Ginsburg, Léonard, et al. 1966. Empreintes de pas de vertébrés tétrapodes dans les séries continentales à l'Ouest d'Agadès (République du Niger). Acad. Sci., C. R., Sér. D, 263:28–31.

Heilmann, Gerhard. 1927. The origin of birds. D. Appleton, New York, i–vii, 1–210.

Lull, R. S. 1904. Fossil footprints of the Jura-Trias of North America. Boston Soc. Natur. Hist., Mem., 5:461–557, pl. 72.

———. 1915. Triassic life of the Connecticut Valley. Conn. Geol. Natur. Hist. Surv., Bull., 24:1–285, 12 pls., map.

———. 1953. Triassic life of the Connecticut Valley. Conn. Geol. Natur. Hist. Surv., Bull., 81:1–336, 12 pls.

McKee, E. D. 1979. Ancient sandstones considered to be eolian. *In* E. D. McKee, ed., A study of global sand seas. U. S. Geol. Surv., Prof. Paper, 1052:187–238.

Nopcsa, Franz. 1923. Die Familien der Reptilien. Fortschr. Geologie und Paläontologie, 2:1–210, 6 pls.

Olsen, P. E., and Galton, P. M. 1977. Triassic-Jurassic tetrapod extinctions: are they real? Science, 197:983–86.

Schaeffer, Bobb. 1941. The morphological and functional evolution of the tarsus in amphibians and reptiles. Amer. Mus. Natur. Hist., Bull., 78:395–472.

Walker, A. D. 1968. *Protosuchus, Proterochampsa,* and the origin of phytosaurs and crocodiles. Geol. Mag., 105(1):1–14.

A DIPLODOCID SAUROPOD FROM THE LOWER CRETACEOUS OF ENGLAND

by Alan J. Charig

Diplodocus is one of the most familiar of all dinosaurs to the general public. The name (Marsh, 1878) means double beam in Greek and refers to the highly characteristic form of the chevron bones in the middle part of the tail.

Chevron bones (otherwise haemal arches or haemapophyses) are essentially modified intercentra, each fitting between the bevelled edges of two successive caudal centra. Where chevrons are most typically developed, as in primitive reptiles, the intercentrum bears a pair of bony rods projecting downwards and a little backwards; these converge to meet each other in the mid-line, together forming a structure which is Y-shaped in anterior or posterior view (Figure 13.1B) and which is attached to the original intercentrum by both limbs of the Y. The two limbs of the Y and the "bridge" connecting them above enclose the haemal canal, within which run the major blood vessels supplying the tail. In more advanced reptiles this "bridge"—which forms the greater part of the original intercentrum—is often reduced; it may even disappear entirely in at least some of the chevrons, so that the two limbs of the Y articulate separately with the overlying centra.

The primary function of such chevrons is to provide additional points of origin and insertion for the hypaxial musculature of the tail, especially important in those animals where lateral movements of the tail are used in locomotion and in self-defense. They may also serve to protect the caudal artery and vein to some extent.

In the mid-caudal chevrons of *Diplodocus*, however, the greater part of the Y is directed more horizontally, backwards rather than downwards, and there is also a pair of forwardly projecting processes which—at their maximum development, on caudal vertebrae numbers 15–19 inclusive in the type-species *D. longus*—likewise converge to meet each other. Thus the entire element assumes a rather boat-like shape, roughly symmetrical in lateral view, with an elongated median longitudinal (sagittal) slit in the bottom and with two upwardly projecting processes attaching it to the overlying vertebrae. (Figure 13.2 shows two such chevrons. B belongs to *Diplodocus carnegii*; A is smaller but similar, and is seen also in two views in Figure 13.4 A and B.) The transition from the "normal" chevrons at the base of the tail to these aberrant boat-shaped structures is gradual (Hatcher, 1901), likewise their progressive reduction and simplification in the more distal parts of the tail. The particular function of these strange bones can only be guessed at, but they would obviously be less suitable for the attachment of muscles, which in the more distal part of a whip-like tail would not be especially large; conversely, with their skid-like shape, they would clearly afford greater protection to the blood vessels running between them and the vertebrae than would the "normal" chev-

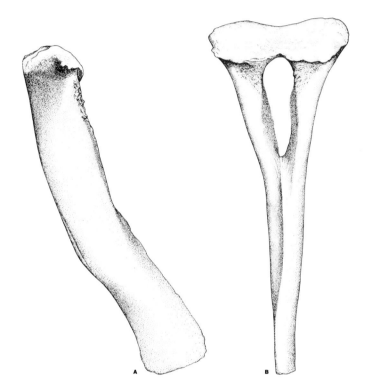

Figure 13.1 A. Proximal caudal chevron (fourth?) of holotype of *Cetiosauriscus stewarti* sp. nov. (see text), of "normal" shape. Original partial skeleton R.3078 in British Museum (Natural History); from the Oxford Clay of Peterborough, Cambridgeshire. From left side. The angle which this chevron as drawn makes with the horizontal is intended to approximate the angle in the articulated skeleton between the chevron and the vertebral column (anterior and to the left). B. The same, from behind. The articulating surface is at the top. One-third natural size.

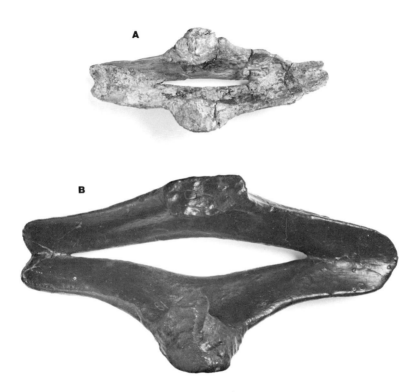

Figure 13.2 A. Mid-caudal chevron of diplodocid sauropod. Isolated bone R.8924 in British Museum (Natural History); from the Wealden Formation, near Grange Chine, Isle of Wight. From above. Length 19.6 cm. Collected and presented by Mr. Stephen Hutt. B. Mid-caudal chevron, lying between 18th and 19th caudal vertebrae, of *Diplodocus carnegii* Hatcher. Copy in British Museum (Natural History) of plaster restoration in Carnegie Museum, Pittsburgh, based on specimens from the Morrison Formation of Wyoming. From above. Length 30.5 cm.

rons, and this would be more necessary in a part of the tail where thick musculature was absent and which may well have been dragged upon the ground or used as a weapon. Indeed, it was long ago pointed out (Hatcher, 1901, p. 37) that "this is just that region of the caudal series which would come in contact with the ground," although the author in question believed that the living animal assumed a tripodal position and that the purpose of the modification was "better to resist the impact brought to bear at this point by the superimposd weight of the tail and body."

There seems to be a general assumption by almost everyone not personally concerned with research on this group that this extraordinary feature was unique to the genus *Diplodocus*, of which the four species described are all from the Morrison Formation (Upper Jurassic) of Colorado, Wyoming, and Utah (Steel, 1970). The double-beam chevrons occur in the type-species *D. longus* (Marsh, 1878; Osborn, 1899; Hatcher, 1901), and in the equally well-known *D. carnegii* (Hatcher, 1901); we may only guess at their occurrence in the other two species of the genus, which were described on skull material alone. In fact, however, mid-caudal chevrons like those of *Diplodocus* are known also in five other sauropod genera. Two of these, *Apatosaurus* (see Riggs, 1903) and *Barosaurus* (see Lull, 1919), were sympatric with *Diplodocus* in the Morrison Formation of the United States; the other three, *Cetiosauriscus* (see Woodward, 1905; my Figure 13.3A, B), *Mamenchisaurus* (see Young, 1954a, b, Young and Chao, 1972) and *Dicraeosaurus* (see Janensch, 1929) are also of Late Jurassic age, occurring in England, China, and Tanzania respectively. Another species of sauropod, *Gigantosaurus africanus* (E. Fraas, 1908), occurs with *Dicraeosaurus* in Tanzania and has been tentatively referred (Janensch, 1922) to the genus *Barosaurus* as *B. africanus;* chevrons found in the same excavation as *Dicraeosaurus* were claimed to belong to *Barosaurus* because of their "*Diplodocus*-artige form" (Janensch, 1929) but could presumably have belonged to either of those genera. As for *Mamenchisaurus*, there are two species of that genus, *M. constructus* (Young, 1954b) (type-species) and *M. hochuanensis* (Young and Chao, 1972). Both possess these *Diplodocus*-like chevrons. It should be noted that the beds in which they were found, the Red Beds of the Szechuan Basin, were dated at the time as "Upper Jurassic or at most lowest Cretaceous in age" (Young, 1954b, p. 501) simply by correlation with *Diplodocus* on the basis of the type of chevron which both genera possess. Nevertheless it cannot be denied that the dinosaurs of Szechuan show an impressive similarity to the dinosaur faunas of the Morrison Formation in the western United States and of Tendaguru in Tanzania, both of which include the largest land animals of all time and both of which are generally considered to be of Kimeridgian age.

A new development in this matter took place on 21 May 1975, when an amateur collector found an isolated bone in the Wealden Formation (Lower Cretaceous) of the Isle of Wight, near Grange Chine on the southwest coast of the island. The finder, Mr. Stephen P. Hutt, repaired his fragmentary discovery and later presented it to the British Museum (Natural History); it has been registered as R.8924 (Figure 13.2A, 13.4B, C). It appears to be a well-preserved, characteristic mid-caudal chevron of diplodociform type, 19.6 cm long and with a maximum width of 8.1 cm. From the length of the new chevron the length of the animal to which it belonged has been estimated on a simple proportional basis at about 16.6 m (54

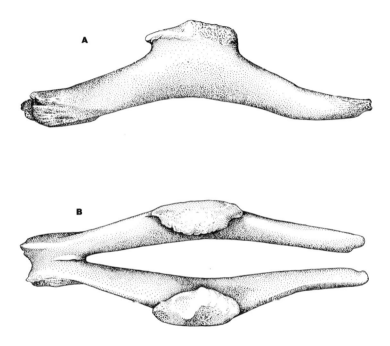

Figure 13.3 A. Mid-caudal chevron of holotype of *Cetiosauriscus stewarti* sp. nov. (see text). Original partial skeleton R.3078 in British Museum (Natural History); from the Oxford Clay of Peterborough, Cambridgeshire. From left? side. B. The same, from above. The articulating surfaces are in the center. One-half natural size.

feet); corresponding measurements for *Diplodocus carnegii* are 30.5 cm (chevron between eighteenth and nineteenth caudal vertebrae) and 25.8 m.

This discovery, incidentally, attracted a great deal of attention from the press. Not only was it mentioned in local newspapers and in "The Job," the organ of the Metropolitan Police (in which Mr. Hutt was then serving as a constable), but there was also a remarkably inaccurate article on the front page of the "Sunday Times" (13 February 1977). It all culminated in a six-page color feature in the "Telegraph Sunday Magazine" (20 February 1977), with a photograph of S. P. Hutt on the cover. The only previous scientific account, however, is a brief article, contributed by myself and published anonymously in 1978, on pp. 20–21 of the *Report on the British Museum (Natural History) 1975–1977*.

The new chevron has been compared with the corresponding elements of *Diplodocus carnegii* (cast; Figure 13.2B) and of *Cetiosauriscus stewarti* (original; Figure 13.3A, B); comparison is difficult because there is so much variation between neighboring chevrons in both those species, but the following general observations seem to apply:

1. The "horizontal" part of the chevron is more solidly constructed in *Cetiosauriscus*, where it is almost rod-like. The processes are more plate-like in *Diplodocus* and even more so in the Wealden specimen, where they are thin and fragile.

2. The dorsal part of the chevron which articulates with the vertebrae seems to be longer and heavier in *Diplodocus* than in the other two forms. In *Diplodocus* it is a fairly distinct separate process; in *Cetiosauriscus* (Figure 13.3A) such a process is hardly distinguishable; and in the lateral view of the Wealden specimen (Figure 13.4A) it is no more than a convexity of the dorsal margin.

3. The tendency of the chevron to split into two separate elements, left and right, is greatest in *Cetiosauriscus*, where all but one of the "straight" chevrons preserved (i.e., those with the anterior and posterior processes directed horizontally rather than obliquely downwards) are already completely separated. The only exception is the chevron illustrated (Figure 13.3B) where there is still a firm connection at one (anterior?) end. In *Diplodocus* this tendency is much less than in *Cetiosauriscus*, but the median slit is still relatively much longer (19.5 cm, i.e., 64 percent of the total length) than in the Wealden specimen (6.9 cm, i.e., 35 percent).

The chevrons of the holotype of *Diplodocus longus* (Marsh, 1878), the specimen to which the generic name originally referred, were not available for a direct comparison. The several chevrons discovered in that specimen were all double-beamed, and one, found attached to the eleventh caudal vertebra, was figured by Marsh (1878, plate 8, figure 3). The anterior and posterior processes of the figured chevron are both directed slightly downwards at their ends, whereas in the Wealden specimen they both project more horizontally.

Comparison with other diplodociform chevrons was even more difficult. Apart from the extensive variation along the tail of each individual, complicating factors included the fewness of such chevrons even mentioned in the literature (*Barosaurus, Dicraeosaurus*), uncertain association of chevrons with the rest of the skeleton (*Dicraeosaurus*), inadequacy of description and illustration (*Apatosaurus*) and, in one case, publication of the description in Chinese only without one word of any other language (*Mamenchisaurus hochuanensis*, Young and Chao, 1972). None of the descriptions or illustrations published indicates that any of the chevrons in any

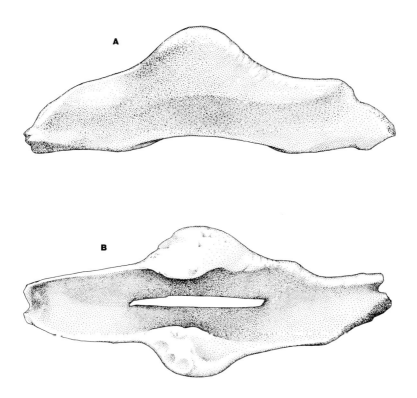

Figure 13.4 A. Mid-caudal chevron of diplodocid sauropod. Isolated bone R.8924 in British Museum (Natural History); from the Wealden Formation, near Grange Chine, Isle of Wight. From side. Length 19.6 cm. Collected and presented by Mr. Stephen Hutt. B. The same, from above.

of these genera were as closely comparable to the three mentioned above as were those three to each other. In the *Barosaurus* chevrons the two arms of the Y are expanded at their proximal (articular) end to touch each other, and there is no median opening in the horizontal part of the chevron, only "an elongated, alveolus-like depression." In *Dicraeosaurus* the processes connecting the horizontal part to the vertebrae are very much heavier, and the entire chevron is not nearly so symmetrical in lateral view. In *Mamenchisaurus constructus* not only are the two arms of the Y connected above into a single, obliquely rectangular articular facet but also the anterior and posterior processes are directed, not horizontally, but obliquely downwards. No useful information is available on the chevrons of *Apatosaurus*. However, it is quite possible—even likely—that undescribed chevrons more closely comparable to those of *Diplodocus* itself were present in other parts of the tail in some or all of those genera.

It is not often that chevrons of the genera concerned have been found in articulation with the overlying vertebrae (as they were in *Diplodocus longus*, likewise three in *Cetiosauriscus*) but in other cases they have been arranged in a series according to diminishing size (*Mamenchisaurus constructus*). Where the chevrons can be placed in their proper positions it seems that the diplodociform chevrons are developed always in the same part of the tail, being at their maximum development in the region of the twentieth caudal.

Several other supposed sauropod specimens are already known from the Wealden of the Isle of Wight, something like four new genera and eight new species having been based upon them (mostly between 1870 and 1890). All, however, are of an extremely fragmentary nature, comprising (for example) one tooth or one imperfect vertebral centrum; indeed, even the best consists of no more than two ischia and a pubis (*Ornithopsis eucamerotus* Hulke 1882). Most of these type specimens would be considered indeterminate by modern taxonomists, so that the generic and specific names given to them would accordingly be regarded as *nomina dubia;* and, although the taxa concerned have sometimes been synonymized (see Steel, 1970) with each other and with sauropod taxa founded upon material from the Wealden of the English mainland (usually Sussex), such synonymies are generally based upon inadequate evidence. The new chevron, of course, could well have belonged to the same species as some of these previously described remains, from the mainland or the Isle of Wight, but this suggestion cannot be tested at present.

On 12 February 1976 yet another sauropod bone was collected from the Wealden beds of the southwest coast of the Isle of Wight (Brighstone Bay). This is a large caudal vertebra, 17.0 cm long; it comprises a fairly well-preserved, elongated centrum (somewhat obliquely crushed by lateral pressure) and an incomplete neural arch. It is similar in shape to the middle-to-distal caudal vertebrae of *Diplodocus carnegii*. The best comparison is with the thirty-third caudal of that species, though the latter is rather more elongated and altogether a little larger, reaching a total length of 22.3 cm. The new vertebra, of course, is too indeterminate on its own to be referred to any particular species or even genus of sauropod. The finder, Senhor Espedito Cordeiro da Silva, has generously donated the specimen to the British Museum (Natural History), where it has been registered as R.9224.

The type of chevron bone described in this article, unique in its form and (as far as we know) in its distribution within the middle part of the tail, is a feature which one would hardly expect to be a product of parallel evolution, i.e., to have evolved simultaneously and independently in two or more lines of the sauropod radiation. It might therefore be regarded as a shared derived character of taxonomic value, a synapomorphy, indicating a close phylogenetic relationship between the animals which possessed it. Berman and McIntosh (1978, p. 33) seem to have come to the same conclusion; they placed all such animals into a single taxon defined on a suite of such characters, one of which is "midcaudal chevrons having distal-, fore- and aft-directed processes." The distribution of these diplodociform caudal chevrons is congruent with the distribution of the other characters employed in the defini- tion; this suggests that the character in question is indeed a valid synapomorphy and not merely a parallel development, a similar adaptation to a common func- tional requirement. The taxon thus defined included the seven genera *Apatosaurus*, *Diplodocus*, *Barosaurus*, *Cetiosauriscus*, *Mamenchisaurus*, *Dicraeosaurus*, and *Nemegto- saurus*. All except the last are known to have possessed diplodociform chevrons. *Nemegtosaurus* occurred much later than the others, being represented only by skull material from the uppermost Cretaceous of Mongolia (Nowiński, 1971). The skull and teeth, however, give distinct indications of close affinities with the other gen- era mentioned. Berman and McIntosh considered that the taxon comprising all these genera was deserving of familial separation and should be called Diplodocidae (Marsh, 1884); older names (Atlantosauridae, Marsh, 1877; and Amphicoeliidae, Cope, 1877) were rejected as having been based on indeterminate type genera, generic names which were themselves based on indeterminate type material and should consequently be regarded as *nomina dubia*.

If it be assumed that the animals which possessed diplodociform chevrons all belong to the same family, then their occurrence in the Upper Jurassic rocks of western North America, England, China, and East Africa provides additional con- firmation of the existence of migratory pathways for large terrestrial animals be- tween those areas during the Mesozoic—more specifically the Jurassic—and of the generally cosmopolitan nature of dinosaur faunas (Charig, 1971). There is good evidence for such a pathway between Europe and North America in Triassic times; for example, the Upper Trias of South Wales has yielded a slab covered in assorted dinosaur footprints which, it now appears, are just like those in contem- porary rocks of the Connecticut Valley (Bassett and Owens, 1974). Charig (1973) listed six dinosaur families—including three sauropod subfamilies—which are com- mon to Upper Jurassic of Europe and North America, while families which seem to be restricted to one continent or the other are comparatively few in number (one in each). The corresponding figures for Lower Cretaceous families are given as eight or nine common to both continents, with two restricted to each.

It has even been claimed, on several occasions, that the same *genus* of dinosaur— though not the same species—is to be found on both sides of the North Atlantic; in every case, however, the remains of at least one of the allegedly congeneric forms are so fragmentary as to cast serious doubt on the matter. Some of these claims date back nearly a century, including a few which relate to the scrappy sauropod material from the Wealdon of the Isle of Wight (see above) and Sussex; other such claims were made only a couple of years ago. For example, Lydekker (1889) named two English femora, one from the Upper Jurassic and one from the Lower Creta- ceous, as two new species of the American ornithopod *Camptosaurus*; he called them *C. leedsi* and *C. valdensis* respectively. Galton (1975) accepted the former, rather doubtfully, as *Camptosaurus*(?); but he rejected the latter (1974) and referred the femur in question to *Hypsilophodon foxii*. Working in the opposite direction, Galton himself (1976a) placed the Morrison theropod *Stokesosaurus* from the uppermost Jurassic of Utah in the synonymy of *Iliosuchus*, based on an isolated ilium from the Middle Jurassic (Bathonian) of Oxfordshire; later, however, he changed his mind about this (Galton and Jensen, 1979). He also recognized (Galton and Jensen, 1975) the English Wealden genera *Hypsilophodon* and *Iguanodon* in the Lower Cretaceous of North America, and even went so far as to base new species (Galton and Jensen, 1978) on the material in question—*H. wielandi* on a femur from South Dakota and *I. ottingeri* on a piece of maxilla from Utah, with teeth. As a final example of this "transatlantic congenericity," Galton claimed (1975) the presence of the American Upper Jurassic genus *Dryosaurus* in the Lower Cretaceous of England, proposing the new species *Dryosaurus? canaliculatus* on yet another femur from the Wealden of the Isle of Wight. In a slightly later publication (1976b) the "?" had disappeared, and with it, presumably, his doubts. Yet two years later (1978, personal communication) he told me that he intended to erect a new genus for the species in question.

To be blunt, it seems to me that the endless, frequently repeated discussion as to whether two similar fossils from different continents represent the same species, the same genus, the same genus with a question mark, closely related genera, or merely members of the same family is, in most cases, a complete waste of time. It is not difficult to demonstrate the similarities or differences between two related forms, geographically widely separated, when the material of both is fairly abundant (as with the American *Dryosaurus* and the East African *Dysalotosaurus*, see Galton, 1977; Shepherd, Galton, and Jensen, 1977). Even when the material of one or both is very restricted in extent, a convincing case may often be made (provided there is some degree of overlap). But to assess the taxonomic significance of those similarities and differences can present a much greater problem. Bearing in mind that the limits of paleontological taxa are matters of opinion, that well-known forms may show considerable intrageneric and intraspecific variation (the latter especially with age and size), and that structures like teeth and vertebrae can vary greatly within a single individual, some of the conclusions of Galton and his fellow workers seem to be based on insufficient evidence and some are inconsistent. As already mentioned, Galton and Jensen (1975) referred an isolated femur from the Lower Cretaceous of South Dakota to the English genus *Hypsilophodon;* that, perhaps, is not unreasonable, although it might be thought unlikely that there would be great differences in the form of the femur between the several genera of small ornithopods. But later (1978) Galton and Jensen placed their American femur in a new species, *H. wielandi*, on the grounds that "the ratio of the minimum distance between the proximal end of the femur and the distal edge of the fourth trochanter [ratio to what? presumably to the total length of the femur] is 0.45 (0.43 in *H. foxii*), and distally there is a slight anterior intercondylar groove." No other differences were specified. In fact, the measurements they quoted (table 1) yield ratios of 0.426, 0.433, and 0.435 for the three specimens of *H. foxii* cited and 0.426 for the much larger American specimen (35 percent longer than the biggest *H. foxii*, 167 percent longer than the smallest). I cannot see that this proves anything. The same applies to Galton and Jensen's proposal of *Iguanodon ottingeri* as a new species. (Both *Hypsilophodon* and *Iguanodon* occur abundantly—in the case of *Hypsilophodon,* virtually exclusively—in the Wealden of the Isle of Wight.) Yet Galton was critical (1977) of past attempts to claim the presence of genera from the Upper Jurassic of North America in the Upper Jurassic of Tendaguru (Tanzania), on precisely the same grounds on which I criticize him here. The genera he mentioned were *Diplodocus, Barosaurus, Brachiosaurus, Allosaurus?,* and *Ceratosaurus?;* he, himself, claimed the presence of *Dryosaurus* in the same work.

As I see it, to attempt the definition of new taxa upon indeterminate material is highly undesirable. It is far better to refer to inadequate specimens like those mentioned above by such expressions as "indeterminate hypsilophodontid," "indeterminate iguanodontid," and to leave it at that. Their paleobiogeographical significance would be no less. Indeed, such tenuous evidence of genera present in areas A and B, absent in area C, ought not to be used as a basis for paleogeographical conclusions, for it could well mislead; for such purposes it is surely better to use only family linkages, of which we can be much more certain and upon which we can therefore rely. In any case, all distribution data—if they are to have any paleogeographical significance—must include some *negative* evidence, i.e., evidence of apparent absence of the taxa concerned from particular strata or regions; and the possibility always exists that such negative evidence might be disproved.

It follows from the above that the presence of a diplodociform chevron in a given deposit cannot of itself be taken as indicating the presence of the genus *Diplodocus,* even though it might be morphologically closer to *Diplodocus* than to any other genus in the group. As already noted, such chevrons are known to occur in several genera; their precise form varies widely within the individual, so that one single isolated chevron is more easily assigned to an approximate position in the tail than to a particular genus. In any case, the limits of paleontological genera are highly subjective. The Isle of Wight sauropod represented by the isolated chevron may therefore be referred to the family Diplodocidae, but it cannot be determined more precisely at present and should not be given a new name.

The new chevron from the Isle of Wight does not seem to be derived, for it shows no sign of abrasion and, though fragile, is remarkably well preserved. It is therefore justifiable to regard it as of Wealden origin, affording our first positive indication that sauropods with diplodociform chevrons survived into the Early Cretaceous. Thus, because of the antiquity of *Cetiosauriscus* (which is probably the *oldest* of those reptiles), the English material alone gives them a range from the Callovian to the Neocomian. (Nevertheless this extended range is hardly surprising if it be accepted that the family Diplodocidae occurred also in the latest Cretaceous, where it is represented by the skull of *Nemegtosaurus.*) Since it was already known that Late Jurassic sauropods possessing these peculiar chevrons were widely distributed in both Mesozoic supercontinents, likewise in England as well as in North America, China, and East Africa, it is clear that the real importance of the new find is stratigraphical rather than geographical. The new chevron, incidentally, is indeed morphologically closer to *Diplodocus* itself than to any other genus in the group; this appears all the more remarkable when the difference in stratigraphical levels is taken into account.

One of the genera mentioned above as belonging to the Diplodocidae is *Cetiosauriscus.* The genus *Cetiosauriscus* was unequivocally based by von Heune (1927) upon a single specimen, R.3078 of the British Museum (Natural History), a partial sauropod skeleton formerly exhibited in the recently closed Dinosaur Gallery. It is that specimen which possesses the diplodociform chevrons discussed in this article. It formed part of the famous Leeds Collection from the Oxford Clay of Peterborough and was described originally by Woodward (1905) as another specimen of *Cetiosaurus leedsi* (Hulke, 1887), itself based upon the unique holotype B.M.(N.H.) R.1988 from the same geological formation and locality. (Hulke, in fact, had placed his new species in the genus *Ornithopsis*, and only later was it tentatively suggested [Seeley, 1889] that it might be congeneric with *Cetiosaurus oxoniensis*, a suggestion which was generally accepted. *O. leedsi* is not, of course, the type-species of either of those genera.) Von Huene (1927) uncritically accepted R.3078 as belonging to the species *leedsi*. Unfortunately, however, the only element preserved in both R.3078 and R.1988 is the ilium, and it was written of R.1988 (Seeley, 1889, p. 394) that "From the imperfect fragment preserved it is impossible to judge of the form of the ilium." A recent examination of both specimens confirms that a direct comparison cannot be made. In reality, therefore, there was nothing on which to base the reference of R.3078 to *C. leedsi* (which reference, in any case, would have been purely subjective) and there is no specific name available for the former. I consequently designate R.3078 herewith as the holotype of *Cetiosauriscus stewarti* sp. nov., defined on the same characters as those used (von Huene, 1927) to define the genus of which it is the type-species, *Cetiosauriscus*. (The new specific name is intended as a tribute to Sir Ronald Stewart, Bt., the recently retired Chairman of the London Brick Company Limited, in grateful recognition of the generous cooperation which his Company has long afforded the British Museum [Natural History] in recovering many important fossils from the Oxford Clay.) Because of the unjustifiable reference (Seeley, 1889; Woodward, 1905) of this specimen to *Cetiosaurus leedsi*, the erection of the generic name *Cetiosauriscus* (von Huene, 1927) upon the same specimen and the consequent designation (von Huene, 1927) of *Ornithopsis* [*Cetiosaurus*] *leedsi* as the type-species of *Cetiosauriscus*, it is evident that we have here a case of "misidentified type-species," which, according to Article 70(a) of the International Code of Zoological Nomenclature, should be referred to the Commiision for consideration of a change of type-species. This I intend to do.

Acknowledgments

I thank Mr. Stephen Hutt for finding the chevron bone on which this article is based, for permitting me to describe it and for donating it to the national collection; Senhor E. Cordeiro da Silva for finding and presenting the sauropod vertebra; Sir Ronald Stewart for allowing me to bestow his family name upon the proposed type-species of *Cetiosauriscus*; Dr. J. S. McIntosh for letting me see, before publication, the relevant part of a draft of the article by Berman and McIntosh on *Apatosaurus* and *Diplodocus*; Mrs. Li Jin-ling for translating the titles of papers published only in Chinese; Miss Marilyn Holloway for making the drawings; Miss Sandra Chapman for helping in several ways in the preparation of this article; my wife for typing two drafts of the article; and last but by no means least, Dr. H. W. Ball and Dr. Angela Milner for reading and criticizing my manuscript.

Addendum

At the end of my paper I should like to seize this opportunity of adding a small transatlantic tribute to Ned Colbert, mentor and friend, who has done more than anyone else alive today to extend and disseminate our knowledge of dinosaurs. His innumerable works have ranged from highly erudite and esoteric research papers at one end of the spectrum to best-selling popular books at the other, the books more easily digestible by the layman but no less scholarly than the papers.

Nor have I forgotten Ned's valiant efforts in the field, laboratory, and museum gallery, in the lecture hall, and on television.

I had hoped that I myself might produce, for the Colbert Festschrift, an interesting and fundamental contribution to dinosaur studies (on "The Origin of Dinosaurs") as a mark of my esteem and affection for Ned. Alas, the project grew out of hand until it seemed that there would be neither sufficient time nor space to present it properly; then, when I became seriously ill shortly before the deadline, I was forced to admit defeat and submit a work of more modest proportions (reserving the longer work for another Festschrift, to be assembled later this year). I bear my disappointment and ask Ned to forgive my failure. However, the lesser work submitted—trivial though it might be—does have symbolic value in that it documents the discovery in England of what we always think of as a famous and characteristically American group of dinosaurs. England, my own native land, is where dinosaurs were first discovered; the United States, Ned's native land, is where they have been found in the greatest abundance. The Diplodocidae forge yet another link between us.

Literature Cited

Bassett, M. G. and Owens, R. M. 1974. Fossil tracks and trails. Amgueddfa, Bull., 18:1–18.

Berman, D. S. and McIntosh, J. S. 1978. Skull and relationships of the Upper Jurassic sauropod *Apatosaurus* (Reptilia, Saurischia). Carnegie Mus., Bull., 8:1–35.

Charig, A. J. 1971. Faunal provinces on land: evidence based on the distribution of fossil tetrapods, with especial reference to the reptiles of the Permian and Mesozoic. *In*, F. A. Middlemiss, P. F. Rawson and G. Newall (eds.), Faunal provinces in space and time. Geol. J., special issue no. 4. Seel House Press, Liverpool, pp. 111–28.

———. 1973. Jurassic and Cretaceous dinosaurs. *In* A. Hallam, (ed.), Atlas of paleobiogeography. Elsevier, Amsterdam, pp. 339–52.

Cope, E. D. 1977. On *Amphicoelias*, a genus of saurians from the Dakota epoch of Colorado. Paleontol. Bull., 27:1–5.

Fraas, E. 1908. Ostafrikanische Dinosaurier. Palaeontographica, 55(2):105–44.

Galton, P. M. 1974. The ornithischian dinosaur *Hypsilophodon* from the Wealden of the Isle of Wight. Brit. Mus. (Natur. Hist.), Bull., Geol., 25(1):1–152c.

———. 1975. English hypsilophodontid dinosaurs (Reptilia: Ornithischia). Palaeontology, 18:741–52.

———. 1976a. *Iliosuchus*, a Jurassic dinosaur from Oxfordshire and Utah. Palaeontology, 19:587–89.

———. 1976b. The dinosaur *Vectisaurus valdensis* (Ornithischia: Iguanodontidae) from the Lower Cretaceous of England. J. Paleontol., 50:976–84.

———. 1977. The ornithopod dinosaur *Dryosaurus* and a Laurasia-Gondwanaland connection in the Upper Jurassic. Nature, 268:230–32.

———. 1978. Remains of ornithopod dinosaurs from the Lower Cretaceous of North America. Brigham Young Univ. Geol. Stud., 25(3):1–10.

———. 1979. A new large theropod dinosaur from the Upper Jurassic of Colorado. Brigham Young Univ., Geol. Stud. 26(2):1–12.

Galton, P. M., and Jensen, J. A. 1975. *Hysilophodon* and *Iguanodon* from the Lower Cretaceous of North America. Nature, 257:668–69.

Hatcher, J. B. 1901. *Diplodocus* (Marsh): Its osteology, taxonomy, and probable habits, with a restoration of the skeleton. Carneg. Mus. Mem., 1(1):1–63.

Huene, F. von. 1927. Sichtung der Grundlagen der jetzigen Kenntnis der Sauropoden. Eclogae. Geol. Helv., 20(3):444–70.

Hulke, J. W. 1882. Note on the os pubis and ischium of *Ornithopsis eucamerotus*. Geol. Soc. London, Quart. J., 38:372–76.

————. 1887. Note on some dinosaurian remains in the collection of A. Leeds, Esq., of Eyebury, Northamptonshire. Geol. Soc. London, Quart. J., 43:695–702.

Janensch, W. 1922. Das Handskelett von *Gigantosaurus robustus* und *Brachiosaurus brancai* aus den Tendaguru-Schichten Deutsch-Ostafrikas. Zentralbl. Miner. Geol. Paläont., 15:464–8l.

————. 1929. Die Wirbelsäule der Gattung *Dicraeosaurus.* Palaeontographica, Suppl. 7, Reihe 1, 2(1):35–133.

Lull, R. S. 1919. The sauropod dinosaur *Barosaurus* Marsh. Conn. Acad. Arts Sci. Mem., 6:1–42.

Lydekker, R. 1889. On the remains and affinities of five genera of Mesozoic reptiles. Geol. Soc. London, Quart. J., 45:41–59.

Marsh, O. C. 1877. Notice of new dinosaurian reptiles from the Jurassic formation. Amer. J. Sci., 14(3):514–16.

————. 1878. Principal characters of American Jurassic dinosaurs, Part I. Amer. J. Sci., 16(3):411–16.

————. 1884. Principal characters of American Jurassic dinosaurs, Part VII, Diplodocidae, a new family of the Sauropoda. Amer. J. Sci. 27(3): 161–68.

Nowiński, A. 1971. *Nemegtosaurus mongoliensis* n. gen., n. sp. (Sauropoda) from the uppermost Cretaceous of Mongolia. *In* Results of the Polish-Mongolian Palaeontological Expeditions, part 3. Palaeontol. Polonica, 25:57–81.

Osborn, H. F. 1899. A skeleton of *Diplodocus.* Amer. Mus. Natur. Hist., Mem., 1:191–214.

Riggs, E. S. 1903. Structure and relationships of opisthocoelian dinosaurs, Part I, *Apatosaurus* Marsh. Field Mus. Publ., Geol. Ser., 2:165–96.

Seeley, H. G. 1889. Note on the pelvis of *Ornithopsis.* Geol. Soc. London, Quart. J., 45:391–97.

Shepherd, J. D.; Galton, P. M.; and Jensen, J. A. 1977. Additional specimens of the hypsilophodontid dinosaur *Dryosaurus altus* from the Upper Jurassic of western North America. Brigham Young Univ. Geol. Stud., 24(2):11–15.

Steel, R. 1970. Saurischia. *In,* O. Kuhn (ed.), Handbuch der Paläoherpetologie, Band 14. Gustav Fischer Verlag, Stuttgart & Portland U.S.A., v + 87 pp.

Woodward, A. S. 1905. On parts of the skeleton of *Cetiosaurus leedsi,* a sauropodous dinosaur from the Oxford Clay of Peterborough. Zool. Soc. London, Proc., 1905:232–43.

Young, C.-C. 1954a. A new sauropod from Yiping, Szechuan Province. (In Chinese.) Acta Palaeontol. Sin., 2:355–69.

————. 1954b. On a new sauropod from Yiping, Szechuan, China. Scientia Sin., 3:491–504.

Young, C.-C., and Chao X.-J. 1972. *Mamenchisaurus hochuanensis* sp. nov. (In Chinese.) Inst. Vertebr. Palaeontol. Palaeoanthropol., Peking, Mem., 8:1–30.

COELURUS AND ORNITHOLESTES: ARE THEY THE SAME?

by John H. Ostrom

Introduction

A century ago, the name *Coelurus* was proposed by O. C. Marsh for several bones of a small reptile collected from the *"Atlantosaurus* Beds" at Como Bluff, Wyoming. A quarter century later, H. F. Osborn created the name *Ornitholestes* for a small "compsognathoid dinosaur," which he based on a partial skeleton from the same strata at the famed nearby (eight–ten miles) "Bone Cabin Quarry" in southeastern Wyoming. Over the last half century, beginning with Gilmore (1920), there has been a tendency to equate these two—even though no one apparently has ever actually compared the specimens on which these taxa are based. In fact, to this day, there is no complete published record of just exactly what material does exist that has been, or can be, referred to either.

The purpose of this paper is two-fold: first, to provide a public record of the material and the data pertaining to these two genera, and second, to offer a preliminary assessment that *Coelurus* and *Ornitholestes* are not the same.

The Genus *Coelurus*

In 1879, O. C. Marsh proposed a new genus and species, *Coelurus fragilis,* for an unspecified number and kinds of bones from the *"Atlantosaurus* Beds" (Morrison Formation) of the Como Bluff area of Wyoming Territory, which he identified as a "very small reptile, apparently a Dinosaur." He described certain distinctive features of dorsal, lumbar, and anterior caudal vertebrae (especially their extreme hollowness), but did not mention any other elements, nor did he include any illustrations.

In 1881, Marsh published a more informative note on *C. fragilis* and included illustrations of three vertebrae. Here he concluded that *Coelurus* could not be placed in any known order, and proposed a new order Coeluria for this new reptile. The very next month, Marsh proposed the dinosaurian order Theropoda, but he placed *Coelurus* under the suborder Coeluria, which he listed as "Dinosauria?"

The illustration in Marsh's 1881 paper shows a cervical vertebra in anterior and lateral views, together with a mid-length cross section, and similar views of "dorsal" and anterior caudal vertebrae (see Figure 14.1). The cervical vertebra illustrated is YPM (Yale Peabody Museum) 1993, the "dorsal" is YPM 1991, and the caudal is YPM 1992. Gilmore (1920, p. 128) mentions these separate cataloguings, noting that YPM 1991 is the only one of the three that is marked in Marsh's handwriting as "type," and designating the others as only "plesiotypes." More will be said about this later, but it is important to record here that the three vertebrae in question were received in New Haven at different times. However, they all orig-

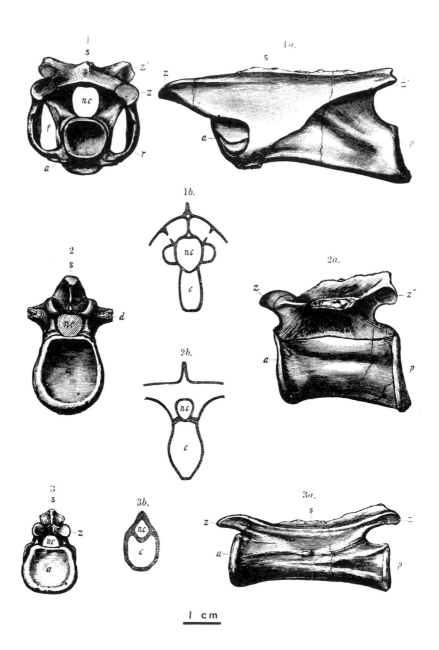

1 cm

Figure 14.1. Reproduction of Plate X from Marsh, 1881. 1, 1a and 1b = cervical vertebra of *Coelurus fragilis* Marsh in anterior and lateral views and in transverse section. This specimen is YPM 1993 (Accession Number 1395). 2, 2a and 2b = proximal caudal (*not* a dorsal) vertebra of *Coelurus fragilis* Marsh in anterior and lateral views and in transverse section. This specimen is YPM 1991 (Accession Number 1282). 3, 3a and 3b = mid-caudal vertebra of *Coelurus fragilis* Marsh in anterior and lateral views and transverse section. This specimen is YPM 1992 (Accession Number 1416).

inated from the same Como Bluff quarry— Quarry 13. Details from the Peabody Museum accession catalogues concerning these specimens are given in Appendix A.

In 1884, a third paper by Marsh appeared in which he proposed a second species of *Coelurus*, *C. agilis*, based on a pair of pubes, which Marsh compared with those of *Ceratosaurus* and *Allosaurus* as examples of his new order Theropoda. Illustrations of these pubes in lateral and anterior view were included (see Figure 14.2). Marsh stated that *C. agilis* "was at least three times the bulk of the type" of *C. fragilis*—a conclusion that certainly is incorrect. No other remains were mentioned by Marsh, nor were other descriptions provided.

Unfortunately, what has not been adequately recorded in the literature is that the type specimen of *C. agilis* consists of a great deal more than the two pubes. Gilmore (1920, p. 129) did quote a letter from R. S. Lull that it consisted of "pubes, femur, tibia, fibula, humerus, radius, ulna, coracoid, an ungual, several podial bones and several vertebrae," but he attached no importance to this. In fact, that quote is followed by this astonishing statement by Gilmore (1920, p. 129):

> At present the distinction of this species [*C. agilis*] from *C. fragilis* rests on Marsh's statement of its larger size. Since *C. fragilis* is based on vertebrae alone and there *are no vertebrae present* (my emphasis) in the type of *C. agilis*, I question the authenticity of this species.

Despite the obvious contradiction of the above statements, apparently it was this erroneous statement by Gilmore leading him to doubt the authenticity of *C. agilis*, coupled with his assertion that *C. fragilis* was based on a single "type" vertebra, that has led to the present tendency of equating *Coelurus* and *Ornitholestes*. That equation has not included all relevant factors.

In addition to the pubes, YPM 2010 includes at least eleven vertebrae, plus a number of neural arches, including elements from the cervical, dorsal, and caudal regions. A full listing of the elements comprising YPM 2010, the type of *C. agilis*, is given in Appendix A. It must be emphasized here how extremely important these additional materials of YPM 2010 are. They permit direct comparison with the type (YPM 1991) and "plesiotypes" (YPM 1992, 1993) of *C. fragilis*—something that Gilmore explicitly stated was not possible. Paradoxically, such comparisons confirm Gilmore's suspicion about the validity of *C. agilis*.

A third species of *Coelurus*, *C. gracilis*, was erected by Marsh in 1888 on the basis of thin-walled and elongated "metapodials" and an ungual phalanx. These were collected by John Hatcher, not from the Morrison Formation of Wyoming, but from the Potomac Formation (Lower Cretaceous) of Maryland. At present, only the ungual, USNM (United States National Museum) 4973, can be located. It was figured (Plate 36, Figure 4) by Gilmore in his 1920 monograph on carnivorous dinosaurs. *Coelurus gracilis* is here considered a nomen dubium, inadequately founded, and will not be discussed further.

There are additional materials in the Yale Peabody Museum collections that have been referred to *Coelurus*, most of which have never been reported or described. These are listed in Appendix A, following the registry of type specimens. Welles and Long (1974) described a right astragalus, YPM 9163 (they cite the Yale accession number, 1252), which is possibly referrable to *Coelurus*, although it is not known to have been associated with other *Coelurus* materials, and came from a different quarry, Quarry 9. It is very important to note that with the exception of

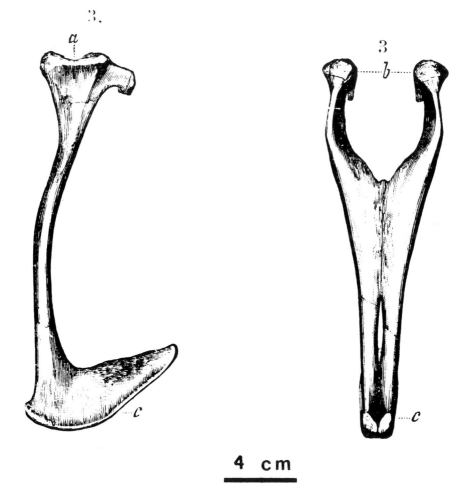

Figure 14.2. Reproduction of part of Plate XI from Marsh, 1884. The pubes of *Coelurus agilis* Marsh in lateral (left) and anterior views. This specimen is YPM 2010 (Accession Number 1483).

this astragalus and three other specimens, all of the Yale *"Coelurus"* materials were recovered from a single quarry, Quarry 13 at Como, Wyoming. The exceptions: YPM 1997, a well-preserved anterior cervical and one caudal; YPM 1996, a well-preserved cervical very similar to YPM 1993; and YPM 1933, an isolated tooth that will be discussed later.

Students of the Theropoda know that *Coelurus* has been cited in numerous papers and texts since Marsh's 1888 paper and his classic *Dinosaurs of North America* in 1896, but no detailed analysis of the Yale *Coelurus* material has been published. Such an analysis is now underway by this author.

The Genus *Ornitholestes*

The type specimen of *Ornitholestes,* AMNH (American Museum of Natural History) 619, was collected from the Morrison Formation at the famed Bone Cabin Quarry approximately eight or nine miles from Como, in 1900. Its discovery was announced by H. F. Osborn in 1903 in a brief paper in which he proposed the name *Ornitholestes hermanni.* The specimen was there described as consisting of a skull; 45 vertebrae, including three cervicals, eleven dorsals, a complete sacrum, twenty-seven caudals; the complete pelvic girdle; a representative portion of both fore and hind limbs—all belonging to one individual. A second specimen (AMNH 587) consisted of most of a left manus. A complete listing of the remains of *Ornitholestes* is provided in Appendix B.

In establishing this new taxon, Osborn noted that it was distinct from *Coelurus* in having non-serrated teeth, relatively short cervical vertebrae, and by the less extreme hollowness of all vertebrae. More will be said of the dental distinction later.

Osborn's original paper included two illustrations of the manus (AMNH 587), which was not the type (reproduced here in Figure 14.3), but only a line drawing reconstruction of the type skeleton (see Figure 14.4). In a later paper, Osborn (1917) offered a modified reconstruction of the hand based on information from the type specimen, plus an outline drawing of the skull and jaws. In this paper, Osborn added a few anatomical details to the description provided in his 1903 paper, but most of these details were offered in comparison with anatomical details of *Struthiomimus* and *Tyrannosaurus.* There is no further mention of *Coelurus* in the 1917 paper.

As with *Coelurus, Ornitholestes* has been mentioned and figured in numerous papers and books since Osborn's 1917 paper, but a thorough analysis of the specimens on which *Ornitholestes* was established has never been made. That study is now in progress by this author.

The Current Status of *Coelurus* and *Ornitholestes*

White (1973), in his catalogue of dinosaur genera, declared *Coelurus fragilis* a nomen dubium, but considered *Ornitholestes hermanni* to be valid. Steele (1970) considered the genera to be synonymous and listed four species of *Coelurus,* including *C. hermanni.* Romer in 1956 and again in 1966, listed *Ornitholestes* as a junior synonym of *Coelurus.* Huene (1956) believed the two genera were distinct, whereas Colbert (1961, 1969) apparently equated *Coelurus* and *Ornitholestes* but discussed only *Ornitholestes.* Swinton (1970) and Stahl (1974) both equated the two, but cite *Coelurus* as the proper name.

As suggested earlier, it is quite clear that Gilmore, with information provided

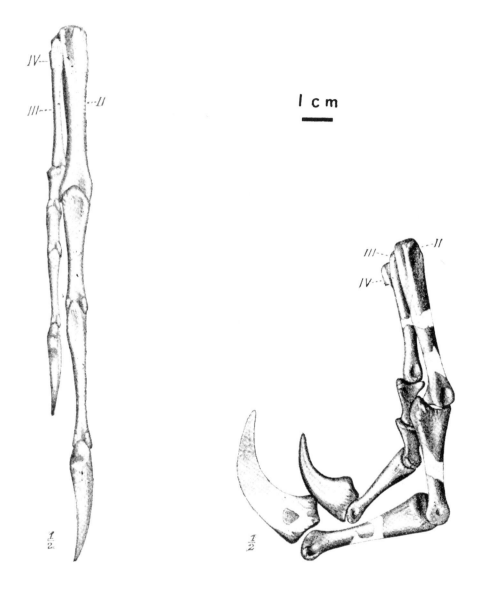

Figure 14.3. Reproduction of Figures 2 and 3 from Osborn in palmar (left) and medial views. This specimen is AMNH 587.

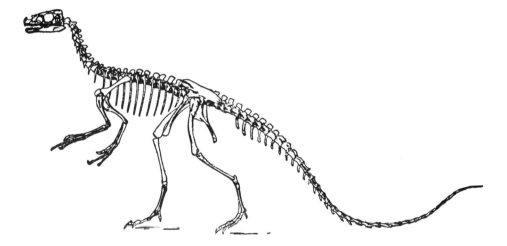

Figure 14.4. Skeletal reconstruction of the type specimen of *Ornitholestes hermanni* Osborn, reproduced from Osborn, 1903. This specimen is AMNH 619.

by R. S. Lull, is responsible for the present state of affairs. Gilmore (1920), commenting on the distinctions between *Ornitholestes* and *Coelurus* cited by Osborn, observed that:

> although the genus *Coelurus* was established in 1879, no mention was made of the presence of teeth until 1896 when Marsh figured a [serrated] tooth ascribed to *Coelurus fragilis*.

Gilmore then quoted a letter from Lull (June 28, 1915):

> The tooth, Figure 1 [Fig. 4, Pl. 34], was entirely disassociated from the rest of the material and bears the accession number [1271], being collected by Reed in 1879 from Como, Wyoming; but I cannot tell from which quarry; but as you will see, the tooth came in some time in advance of the rest of the material. Hence there is no indication of the association. As the original description is based on the vertebrae, the association of the serrated tooth is highly conjectural.

Lull's and Gilmore's assessment of this non-association is entirely correct, and thus the dental distinction between *Coelurus* and *Ornitholestes* claimed by Osborn is unproven, indeterminate, and irrelevant to the question before us.

Gilmore then challenged the other distinctions cited by Osborn (relatively short cervical vertebrae and less extreme hollowness of all vertebrae). He noted that only "two" cervicals were preserved in *Ornitholestes* (actually there are three, as Osborn noted) and since relative lengths of cervicals vary with position in the series, difference in cervical length had no merit. He noted further that the cervical figured by Marsh did not pertain to the type specimen anyway—a point that I will speak to in a moment. Gilmore then claimed:

> The extreme hollowness of the vertebrae is a difference of degree and certainly not a character upon which to base generic distinctions.

On these grounds Gilmore maintained that generic distinctness had not been established. I take this to mean that Gilmore believed the two specimens of *Ornitholestes hermanni* were referrable to *Coelurus fragilis,* but he never says that.

I believe that Lull and Gilmore were wrong in their interpretation that only the "dorsal" vertebra YPM 1991, figured by Marsh in 1881 and again in 1896, is the one and only "type" specimen and that the other two vertebrae illustrated in the same plate are only "plesiotypes." We don't know when or by whom the word "type" was written on the label to YPM 1991, but we do know that Marsh figured all three vertebrae and identified all three as belonging to *Coelurus fragilis* in the first paper in which any material was illustrated. Marsh's intention is clear.

The Yale Type Specimens

YPM 1991: The "dorsal" vertebra figured by Marsh in 1881 and catalogued under this number is not a dorsal, but rather is a proximal caudal. This is established by the small diameter of the neural canal; the high angle pitch of the zygapophyseal facets and their placement close to the mid-line ("The zygapophyses are near together, and stand nearly vertical." Marsh, 1879, p. 505); the transverse process is a simple, nearly horizontal, blade-like lateral projection; and there is no sign of parapophyses. True dorsal vertebrae of theropods have widely spaced, nearly horizontal zygapophyseal facets; complex, strut-buttressed diapophyses; and conspic-

uous parapophyses; as well as enlarged neural canals. Furthermore, the perfora-
tions (= pleurocoels) into the centrum of a true dorsal are of moderate or large
size, whereas in YPM 1991 they are very tiny.

An important fact that has not been reported before is that YPM 1991 consists of
more than the single "dorsal" figured by Marsh. The additional material consists of
another nearly complete proximal caudal of the same size, a third caudal centrum,
and a caudal neural arch which appears not to belong to the centrum. Why these
were not reported by Lull in his letter to Gilmore is not known. At this date, 100
years after their collection, there is no way to certify that these four objects were
found in close association in the quarry, but the fact that they came from the same
quarry, were sent together in the same shipment, and obviously are very much
alike and of the same size, make it most probable that they belong to a single
individual. YPM 1991 was collected from Quarry 13 East in September 1879.

YPM 1993: This specimen also consists of more than the single cervical vertebra
figured by Marsh, a fact not published before. Catalogued under this number is a
second cervical that matches the one figured by Marsh (in fact, Marsh's 1881,
Figures 1 and 1a of Plate X, might well be a composite representation of the two),
plus a neural arch. This last item is important because it matches those of the
proximal caudals of YPM 1991. YPM 1993 was collected from Quarry 13 Middle
(= 13 East) by Marsh himself in September 1880.

Although Lull and Gilmore seem to attach importance to this, the fact that a
year lapsed between the collection of YPM 1991 and 1993 is not significant when
one realizes that quarrying operations at Quarry 13 were carried out intermittently
by one, two, or three men, with interruptions of several months duration over the
period from August 1879 to the fall of 1887 (Ostrom and McIntosh, 1966). The
facts that these specimens came from the same quarry, and that both include
matching proximal caudal neural arches are sufficient grounds for equating these
specimens.

YPM 1992: As with the two preceding type specimens, YPM 1992 also consists
of more than the single "anterior" caudal vertebra figured by Marsh in 1881. This
specimen includes an uninterrupted (but now disarticulated) series of eight mid-
caudals. The one figured by Marsh appears to be the seventh in the series. The
largest of the series is represented by the posterior half only. The next largest of the
series is only slightly longer, and has vertical and transverse centrum diameters
slightly less than those of the shortest (most distal) of the three proximal caudal
centra of YPM 1991. In other words, there is reason to believe that these two
specimens are from similar sized individuals, if not the very same skeleton.

The caudals of YPM 1992 were collected from Quarry 13 East in December
1880, just three months after YPM 1993. Again, I attach no significance to the time
interval separating their collection. That they were catalogued separately was
entirely proper, but it should not mislead us, and should not prevent us from
comparing the materials—or, in my judgment, equating them as long as proper
qualification is included. I suggest that YPM 1991, 1992, and 1993 probably came
from the same individual skeleton—one that was partially disarticulated and somewhat scattered in the eastern part of Quarry 13. This is beyond proof now, but the
following specimen seems to support that conclusion.

YPM 2010: This is the type specimen of *Coelurus agilis,* which I noted above
consists of much more than the pubes that Marsh figured in his original announce-

ment of 1884. It includes at least fifty-four nonduplicated elements including a minimum of eleven vertebrae consisting of cervicals, dorsals, and one caudal. The cervicals correspond closely in size and morphology with those of YPM 1993, and the dorsals are of appropriate size to go with those cervicals, as well as with the caudals of YPM 1992 and 1991. YPM 2010 was collected during the summer of 1881 from Quarry 13 East!

The fact that all four of the Yale type specimens of *Coelurus* were collected from the *same part* of the *same quarry*—and that no part of any of these type specimens duplicates any part of another—adds up to only one reasonable conclusion; these four specimens most probably represent the remains of a single individual.

Are *Coelurus* and *Ornitholestes* the Same?

If one accepts the above conclusion, a much more extensive comparison can be made of the osteology of *Coelurus* and *Ornitholestes*. As noted earlier, this comparison is in progress. My preliminary findings from this comparison are as follows:

1. The cervical vertebrae of the two are different, with those of *Coelurus* much longer and more complex.

2. The metapodials of *Coelurus* are at least fifty percent longer than those of *Ornitholestes*, even though femora are of nearly equal lengths.

3. There are clear differences in the morphology of the dorsal centra and arches.

4. I consider the difference in degree of hollowness of the vertebrae significant, and not simply a matter of individual variation.

5. Of questionable value is the extreme elongation of the caudal prezygopophyses in *Ornitholestes*, which appears not to have been true of *Coelurus*.

Conclusions

Coelurus and *Ornitholestes* are not the same.

Literature Cited

Colbert, E. H. 1961. Dinosaurs: Their discovery and their world. E. P. Dutton, New York, 300 pp.

———. 1969. Evolution of the vertebrates. Second edition. John Wiley & Sons, New York, 535 pp.

Gilmore, C. W. 1920. Osteology of the carnivorous Dinosauria in the United States National Museum, with special reference to the genera *Antrodemus (Allosaurus)* and *Ceratosaurus*. U.S. Nat. Mus. Bull., 110:1–159.

Huene, F. von. 1956. Palaontologie und Phylogenie der Niederen Tetrapoden. Gustav Fischer Verlag, Jena, 716 pp.

Marsh, O. C. 1879. Notice of new Jurassic reptiles. Amer. J. Sci., (3) 18:501–5.

———. 1881. A new order of extinct Jurassic reptiles (Coeluria). Amer. J. Sci., (3) 21: 339–40.

———. 1884. Principal characters of American Jurassic dinosaurs Part VIII: The order Theropoda. Amer. J. Sci., (3) 27:329–40.

———. 1888. Notice of a new genus of Sauropoda and other new dinosaurs from the Potomac Formation. Amer. J. Sci., (3) 35:93–94.

———. 1896. The dinosaurs of North America. U.S. Geol. Surv., 16th Annual Report, pp. 133–415.

Osborn, H. F. 1903. *Ornitholestes hermanni*, A new compsognathoid dinosaur from the upper Jurassic. Amer. Mus. Natur. Hist., Bull., 19:459–64.

———. 1917. Skeletal adaptations of *Ornitholestes, Struthiomimus, Tyrannosaurus*. Amer. Mus. Natur. Hist., Bull., 35:733–71.

Ostrom, J. H., and McIntosh, J. S. 1966. Marsh's dinosaurs: The collections from Como Bluff. Yale Univ. Press, New Haven, 388 pp.

Romer, A. S. 1956. Osteology of the reptiles. Univ. Chicago Press, Chicago, 772 pp.

———. 1966. Vertebrate paleontology. Third edition. Univ. Chicago Press, Chicago, 468 pp.

Stahl, B. J. 1974. Vertebrate history: Problems in evolution. McGraw-Hill, New York, 594 pp.

Steele, R. 1970. Saurischia. *In* Handbuch der Paläoherpetologie, Teil 14:1 87.

Swinton, W. E. 1970. The dinosaurs. John Wiley & Sons, New York, 331 pp.

Welles, S. P., and Long, R. A. 1974. The tarsus of theropod dinosaurs. S. Afr. Mus., Ann., 64:191–218.

White, T. E. 1973. Catalogue of the genera of dinosaurs. Carnegie Mus., Ann., 44:117–55.

Appendix A
The Yale Specimens

Specimen: YPM 1991 (Part of type of *Coelurus fragilis*)
  Accession Number: 1282
  Date Received: September 30, 1879
  Locality: Quarry 13 East, Como, Wyoming
  Material: Two proximal caudal vertebrae; one proximal caudal centrum; one proximal caudal neural arch.

Specimen: YPM 1992 (Part of type of *Coelurus fragilis*)
  Accession Number: 1416
  Date Received: December 23, 1880
  Locality: Quarry 13 East, Como, Wyoming
  Material: Eight mid-caudal vertebrae; one partial centrum of a mid-caudal vertebra.

Specimen: YPM 1993 (Part of type of *Coelurus fragilis*)
  Accession Number: 1395
  Date Received: September 27, 1880
  Locality: Quarry 13 Middle (= East), Como, Wyoming (Also termed Rock Creek Quarry at this time)
  Material: One cervical vertebra; one proximal caudal neural arch.

Specimen: YPM 2010 (Type specimen of *Coelurus agilis*)
  Accession Numbers: 1483, 1485, 1489
  Date Received: July 28, 1881; July 29, 1881; August 16, 1881
  Locality: Quarry 13 East, Como, Wyoming
  Material: Three cervical vertebrae; two dorsal vertebrae; five dorsal centra; six dorsal neural arches; two indeterminate neural arches; one proximal caudal vertebra; left and right ulnae; distal ends of both radii; left humerus; left femur; proximal end of right femur; both pubes; fragment of an ilium?; distal three-quarters of left tibia; proximal end of left fibula; right scapula; distal end of metatarsal III; metatarsal II or IV; left and right radiale; left and right metacarpal II; right metacarpal III; fragment of right metacarpal IV; seven phalanges of the manus; many unidentified fragments.

Specimen: YPM 1933 (Doubtfully referrable to *Coelurus*)
  Accession Number: 1271
  Date Received: September 10, 1879
  Locality: Robber's Gulch (the same as Robber's Roost Quarry, which is Quarry 12), Como, Wyoming. (Quarry 12 is more than nine miles west of Quarry 13.)
  Material: A single tooth (figured by Marsh in 1896).

Specimen: YPM 1994
  Accession Number: 1395
  Date Received: September 27, 1880
  Locality: Quarry 13 East, Como, Wyoming
  Material: One caudal centrum.
  Comment: Received together with YPM 1993.

Specimen: YPM 1995

Accession Number: 1394
Date Received: September 18, 1880
Locality: Quarry 13 East, Como, Wyoming
Material: One caudal vertebra, plus fragments.

Specimen: YPM 1996
Accession Number: 1394
Date Received: September 18, 1880
Locality: Quarry 9, Como, Wyoming
Material: One cervical vertebra.
Comment: Received together with YPM 1995.

Specimen: YPM 1997
Accession Number: 1304
Date Received: November 5, 1879
Locality: Quarry 9, Como, Wyoming
Material: One anterior cervical vertebra; one caudal vertebra.

Specimen: YPM 9162
Accession Number: 1453
Date Received: May 11, 1881
Locality: Quarry 13 East, Como, Wyoming
Material: One partial sacral vertebra.

Specimen: YPM 9163
Accession Number: 1252
Date Received: August 1, 1879
Locality: Quarry 9, Como, Wyoming
Material: One astragalus.

Appendix B
The American Museum Specimens

Specimen: AMNH 619
Date Received: Summer, 1900
Locality: Bone Cabin Quarry, eight miles north of Como, Wyoming
Material: Skull; both mandibles; three cervical vertebrae; eleven dorsal vertebrae; four sacral vertebrae; twenty-seven caudal vertebrae; both ischia; left ilium; both pubes missing the distal ends; incomplete left femur; proximal end of left fibula; both humeri; right metatarsal II, III and IV; left metatarsal IV; four phalanges of the pes; two pedal unguals; one right tarsal; right metacarpal II or III; fragments of two other metacarpals; two fragments of manus phalanges; one ungual of the manus; numerous fragments.

Specimen: AMNH 587
Date Received: Summer, 1900
Locality: Bone Cabin Quarry, eight miles north of Como, Wyoming
Material: Left metacarpal II; left metacarpal III; three phalanges of left manus digit II (impression of ungual II); four phalanges of left manus digit III.

REFLECTIONS OF THE DINOSAURIAN WORLD

by Dale A. Russell

Dedicated, in gratitude, to my former professor,
Dr. Edwin H. Colbert.

Introduction

Paleontologists are usually supported by institutions, and are frequently immersed in the day-to-day preoccupations of human society. Alone in a residential area where a few winter stars are visible above the glow from the city, it is easy to wonder why we have chosen to study dinosaurs—an occupation as remote from the cares of our neighbors as the Age of Reptiles at that moment may seem to us. Indeed, there have been only a few fleeting occasions when, for me, the presence of the Mesozoic world became almost tangible: the sight of the pelvic region of a hadrosaur weathering out of gray shales on a gray overcast day in the northern prairies; petrified wounds from the teeth of a mosasaur in bones from the foreflipper of another of its kind, in a cabinet in the basement of Yale Peabody Museum; water dripping from leaflets of giant podocarps in a rainstorm in Urewera National Park, New Zealand; overbank silts thoroughly plowed by the feet of large terrestrial vertebrates in South Luangwa National Park, Zambia; looking downdip across the undulating surfaces of sauropod bones oriented by cascading streams which flowed through Dinosaur National Monument an incomprehensibly long time ago; and a waterfall in a scrub forest of primitive conifers and archaic angiosperms in New Caledonia.

An aura of mystery thus surrounds the dinosaurs. Stimulated by it, and sensing a similar although less well-defined public interest, for a century and a half paleontologists have scanned the debris of ancient ecosystems in order to transport our only time machine, the human mind, into worlds we will never see. Our research orientations have diversified as our insights have become more sharply defined. However, the process of specialization can obscure an appreciation for the whole. This, in turn, may obscure notions of what the world was like during the Age of Reptiles, and discourage popular support for research programs. In order to promote a balance of approach, the spectrum of current research endeavors is here summarized within five categories. All are equally valid but some are less well explored than others. Listed in increasing order of abstraction, they include:

A primary level of questioning—dinosaurian morphology. What did they look like?
A secondary level of questioning—dinosaurs in ecosystems. What did they do?
A third level of questioning—evolution in dinosaurian ecosystems. How did they change?

A fourth level of questioning—delimiting the dinosaurian era. How did dinosaurs come to be and what became of them?

A fifth level of questioning—the broader implications. What do dinosaurs mean to us?

Dinosaurian Morphology

Size is an important physical attribute of dinosaurs. They apparently varied in weight between 3.5 kg (Ostrom, 1978, *Compsognathus longiceps*) and 55 metric tons (my estimate for *Brachiosaurus atalaiensis*, Lapparent and Zbyszewski, 1957). In addition to its defensive value, large body size in terrestrial metazoans also implies lower per unit weight metabolic and transportation costs (Bennett and Dawson, 1976; Fedak and Seekerman, 1979; the concepts are currently being revised by B. K. McNab). However, beyond certain limits of size these per unit weight economies are more than offset by a large organism's increased foraging demands, perhaps due to a combination of restrictive food preferences and increasingly heterogeneous range (Harestad and Bunnell, 1979, pp. 396, 397).

Transportation costs are similar in reptiles and mammals of comparable body weights (Bennett and Ruben, 1979, p. 651). One may postulate that, during the Mesozoic, terrestrial environments were about as productive and heterogeneous as they now are (cf. Farlow, 1976, p. 851). Reptilian metabolic rates would allow the weights of the largest dinosaurs to surpass those of the largest living terrestrial mammals, for range apparently places limits on individual foraging (metabolic) requirements, not body size (cf. Harestad and Bunnell, 1979). The greater bulks of dinosaurs relative to those of terrestrial mammals would then imply relatively lower per-unit-weight metabolic requirements. Alternatively, one may postulate that terrestrial environments were more uniform during the Mesozoic, but were as productive, and the individual metabolic requirements of endothermic vertebrates of dinosaurian proportions could have been sustained. However, few Cenozoic mammals reached the dimensions of moderately large sauropods, even in highly productive, lowland tropical environments. The largest known land birds only weighed on the order of 450 kg (Wetmore, 1967).

The shapes of dinosaurs were also peculiar. A much larger proportion were bipedal than is the case in modern terrestrial vertebrates. This may be interpreted as a result of bipedal ancestries and identical transportation costs in bipedal and quadrupedal forms (Fedak and Seekerman, 1979). In a general way their silhouettes seem to have foreshadowed the more familiar shapes of modern birds and mammals. Thus one can readily appreciate the ecological possibilities of dinosaurian rhinos (ceratopsians) and dinosaurian ostriches (ornithomimids). The bodies of others appear as curious blends of familiar animals, such as elephant-giraffes (brachiosaurs) and moa-wolves (tyrannosaurs). Some, such as *Therizinosaurus*, known only from elements of giant 2.4 m long arms terminating in scythe-like claws (Barsbold, 1976), and *Segnosaurus*, characterized by a bizarre combination of primitive and advanced saurischian features (Perle, 1979), are nearly incomprehensible. They underscore our ignorance of the ecological possibilities open to terrestrial vertebrate life.

Unexpected dinosaurian shapes continue to be described. Among these may be cited the small, highly raptorial *Deinonychus* (Ostrom, 1976a, figure 2), the fin-backed ornithischian *Ouranosaurus* (Taquet, 1976, figure 73), and the ponderous sauropod *Opisthocoelicauda* (Borsuk-Bialynicka, 1977, figure 19) which held its short tail far off the ground. The largest sample of dinosaurs collected from one area (Dinosaur Provincial Park, Alberta, cf. Béland and Russell, 1978) includes the associated remains of approximately 300 specimens belonging to at least 30 different species. The number of genera of large dinosaurs identified here (14) might lead one to suspect that the diversity of large dinosaurs living within the northern interior of our continent at that time was well sampled. However, a highly distinctive horned hadrosaur (*Maiasaura*, Horner and Makela, 1979) was recently described from sediments of about the same age, but containing elements of a different faunal facies, which outcrop only 250–350 km to the southwest. Significant faunal changes over smaller horizontal and vertical intervals are known in western Canada and Tanzania (Russell, et al., 1980), suggesting that continental dinosaur faunas may well have been quite diverse by modern standards.

The diversity of dinosaurs was gradually compounded as the continents slowly separated during the course of Mesozoic time. South American Late Cretaceous forms appear to be somewhat archaic and morphologically removed from their relatives in the northern hemisphere (cf. Bonaparte, 1978; Brett-Surman, 1979; Bonaparte and Powell, 1980). If the dinosaurs which inhabited the increasingly disjunct fragments of Gondwanaland paralleled the southern trees in their biogeographic development (cf. Raven and Axelrod, 1972), then a suite of Austral forms may have appeared, ecologically analagous to but taxonomically distinct from Holarctic dinosaurs.

Badlands eroded in continental Mesozoic strata in the arid regions of the Earth have not all been prospected with equal intensity. The morphology, diversity, provinciality, and biogeography of dinosaurs indicate that we will continue to be amazed by the appearance of newly discovered skeletons. What did *Torvosaurus* (Galton and Jensen, 1979) look like? What titanic sauropod made the spectacular trackway recently discovered in Morocco (Dutuit and Ouazzou, 1980)? What are the affinities of the animal which produced the long, flamingo-like jaw from Uzbekistan (A. K. Rozhdestvensky, personal communication)? How large, indeed, were "Super-" and "Ultrasaurus" (Jensen, 1979)?

Dinosaurs in Ecosystems

Many physiological variables can be expressed as a function of body weight in terrestrial metazoans. Body weight can therefore be employed as a means of gaining insight into certain behavioral attributes of dinosaurs. It may didactically be useful to combine the relationships presented in Table 15.1 (where all equations cited below are listed) with ichnological, physiological, and osteological information in order provisionally to reconstruct the natural history of an "average" sauropod weighing 20 metric tons.

Because sauropods were about as encephalized as are living crocodilians (Hopson, 1977, Russell, et al., 1980), courtship would probably have been a relatively simple procedure involving proximity, bumping, and scratching. Mating may have occurred in a lateral position. The male organ in dinosaurs was probably single, as in their nearest surviving relatives, crocodiles and birds (squamatans have two hemipenes). In crocodiles the organ extends laterally and upwards. It is doubly recurved when extended in elephants (weight 27 kg, length 1.5 m), where sauropod dimensions may be approached. Neither elephants, crocodiles, nor dinosaurs

TABLE 15.1

Physiological parameters (Y) expressed as a function of body weight (X, gms) in the form of $Y = aX^b$.[b]

Y	a	b	Notes
FOR REPTILES			
1. Egg weight (gms)	0.42	0.59	After Case, 1978, Figure 2.
2. Number of eggs per clutch	6.8×10^3	−0.44	Modified from Seymour, 1979: maximum oxygen consumption per nest divided by Seymour's equation 8, egg weight expressed as a function of body weight.
3a. Total number of eggs laid	2.4×10^{-2}	0.41	Body weight divided by egg weight expressed as a function of body weight, multiplied by reproductive mass as percent of body weight, after Case, 1978.
3b. Total number of eggs laid	7.1×10^{-2}	0.41	As above, reproductive mass after Seymour, 1979.
4. Growth rate to sexual maturity (gms/day)	4.7×10^{-3}	0.67	Case, 1978.
5. Cost of transport (kcal/km)	1.9×10^{-2}	0.72	Fedak and Seekerman, 1979, Figure 2, 1 liter oxygen = 4.8 kcal.
6. Maximum metabolic rate at 30° C (kcal/hr)	6.72×10^{-3}	0.94	Bennett and Dawson, 1976, equation 26, exponent after Prange and Jackson, 1976.
7. Intake (kcal/day)	5.6×10^{-2}	0.82	Béland and Russell, 1980, Figure 2.
8. Resting metabolic rate at 30° C (kcal/hr)	1.33×10^{-3}	0.77	Bennett and Dawson, 1976, equation 23, 1 liter oxygen = 4.8 kcal.
9. Home range (km²)	1.6×10^{-7}	1.21	Assumes mammalian intake 0.791 reptilian intake (Bennett and Dawson, 1976, p. 170), solving for weight of mammal ingesting this quantity (Béland and Russell, 1980, Figure 2) and transferring value into equation for herbivores (Harstad and Bunnell, 1979, Figure 1).
10. Thermal conductance (kcal/hr ° C)	1.33×10^{-3}	0.77	Set equal to resting metabolic rate, yielding cooling constants as in large reptiles (cf. McNab and Auffenberg, 1976).
11. Secondary productivity/ standing crop	3.16	−0.29	Béland and Russell, 1980, Figure 3, line for carnivores.
12. Skeletal mass (gms)	3.3×10^{-2}	1.09	Anderson, et al., 1979, Table 1.
FOR MAMMALS			
13. Resting metabolic rate (kcal/hr)	1.63×10^{-2}	0.75	McNab, 1974.
14. Intake (kcal/day)	1.84	0.69	Béland and Russell, 1980, Figure 2.
15. Maximum growth rate(gms/day)	1.51	0.43	Case, 1978.
16. Secondary productivity/ standing crop	7.28	−0.29	Béland and Russell, 1980, Figure 3, line for herbivores.

are known to have possessed a baculum (cf. Neill, 1971; Sikes, 1971; Douglas-Hamilton and Douglas-Hamilton, 1975).

Between twenty-four and seventy eggs, weighing 8.5 kg each, were laid in clutches of four (equations 1–3). The physiological properties of the shells of sauropod eggs indicate they were buried in nests of earth and decaying vegetation (Seymour, 1979; see also Horner and Makela, 1979). Assuming an average incubation period for reptiles, young sauropods weighing 7.7 kg (hatchling weight = 0.9 egg weight, Case, 1978) would emerge from the nest after about seventy days (Seymour, 1979). They would soon begin to grow at a rate of 0.366 kg per day (equation 4), achieving sexual maturity at a weight of 6 metric tons and age of forty-five years (sexual maturity coincides with attainment of two-thirds body length in reptiles, Case, 1978). Hatchlings were vulnerable to extremes of ambient temperature (Thulborn, 1973), and may have preferred densely vegetated, wet areas where diurnal temperature changes were not great and abundant food was within easy reach. In the Glen Rose trackways (Bandera County, Texas; Langston, 1974) the length of the smallest footprints are 43 percent that of the largest, suggesting that young animals of our 20-ton sauropod would have joined adult herds when they weighed about 1.6 metric tons (or were about twelve years old, cf. Case, 1978, p. 323). These fascinating trackways have been cited as evidence that immature sauropods were flanked by larger animals walking on the periphery of the herd (Bakker, 1968, p. 20), although this is not supported by trackway diagrams (cf. Langston, 1974, fig. 9). Interestingly, however, one group of four animals of intermediate size did follow each other in a single file across the ancient mudflats, stepping in each other's footprints (Bird, 1944, p. 65).

The velocity of moving sauropods can be estimated from trackway evidence (velocity in km per hour = 2.8 · stride length in meters $[\lambda]^{1.67}$; hip height in meters $[h]^{1.17}$; cf. Alexander, 1976, with hip height given as 4 · length of footprint). Assuming the largest Glen Rose sauropods weighed 40 metric tons and were walking at a velocity of 3.6 km per hour, as suggested by Alexander, they would have been generating 20,000 kcal per hour (transportation cost, equation 5 · 3.6 km), or 21 percent of their maximum aerobic metabolism (equation 6). These animals were walking in shallow, brackish water (Perkins and Stewart, 1971) and hence were not drinking or feeding, although they may have been thermoregulating. The gigantic sauropod that made the 100 m trackway in Morocco (Dutuit and Ouazzou, 1980) was moving at a speed of 11.6 km per hour, covering the distance in 31 seconds at a cost of 64 percent of the aerobic potential of a 60-ton animal. A small sauropod (approximately 2 tons) left a 60 m trackway in Niger (Ginsburg, et al., 1966), indicating that it was running at about 20 km per hour and exceeding its maximum aerobic metabolism by 235 percent. It ran the preserved portion of its trackway in 11 seconds.

The hypothetical sauropod under discussion would have approached its maximum weight of 20 tons after nearly 300 years, assuming a lifetime growth rate of about half that which preceded sexual maturity (Case, 1978). At this time it would have consumed 54,300 kcal per day (49.4 kg of vegetation, cf. Farlow, 1976, p. 842), of which 13,400 kcal were diverted for basal metabolic needs (equations 7, 8). Given a transportation cost of 3,400 kcal per km (equation 5) it could have walked about 12 km per day on a sustained basis; the distance could be traversed in 50 minutes at maximum levels of aerobic metabolism (equation 6). If the animals could fast for several weeks, while moving during the cooler hours of the diurnal cycle, migrations of 1,000 km could easily be envisaged. Based on the petrology of gastroliths, sauropod displacements of 600 km have been postulated (Janensch, 1929). The home range of feeding animals would have been confined to an area 12 km in diameter (equation 9), in regions of rather lush vegetative growth. Elephants with comparable metabolic needs occupy home ranges smaller than this in the highly productive ground-water forest of Lake Manyara, but are spread over much more extensive ranges in semiarid regions of East Africa (Douglas-Hamilton and Douglas-Hamilton, 1975). Sauropods must have been rather efficient browsers, sweeping their necks over broad areas of low bush without needing to move their bodies.

Balancing the thermal budget is a very important dimension to the existence of large animals living in warm climates. If a 20-ton sauropod could radiate 557 kcal per hour to the environment for each degree that its body temperature exceeded the ambient temperature (equation 10), it would acquire intolerable heat loads at maximum aerobic activity levels within 30 minutes at ambient temperatures of 26° C, and 15 minutes at ambient temperatures of 30° C. Assuming a mammalian metabolism (equation 13), such a sauropod would be immobilized at 30° C, even in shade. By inference, sauropods possessed a reptilian metabolism and avoided strenuous activity at midday. However, a reptilian sauropod might just have been able to support an overhead equatorial sun generating 0.09 kcal per hour per cm^2 (cf. Delany and Happold, 1979, figure 11.15), particularly with special circulatory shunts to protect the brain (Wheeler, 1978). A 20-ton animal may crudely be represented as a sphere with a radius of 168 cm and an illuminated area of approximately 90,000 cm^2. In one hour it would acquire 8,100 kcal. Added to a basal metabolic increment of 557 kcal, the animal would achieve a thermal equilibrium at slightly over 15° C above ambient temperatures. If the latter were 30° C, lethal body temperatures would have been approached but perhaps not exceeded (Heatwole, 1976, Table 1).

Intake rates projected for 20-ton sauropods (equations 7, 14) suggest that the vegetation of a given area would support 3.7 times more animals with a reptilian physiology than it would of those with a mammalian physiology. If dinosaurs were metabolically reptilian, the standing crop (SC) of large animals of all kinds would have been truly spectacular in productive Mesozoic environments (Béland and Russell, 1978). However, the richness was probably fragile. Reptiles grow much more slowly than do mammals, and their secondary productivities (SP) are probably also relatively low (compare equations 4 and 15, 11 and 16). As in living reptiles, dense populations of dinosaurs could not have been exploited to the extent that are mammalian populations (cf. Ehrenfeld, 1974).

At the end of its life our 20-ton sauropod would have left behind a skeleton weighing nearly 3 metric tons (equation 12). However, relative to elephant populations, sauropods would have produced skeletons at a lower rate: a ratio of five elephant skeletons to one sauropod according to longevities (300 versus 60 years); of 3.7 to 1 according to SP/SC (equations 11 and 16, elephant weighing 4 tons); or of 1.6 to 1 (equation 11, assuming the "reptilian" value for SP/SC). The fragile vertebral column and skull would have rapidly disintegrated, but in semiarid environments heavy limb elements would have remained on the ground for more than twenty years (cf. Coe, 1978).

The continuing integration of information concerning the effects of vertebrate morphology on behavior will permit increasingly precise evaluations of the performance "envelope" of various kinds of dinosaurs.

Evolution in Dinosaurian Ecosystems

As noted above, the fragmentation of Pangea probably produced a global increase in dinosaurian diversity. On a smaller scale, this may have been further augmented by increasingly diversified terrestrial floras (Bakker, 1978; Knoll, et al., 1979), as well as by a slow expansion of vegetation toward more arid environments. In order to better understand directionality in dinosaurian evolution, it would be useful to document more fully conditions in their general physical environment. Apparently there was no trend in the proportion of continental crust covered by epeiric seas (Jeletzky, 1978). However, gradual declines may have occurred in the carbon dioxide content of the atmosphere (Budyko, 1977) and in mean temperatures (Salop, 1978, figure 1) from Triassic through Cretaceous time. These would have been accompanied by changes in climate, the carbon cycle and, in the absence of compensatory physiologic improvements, declines in the productivity of terrestrial plants (cf. Garrels, et al., 1976; Gartner and McGuirk, 1979). It has been suggested (cf. Steiner, 1977) that the radius of Earth has increased through Mesozoic and Cenozoic time. If this should prove to be correct, the force of gravity at the surface of Earth would have declined during this interval. It would then be necessary to distinguish between weight and mass of dinosaurs, and accordingly to correct the relationships listed in Table 15.1. Advances in these areas of knowledge will greatly increase our understanding of changes in the general appearance of the terrestrial biosphere during the Mesozoic.

Dinosaurs were not evolutionarily static. They approached their maximum corporal dimensions by the Late Jurassic, decreasing in mean size through Cretaceous time (cf. Stormberg, Morrison, and Oldman faunas). Large sauropods continued to be present throughout the world during the Cretaceous and dominated South American faunas (Bonaparte, 1978). Dinosaur eggs were larger but less numerous than those of modern reptiles (corrected for body size), producing larger hatchlings which were less vulnerable to predation (Case, 1978). Although mammalian levels of encephalization were only just attained prior to their extinction, a general increase in relative brain size is apparent (Hopson, 1977; Russell, in press). A trend toward endothermy (physiological stamina) is evidenced by the derivation of birds from small theropods (Ostrom, 1976b; McGowan, 1979). Morphological evolution is not nearly as well documented in dinosaurs as it is in mammals. However, long-limbed gracile ornithomimids were derived from short-limbed animals like *Elaphrosaurus* in 65 million years (cf. Janensch, 1925, pl. 1; Russell, 1972, figure 4). The same trend is evidenced by a comparison of *Allosaurus* (Madsen, 1976, figure 6) with tyrannosaurids (Russell, 1970, fig. 8). Crudely elephant-like dental mills evolved in hadrosaurs over a span of 40 million years (Rozhdestvensky, 1977). One is left with the impression that morphological differentiation proceeded somewhat more slowly in dinosaurian lineages than in mammalian lineages.

Delimiting the Dinosaurian Era

The Mesozoic may be thought of as the interval in earth history which followed the population of the land by large, thoroughly terrestrial animals, but preceded the dominance of the endotherms (for a dissenting view see Bakker, 1975). It is bracketed biostratigraphically by profound extinction events (cf. Anderson and Cruikshank, 1978, chart 3; Raup, 1979; Russell, 1979). The reign of the dinosaurs did not coincide with Mesozoic time. Dinosaurs were derived from unknown thecodont ancestors late in the Triassic and did not begin to dominate terrestrial faunas until Early Jurassic time (Cox, 1976; Olsen and Galton, 1977). It has been suggested (Robinson, 1971) that dinosaurs, and the somewhat less-advanced thecodonts, were able to displace large therapsids in warming terrestrial environments because of their ability to conserve water by excreting uric acid instead of urea, and their greater tolerance of heat. Therapsids simultaneously gave rise to mammals, which remained ecologically separated from dinosaurs in productive, closed environments until the dinosaurs became extinct (cf. Henkel and Krebs, 1977; Béland and Russell, 1978).

The dinosaurs dominated the land for 130 million years, or approximately twice the duration of subsequent geologic time. Had they not vanished in the extinction event that brought the Mesozoic to a close, how would the resulting development of tetrapod faunas have differed? Such speculation might be sustained by a review of trends observed during the period of dinosaurian ascendancy. The more pervasive bipedality characteristic of dinosaurs would have continued. Diversity would have increased in harmony with continental dispersal and climatic differentiation. Because the morphological radiation of dinosaurs was already so great at the close of the Cretaceous, the number of distinct morphotypes would probably have exceeded those of living terrestrial mammals. As during Jurassic and Cretaceous time, the development of cursorial mammals and birds would probably have been suppressed.

However, dinosaurian body size may have continued to decline because of increased vegetational diversity and a progressive acquisition of endothermy. Feather-like, insulating structures would have covered the bodies of the smaller forms. Encephalization trends established during the Mesozoic would also have continued, culminating in creatures of humanoid brain weight—body weight proportions by perhaps several millions of years ago (Russell, in press). The absence of facial musculature would have been compensated by other behavioral complexities in intraspecific communication. Thus the actors would have been different, the drama would have been moved forward in time but the theme would have probably been the same: the movement of metazoans toward intelligence from a phase of medieval gigantism.

The Broader Implications

After perusing this essay, one might ask how does a knowledge of dinosaurs enrich our lives? Awareness of dinosaurs is widespread, as evidenced by a multi-million dollar market in the United States and Canada for children's books on dinosaurs, often of indifferent quality. Basic research on dinosaurs is only funded to the extent of about $150,000 annually in both countries combined.

Dinosaurs do appeal to a wide range of interests, and can be approached in the context of present-day problems. This can be illustrated by estimating the cost of Flintstone brontoburgers. Cows grow 1.16 times faster than 20-ton sauropods (equations 4, 15) and consume 0.27 times as much food when both are one-quarter grown (equations 7, 14; corrected for the increased assimilation efficiency of the cow). Brontoburger would then cost about four times as much as hamburger, given similar overhead expenses. Another example may lie in the fact that dinosaur bone can contain as much as 1% uranium by weight (R. T. Bell, personal communication). The fossilized skeleton of our 20-ton animal might yield as much as 60 kg of uranium (3-ton skeleton, equation $12 \cdot 2$ fossilization factor $\cdot 0.01$) and 420 gms of U^{235}. Two such skeletons could produce enough U^{235} for one nuclear bomb or to illuminate a city of three hundred thousand for about seven hours (D. S. Russell, personal communication).

Most will agree, however, that the main contribution of dinosaurs to mankind lies in a certain philosophical perspective which can be a useful guide in preparing ourselves for the future. Their evolution appears to suggest a repeatability or directionality in the evolution of large terrestrial organisms. The catastrophe which overtook them was far more severe than the perturbations we are presently introducing into the biosphere (Russell, 1976, 1979), and the biosphere subsequently recovered. Had the small, relatively highly encephalized theropods not become extinct, the evolution of primates might have been preempted and we may never have appeared.

The context of the old Mesozoic is worth probing further. To what extent have evolutionary rates increased since then? Have they been offset by a less than ideal or deteriorating planetary environment? What are the implications for organic evolution elsewhere in the cosmos? Will not the insights gained from the record of the biosphere through immense spans of time reveal a unique dimension to mankind's newly acquired ability to extend the biosphere into space (Billingham, et al., 1979)? An appreciation of the dinosaurian world lends a certain fullness to one's appreciation of life. It is comforting to reflect, upon returning from a walk through a darkened winter neighborhood, that we do not know all we are given to know about dinosaurs.

Acknowledgments

I am deeply grateful to Dr. Edwin H. Colbert for the patience and consideration with which he introduced me to the study of Mesozoic vertebrates. It has proven to be a multifaceted and engrossing vocation. I have also benefited greatly from numerous conversations with Dr. Pierre Béland (Paleobiology Division, National Museums of Canada), shared in widely separated parts of the globe over the past several years, concerning the general subject matter of this presentation. I am grateful to Dr. Luis W. Alvarez (Lawrence Berkeley Laboratory), Dr. Richard T. Bell (Geological Survey of Canada), and Mr. Douglas S. Russell (National Research Council of Canada) for information on the occurrence and possible economic significance of uranium in dinosaur bone. Tina Matiisen (Library, National Museums of Canada) provided data relating to North American sales of children's books on dinosaurs. Drs. Béland, James O. Farlow (Department of Geology, Hope College), C. R. Harington (Paleobiology Division, National Museums of Canada), and Brian K. McNab (Department of Zoology, University of Florida) generously reviewed the manuscript. I must, however, accept responsibility for all errors of fact and judgment.

Literature Cited

Alexander, R. M. 1976. Estimates of speeds of dinosaurs. Nature, 261:129–30.
Anderson, J. F.; Rahn, H.; and Prange, H. D. 1979. Scaling of supportive tissue mass. Quart. Rev. Biol., 54:139–48.
Anderson, J. M., and Cruickshank, A. R. I. 1978. The biostratigraphy of the Permian and the Triassic: Part 5. A review of the classification and distribution of Permo-Traissic tetrapods. Palaeontol. Afr., 21:15–44.

Bakker, R. T. 1968. The superiority of dinosaurs. Discovery, 3:11–22.
————. 1975. Experimental and fossil evidence for the evolution of tetrapod energetics. *In* D. M. Gates and R. B. Schmerl (eds.), Perspectives of biophysical ecology, Ecological Studies 12. Springer-Verlag, New York, 609 pp.
————. 1978. Dinosaur feeding behaviour and the origin of flowering plants. Nature, 274: 661–63.
Barsbold, R. 1976. New data on *Therizinosaurus* (Therizinosauridae, Theropoda). Soviet-Mongolian Paleontological Expedition Transactions, 3:76–92. (in Russian).
Béland, P., and Russell, D. A. 1978. Paleoecology of Dinosaur Provincial Park (Cretaceous), Alberta, interpreted from the distribution of articulated vertebrate remains. Can. J. Earth Sci., 15:1012–24.
Béland, P., and Russell, D. A. 1980. Dinosaur metabolism and predator-prey ratios in the fossil record. American Association for the Advancement of Science Selected Symposium 28:85–102.
Bennett, A. F., and Dawson, W. R. 1976. Metabolism. *In* C. Gans (ed.), Biology of the reptilia, Vol. 5. Academic Press, London, 556 pp.
Bennett, A. F., and Ruben, J. A. 1979. Endothermy and activity in vertebrates. Science, 206: 649–54.
Billingham, J.; Gilbreath, W.; and O'Leary B. (editors). 1979. Space resources and space settlements. National Aeronautics and Space Administration, Special Publication 428. 288 pp.
Bird, R. T. 1944. Did brontosaurus ever walk on land? Natur. Hist., 53:60–67.
Bonaparte, J. F. 1978. El Mesozoico de America del Sur y sus tetropodes. Opera Lilloana 26, Tucuman. 596 pp.
Bonaparte, J. F., and Powell, J. E. 1980. A continental assemblage of tetrapods from the Upper Cretaceous beds of El Brete, north-western Argentina (Sauropoda—Coeluro-sauria—Carnosauria—Aves). Société géologique de France 59 (139):19–27.
Borsuk-Bialynicka, M. 1977. A new camarasaurid sauropod *Opisthocoelicaudia skarzynskii* gen. n., sp. n. from the Upper Cretaceous of Mongolia. Palaeontol. Polonica, 37:5–64.
Brett-Surman, M. K. 1979. Phylogeny and palaeobiogeography of hadrosaurian dinosaurs. Nature, 277:560–62.
Budyko, M. I. 1977. Climatic change. Waverly Press, Baltimore, 261 pp.
Case, T. J. 1978. Speculations on the growth rate and reproduction of some dinosaurs. Paleobiology, 4:320–28.
Coe, M. 1978. The decomposition of elephant carcasses in the Tsavo (East) National Park, Kenya. Journal of Arid Environments, 1:71–86.
Cox, C. B. 1976. Mysteries of early dinosaur evolution. Nature, 264:314.
Delany, M. J., and Happold, D. C. D. 1979. Ecology of African mammals. Longman, New York, 434 pp.
Douglas-Hamilton, I., and Douglas-Hamilton, O. 1975. Among the elephants. Collins and Harvill Press, London, 285 pp.
Dutuit, J. -M., and Ouazzou, A. 1980. Découverte d'une piste de dinosaures sauropodes sur le site d'empreintes de Demnate (Haut-Atlas marocain). Société géologique de France 59 (139):95–102.
Ehrenfeld, D. W. 1974. Conserving the edible sea turtle: can mariculture help? Amer. Sci., 62:23–31.
Farlow, J. O. 1976. A consideration of the trophic dynamics of a Late Cretaceous large-dinosaur community (Oldman Formation). Ecology, 57:841–57.
Fedak, M. A., and Seekerman, H. J. 1979. Reappraisal of energetics of locomotion shows identical cost in bipeds and quadrupeds including ostrich and horse. Nature, 282:713–16.
Galton, P. M., and Jensen, J. A. 1979. A new large theropod dinosaur from the Upper Jurassic of Colorado. Brigham Young Univ. Geol. Studies, 26:1–12.
Garrels, R. M.; Lerman, A.; and Mackenzie, F. T. 1976. Controls of atmospheric O_2 and CO_2: past, present and future. Amer. Sci., 64:306–15.
Gartner, S. and McGuirk, J. P. 1979. Terminal Cretaceous extinction scenario for a catastrophe. Science, 206:1272–76.

Ginsburg, L.; Lapparent, A. F. de; Loiret, B.; and Taquet, P. 1966. Empreintes de pas de vertébrés tétrapodes dans les séries continentales à l'ouest d'Agadès (Republique du Niger). Acad. Sci., C. R., Sér. D, 263:28–31.

Harestad, A. S., and Bunnell, F. L. 1979. Home range and body weight—a reevaluation. Ecology, 60:389–402.

Heatwole, H. 1976. Reptile ecology. University of Queensland Press, St. Lucia, 178 pp.

Henkel, S., and Krebs, B. 1977. Der erste Fund eines Säugetier-Skelettes aus der Jura-Zeit. Umsch. Wiss. Tech. 77:217–18.

Hopson, J. A. 1977. Relative brain size and behavior in archosaurian reptiles. Ann. Rev. Ecol. Syst., 8:429–48.

Horner, J. R., and Makela, R. 1979. Nest of juveniles provides evidence of family structure among dinosaurs. Nature, 282:296–98.

Janensch, W. 1925. Die coelurosaurier und Theropoden der Tendaguru-Schichten Deutsch Ostafrikas. Palaeontographica, Supplement 7, Erste Reihe, 1:1–99.

———. 1929. Magensteine bei Sauropoden der Tendaguru-Schichten. Palaeontographica, Supplement 7, Erste Reihe, 2:135–44.

Jeletzky, J. A. 1978. Causes of Cretaceous oscillations of sea level in western and arctic Canada and some general geotectonic implications. Can., Geol. Surv., Paper, 77-18. 44 pp.

Jensen, J. A. 1979. Ultrasaurus. Science News, 116:84.

Knoll, A. H.; Niklas, K. J.; and Tiffney, B. H. 1979. Phanerozoic land-plant diversity in North America. Science, 206:1400–1402.

Langston, W., Jr. 1974. Nonmammalian Comanchean tetrapods. Geoscience and Man, 8: 77–102.

Lapparent, A. F. de, and Zbyszewski, G. 1957. Les dinosauriens du Portugal. Services Géologiques du Portugal, Mémoire 2. 63 pp.

Madsen, J. H. 1976. *Allosaurus fragilis:* a revised osteology. Utah Department of Natural Resources, Bulletin 109. 163 pp.

McGowan, C. 1979. Selection pressure for high body temperatures: implications for dinosaurs. Paleobiology, 5:285–95.

McNab, B. K. 1974. The energetics of endotherms. Ohio J. Sci., 74:370–80.

———, and Auffenberg, W. 1976. The effect of large body size on the temperature regulation of the Komodo dragon, *Varanus komodoensis.* Comparative Biochemical Physiology, 55A:345–50.

Neill, W. T. 1971. The last of the ruling reptiles. Columbia University Press, New York, 486 pp.

Olsen, P. E., and Galton, P. M. 1977. Triassic-Jurassic tetrapod extinctions: are they real? Science, 197:983–86.

Ostrom, J. H. 1976a. On a new specimen of the lower Cretaceous theropod dinosaur *Deinonychus antirrhopus.* Breviora, 439:1–21.

———. 1976b. *Archaeopteryx* and the origin of birds. Linn. Soc., Biol. J., 8:91–182.

———. 1978. The osteology of *Compsognathus longipes* Wagner. Bayer. Staatssamml. Paläontol. Hist. Geol., Abh., 4:73–118.

Perkins, B. F., and Stewart, C. L. 1971. Stop 7: Dinosaur Valley State Park. Louisiana State Univ., Misc. Publ., 71-1:56–59.

Perle, A. 1979. Segnosauridae—a new family of theropods from the Late Cretaceous of Mongolia. Joint Soviet-Mongolian Paleontological Expedition Transactions, 8:45–55. (in Russian).

Prange, H. D., and Jackson, D. C. 1976. Ventilation, gas exchange and metabolic scaling of a sea turtle. Respiration Physiology, 27:369–77.

Raup, D. M. 1979. Size of the Permo-Triassic bottleneck and its evolutionary implications. Science, 206:217–218.

Raven, P. H., and Axelrod, D. I. 1972. Plate tectonics and Australasian paleobiogeography. Science, 176:1379–86.

Robinson, P. L. 1971. A problem of faunal replacement on Permo-Triassic continents. Palaeontology, 14:131–53.

Rozhdestvensky, A. K. 1977. The study of dinosaurs in Asia. J. Palaeontol. Soc. India, 20:-102–19.

Russell, D. A. 1970. Tyrannosaurs from the Late Cretaceous of western Canada. National Museums of Canada Publications in Palaeontology 1. 34 pp.

———. 1972. Ostrich dinosaurs from the Late Cretaceous of western Canada. Can. J. Earth Sci., 9:375–402.

———. 1976. Mass extinctions of dinosaurs and mammals. Nature Canada, 5:18–24.

———. 1979. The enigma of the extinction of the dinosaurs. Ann. Rev. Earth Planet. Sci., 7:163–82.

———. In press. Speculations on the evolution of intelligence in multicellular organisms. National Aeronautics and Space Administration, Special Publication.

Russell, D. A.; Béland, P.; and McIntosh, J. S. 1980. Paleoecology of the dinosaurs of Tendaguru (Tanzania). Mémoires de la Société géologique de France, N.S., 59 (139):169–75.

Salop, L. I. 1978. Relationship of glaciations and rapid changes in organic life to events in outer space. Int. Geol. Rev. 19:1271–91.

Seymour, R. S. 1979. Dinosaur eggs: gas conductance through the shell, water loss during incubation and clutch size. Paleobiology, 5:1–11.

Sikes, S. K. 1971. The natural history of the African elephant. Weidenfeld and Nicholson, London, 397 pp.

Steiner, J. 1977. An expanding Earth on the basis of sea-floor spreading and subduction rates. Geology, 5:313–18 (see also discussion in Geology, 6:377–83).

Taquet, P. 1976. Géologie et paléontologie du gisement de Gadoufaoua (Aptien du Niger). Centre National de la Recherche Scientifique, Cahiers de Paléontologie. 191 pp.

Thulborn, R. A. 1973. Thermoregulation in dinosaurs. Nature, 245:51–52.

Wetmore, A. 1967. Re-creating Madagascar's giant extinct bird. Nat. Geogr. Mag., 132:488–93.

Wheeler, P. E. 1978. Elaborate CNS cooling structures in large dinosaurs. Nature, 275:441–43.

THE MORPHOLOGY, AFFINITIES, AND AGE OF THE DICYNODONT REPTILE *GEIKIA ELGINENSIS*

by Timothy Rowe

Introduction

In 1886 R. H. Traquair reported the discovery of a dicynodont, tentatively identified as *Dicynodon* sp., from Cuttie's Hillock, near Elgin in northern Scotland (Figure 16.1). This was the first discovery of therapsid reptile remains from western Europe. During the next six years several more specimens were collected from this locality. The first detailed description of this material was by Newton (1893), who described two new dicynodont genera, *Geikia* and *Gordonia,* and a new pareiasaur, *Elginia.* Traquair's specimen was referred to *Gordonia.*

Traquair (1886), Smith Woodward and Sherborn (1890), and Smith Woodward (1898) tentatively considered the Cuttie's Hillock fauna to be of Late Triassic age. Newton (1893) believed that it correlates with Upper Permian or Lower Triassic faunas of India and South Africa. Most later authors (Watson, 1909; Watson and Hickling, 1914; von Huene, 1913, 1940; Westoll, 1951; Romer, 1966) have considered it to be of Late Permian age. Walker (1973), however, believed that the Cuttie's Hillock fauna correlates with a position in the South African Beaufort Series lying between a level high in the *Cistecephalus* Zone (equivalent to the *Daptocephalus* Zone of most recent usage, Kitching, 1970, 1977) to one low in the *Lystrosaurus* Zone, i.e., a position between levels closely bracketing the Permo-Triassic boundary. Walker pointed out that the Permo-Triassic boundary within these almost entirely nonmarine sediments is difficult to correlate with the type marine sections, and consequently the dating of the Cuttie's Hillock fauna is tentative and likely to remain so. An additional problem is that the Cuttie's Hillock fauna consists of only three genera, whose relationships to members of better known faunas is currently unclear, making precise correlation difficult.

At the time of discovery of the Cuttie's Hillock fauna, therapsids of Late Permian to Early Triassic age were known only from South Africa and India. Since then, therapsids of this age have been reported from East Africa, South America, China, Indochina, U.S.S.R., Antarctica; many new forms have also been described from South Africa and India. Despite the many discoveries elsewhere, *Geikia* and *Gordonia* remain the only known therapsids of Late Permian or Early Triassic age from western Europe.

The type-species of *Geikia, G. elginensis* Newton (1893), is known from a single specimen. As will be seen, *Dicynodon locusticeps* von Huene (1942), also known from only one specimen, is here referred as a valid species of *Geikia, G. locusticeps* (see ''Affinities of *Geikia*'' for discussion). The purpose of this paper is to redescribe

Figure 16.1. Location of Cuttie's Hillock.

Geikia elginensis and to reassess its affinities in light of recent developments in our understanding of dicynodont systematics.

Systematics

Class: Reptilia
  Order: Therapsida
    Suborder: Anomodontia (*sensu* Romer, 1966)
      Infraorder: Dicynodontia (*sensu* Romer, 1966)
        Family: Cryptodontidae (*sensu* Toerien, 1953; Haughton and Brink, 1954)
          Genus: *Geikia* Newton (1893)

Generic Diagnosis: Dicynodonts having no tusk or postcanine teeth; highly vaulted palate; anterior palatal ridges of premaxilla reduced or absent; large palatine having rugose palatal surface; palatine having extensive contact with maxilla and premaxilla; length of interpterygoidal vacuity not less than half the length of the interpterygoidal fossa; interpterygoidal vacuity lying entirely within roof of interpterygoidal fossa; well developed maxillary caniniform process having pronounced lateral ridge; sharp occlusal margin of beak; sharp ridge or "keel" developed on ventral edge of maxilla behind caniniform process; septomaxilla having exposure on lateral surface of snout behind external nares; anterior surface of premaxilla flat, oriented vertically, and meeting lateral surface of premaxilla in abrupt "corner"; single, prominent preorbital protuberance.

Type-Species: *Geikia elginensis* Newton (1893)

Holotype: Natural mold of single nearly complete skull and mandible, associated left humerus, and one isolated metapodial or proximal phalanx. The block containing the natural mold is broken into several pieces housed in the Institute of Geological Sciences, London, G.S.M. 90998–91015.

Locality: Cuttie's Hillock, near Elgin in northern Scotland.

Horizon: Upper Permian, Cuttie's Hillock Sandstone; equivalent to *Daptocephalus* Zone of Beaufort Series, South Africa.

Specific Diagnosis: Interorbital region approximately twice as wide as intertemporal region; width across temporal arches greater than sagittal length of skull; orbit expanded such that its greatest horizontal length is more than one-third of the sagittal length of the skull; very prominent preorbital protuberance expanding anteriorly to lie in front of the vertical anterior surface of the premaxilla.

Referred Species: *Dicynodon locusticeps* von Huene 1942, p. 159; *Geikia locusticeps* new combination.

Holotype: Single skull lacking tip of premaxilla, right quadrate, left temporal arch, and mandible; Institut für Geologie und Paläontologie, Tübingen, K. 87.

Locality: Near the village of Kingori, Songea District, Tanzania (see Nowack, 1937, Figure 5, for detailed description of locality).

Horizon: Upper Permian, Kawinga Formation (Charig, 1963); equivalent to *Daptocephalus* Zone of Beaufort Series, South Africa (Haughton, 1932).

Specific Diagnosis: Intertemporal width and interorbital width approximately equal; width across temporal arches less than sagittal length of skull; greatest horizontal length of orbit approximately one-quarter of the sagittal length of skull; anterior surface of premaxilla extends forward beyond anterior surface of preorbital protuberance.

Material

Geikia elginensis is known from a single specimen that occurs as a natural mold in a pebbly sandstone. The block containing the mold is broken into several pieces, but by reuniting various combinations of pieces most parts of the specimen have been cast. The specimen consists of a nearly complete skull, mandible, left humerus, and an isolated metapodial or proximal phalanx. The skull is of an unusual shape which tends to give one the impression that it may have been severely distorted. However, comparing the right and left sides of the specimen shows evidence of only very minor asymmetrical distortion. Natural openings in the skull such as the foramen magnum and the interpterygoidal vacuity are symmetrical about the sagittal plane and show no evidence of crushing. The right and left orbits, temporal fenestrae, and temporal arches are of nearly identical shapes and show evidence of only minor distortion. It seems quite unlikely that distortion has contributed significantly to the shape of this specimen. The specimen was described from polyvinylchloride casts made by Dr. Alick Walker of the University of Newcastle-upon-Tyne. They generally show well preserved bone in fine detail, but in a number of small localized areas the bone left no impression or only a vague impression on the natural mold and detail is lost.

Geikia locusticeps is known from a single skull lacking the tip of the premaxilla, right quadrate, left temporal arch, and mandible. Preservation of the bone is good, although some of the bone surface appears to have been ground away during preparation.

Description of *Geikia elginensis*

The skull of *Geikia elginensis* is unusual in shape compared to most other dicynodonts. *Geikia* is a medium-sized dicynodont, its skull length being about 110 mm. The greatest width of the skull, between the temporal arches, is slightly greater than its sagittal length. The skull is quite deep, its height from the bottom of the quadrate to the top of the parietals is about 67 mm. The orbits are very large, the interorbital region broad, and pronounced preorbital protuberances extend far anterolaterally from the anterodorsal "corners" of the orbits. The extreme degree of development of the protuberances is greater than in any other dicynodont and is one of the most striking features of the skull. The snout is abruptly terminated and down-turned such that the orbits lie very near the front of the skull. The beak is square, delicately built with a sharp slender maxillary caniniform process, and lies almost directly below the orbits. The overall impression is of a rather delicately built, box-like skull with a large deep temporal region, very large orbits, a flat vertical face, and a relatively small beak and mouth.

Skull Roof (Figure 16.2)

The entire skull roof is preserved but some of its detail has been lost, and a seam in the mold has obscured the pineal region. The interorbital region is quite broad, being slightly more than twice as wide as the intertemporal region. The interorbital region, measured along the midline from the front of the vertical anterior surface of the premaxilla to the back of the postorbital bar, is nearly twice as long as the intertemporal region, measured from the back of the postorbital bar to the occipital plate. The dorsal edges of the orbits are raised some 20 mm above the

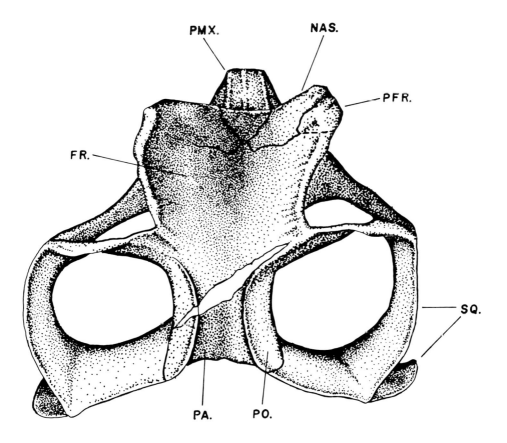

Figure 16.2. *Geikia elginensis,* dorsal view. × 0.85.

midline of the interorbital region, creating a deep trough in the skull roof between the orbits.

The premaxilla extends onto the skull roof as a triangular process. The postero-lateral edges of this triangle meet the nasals. Sutures in this region are not clear, but the nasals do not appear to meet along the midline and the tip of the premaxilla has a short contact with the frontals. Walker (1973, figure 3b) figured the skull roof of *Geikia elginensis* with the nasals meeting along the midline, thus separating the premaxilla from the frontals, but I believe this interpretation to be incorrect.

The nasals have considerable exposure on the anterior border of the skull roof. They fuse with the prefrontals to form a very prominent protuberance, or "boss," which extends obliquely far laterally in front of the orbit. The distal surface of the protuberance overhangs the external nares by nearly 25 mm anterolaterally, and lies 12 mm in front of the vertical anterior surface of the premaxilla. The thickened distal end of the protuberance has a rugose surface which appears to be perforated by several small nutrient foramina. More medially, above the external nares, the bone becomes thin and flat. The preorbital protuberance is one of the most striking features of the skull. A smaller protuberance (or several smaller protuberances) is a common feature among dicynodonts, but the prominent degree to which it is developed in *Geikia elginensis* is unique.

The prefrontal is a small bone confined to the posterior part of the preorbital protuberance and the dorsal border of the orbit. The nasals meet the frontals in a nearly straight transverse suture, similar to the condition found in *Pelanomodon* (von Huene, 1942, plate 28-2). Whether postfrontals and a preparietal are present is uncertain, since no sutures are visible at the back of the interorbital region. The pineal foramen is not visible but this is owing to a seam in the mold that cuts across this part of the parietal region. The parietals join the occipital plate at an angle of nearly 90 degrees, forming an abrupt corner which clearly defines the back of the skull roof. The postorbitals lie against the parietals and extend along the full length of the intertemporal bar, forming the medial borders of the temporal fenestrae. The posterior end of the postorbital overlies the squamosal at the back of the temporal fenestra, and the squamosal and postorbital together project backwards to overhang the occipital plate.

Temporal Region (Figure 16.3)

Little of the structure of the temporal region is visible. The lateral aspect of the braincase is almost completely obscured and only a small portion of the epipterygoid and the lateral part of the prootic can be seen. The epipterygoid slants slightly anteriorly as it rises towards the skull roof, but neither end of the bone is visible. The prootic extends laterally to meet the ventral flange of the squamosal and, dorsolaterally, it meets the supraoccipital in a sutural contact. A wide, shallow, indistinct groove lies on the anterior surface of the prootic, in approximately the same position as a groove in *Kingoria,* interpreted by Cox (1959) as marking the course of the *vena capitis dorsalis.*

The temporal fossa is partially roofed by a narrow lateral extension of the parietal and postorbital. The temporal arches expand far laterally so that the transverse width across the arches exceeds the sagittal length of the skull (beak tip to occiput). In dorsal view the temporal fenestrae are almost rectangular in outline, being slightly wider than long.

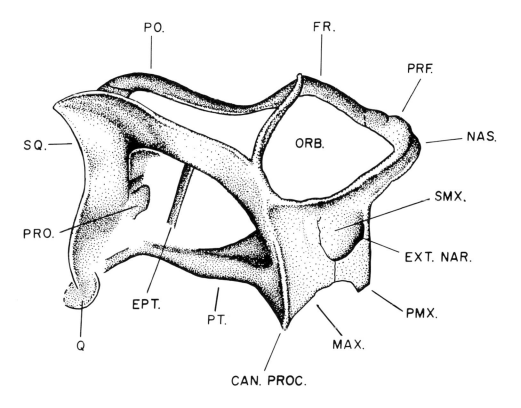

Figure 16.3. *Geikia elginensis,* lateral view, composite illustration from right and left sides. × 0.87.

Snout (Figure 16.3)

The snout is nearly complete and is generally well-preserved, although few of the facial sutures are visible. The occlusal margin of the beak is formed by the maxilla and premaxilla. These bones meet in an indistinct suture that extends vertically from the occlusal margin to the bottom of the external nares. The bone in this region is thin and tapers ventrally to form a sharp cutting edge. The edge remains sharp from the tip of the beak backwards to the caniniform process of the maxilla, and continues behind the caniniform process as a sharp ridge or "post-caniniform keel." The edge of this keel extends posterodorsally onto the ventral edge of the pterygoid bone. In lateral view the occlusal margin forms a smooth uninterrupted curve which rises from the tip of the caniniform process up to the premaxilla. The anteriormost part of the premaxilla curves down and backwards, giving the beak a hooked appearance in lateral view.

The caniniform process of the maxilla is quite slender and is without a tusk. On its lateral surface is a narrow but prominent curved ridge which extends from the tip of the process onto the base of the zygoma below the orbit. A similar condition is seen in *Geikia locusticeps* (von Huene, 1942, figure 3a; Walker, 1973, figure 3d). Newton (1893) pointed out that the absence of a tusk in *Geikia elginensis* might be an indication of immaturity. Newton's comment could also apply to the single known specimen of *Geikia locusticeps*. However, the delicate structure of the caniniform process, together with the rather large size of both skulls (beak tip to occiput = 110 mm in *G. elginensis;* 150 mm in *G. locusticeps*), suggests that the absence of a tusk is an adult character state in *Geikia*. In juvenile specimens of *Lystrosaurus* with unerupted tusks, the caniniform process is robust, clearly foreshadowing the development of tusks. Furthermore, the caniniform process in both species of *Geikia* bears a detailed resemblance to those found in *Oudenodon* (Broili and Schroeder, 1936, figure 1-5) and *Pelanomodon* (von Huene, 1942, plate 28-1), genera in which all known specimens are without tusks. Because the two species of *Geikia* are each known from only a single specimen, the possibility that the absence of tusks is a sexual character cannot be ruled out altogether. Nonetheless, comparison with other dicynodonts strongly suggests that the absence of tusks in *Geikia* is an adult character state, probably common to both sexes.

The front of the snout forms a broad, flat, vertically oriented surface, with a low median ridge extending from the beak tip up to about the level of the external nares. The lateral surface of the snout is also flat and joins the anterior surface at a 60 degree angle. The junction between the flat lateral and anterior surfaces forms a prominent corner which extends vertically from the beak margin up to the level of the top of the nares, where the frontoparietal plane of the skull roof meets the premaxillary plane. The snout thus appears to be box-like, abruptly terminated, and downturned very much like, although not as deep as, the snout of *Lystrosaurus.*

The external nares are located on the lateral surface of the snout, approximately midway between the ventral margin of the orbit and the cutting edge of the beak. Sutures between the bones which form this region are not clearly visible, and difficulty in distinguishing between bone and artifacts in the cast make a detailed interpretation of this region difficult. Judging from the condition found in other dicynodonts, the roof of the narial aperture is formed by the nasal, and the anterior wall and anterior part of the floor are formed by the premaxilla (Cluver, 1971). The posterior part of the floor and the rear wall are formed by the septomaxilla.

On the right side of the specimen the maxilla can be seen overlapping the septo-maxilla behind the nares, but only a small portion of their contact is visible and the precise geometry of the septomaxilla cannot be determined. No sutures are visible on the left side. Posteriorly, the septomaxilla lies exposed on the lateral surface of the snout in a plane with the maxilla, while anteriorly it gradually slopes medially into the narial aperture without forming a distinct posterior narial border. Newton's (1893) figure is therefore incorrect, as he illustrated a distinct posterior narial border. A similar condition to that of *Geikia* is found in *Pelanomodon* (von Huene, 1942, plate 28-1). Whether a septomaxillary foramen is present is uncertain.

The bone of the surface of the maxilla along the occlusal margin of the beak and on the caniniform process, over most of the premaxilla, and on the preorbital protuberances has a rugose, pitted texture. Some of the pitting, particularly on the preorbital protuberances, appears to be the result of perforation of the bone by small nutrient foramina. It has long been recognized that the occlusal margin of the beak in dicynodonts was covered by horn, and that this horny covering extended onto the face (e.g., Watson 1948). The precise extent of the covering varies among species. The distribution of the rugose punctate surface texture on the snout of *Geikia elginensis* indicates that horn covered much of the caniniform process, occlusal margin of the beak, and extended up the premaxilla onto the preorbital protuberance of the skull roof.

The orbit is quite large. Its greatest length, measured along a horizontal line in a parasagittal plane, is 40 mm, which is slightly more than one-third the length of the skull. The greatest vertical height of the orbit is 34 mm. When compared to *G. locusticeps,* much of the orbital expansion appears to be the result of the extreme anterior development of the preorbital protuberances.

Palate and Basicranium (Figure 16.4)

The entire palate and the basicranium back almost to the basioccipital tuberosities are preserved. Only a small part of the premaxillomaxillary suture is visible, and only parts of the borders of the palatine bones can be seen clearly. However, sufficient detail is preserved to permit the general shapes of these bones to be determined. The palate forms a highly arched vault bounded laterally and anteriorly by the sharp occlusal rim of the beak. There are no teeth on the palate. The palatal surface of the anterior half of the premaxilla curves strongly anteroventrally to form a nearly vertical wall at the front of the palate. The premaxilla bears the two anterior palatal ridges that are present in many dicynodont genera (e.g., *Pelanomodon*, Figure 16.8), but they are much reduced, being very short, narrow, and of low relief. A pronounced median ridge is developed on the rear half of the premaxilla in the usual dicynodont fashion. Its anterior end lies midway between the caniniform process and the tip of the beak. Its posterior end meets the vomer at the back of the palate, but no suture is visible.

The maxillae slant steeply ventrolaterally, forming the steep lateral walls of the palate. The back of the palate is formed by the palatine bones, which meet the maxilla and premaxilla in a broad sutural contact. The palatal surface of the palatine slants posteroventrally, curving down and away from the horizontal surface of the posterior half of the premaxilla. The bone on the surface of the palatine has a rugose texture and is perforated by small nutrient foramina. This texture appears

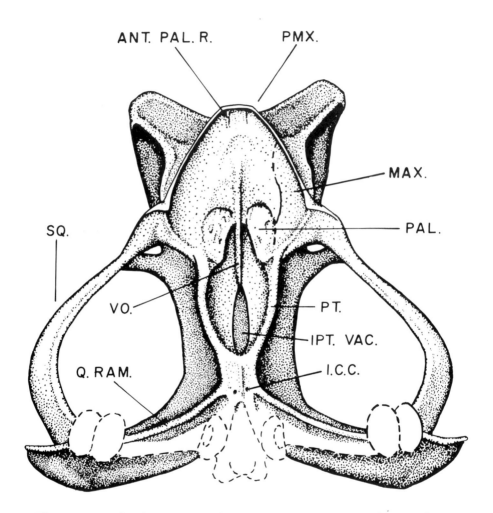

Figure 16.4. *Geikia elginensis,* ventral view, palate and basicranium illustrated from one cast; the remainder of the shaded portion is a composite illustration from several casts; dashed lines surrounding unshaded areas are restorations of unpreserved regions. × 0.96.

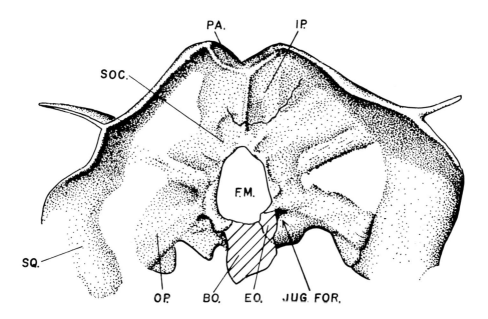

Figure 16.5. *Geikia elginensis,* occipital view. × 0.90.

to extend onto the adjacent parts of the maxilla, but, as no nutrient foramina are visible here, its presence on the maxilla is probably an artifact.

The premaxilla forms the anterior border of the choanae and the palatine forms the anterolateral and lateral borders. The choanae are divided by the vomer, which forms a deep thin septum descending from the roof of the interpterygoidal fossa. The vomer splits posteriorly, forming the anterior and lateral borders of the interpterygoidal vacuity. The interpterygoidal fossa is deep, and the interpterygoidal vacuity lies entirely within its roof. The vacuity is narrow and is two-thirds as long as the fossa. The pterygoids, which form the lateral walls of the fossa, are lightly built and diverge from each other only slightly as they extend forward toward the maxillae. It is not clear whether an ectopterygoid bone is present. The labial fossa is absent, as is the case in most other cryptodontids.

Behind the interpterygoidal fossa, the pterygoids fuse to form a narrow plate with a low median ridge. Posterior to this plate the structure of the basicranium becomes obscure. The basioccipital tuberosities appear to be quite prominent and extend far ventrally, but their distal ends are not preserved. Neither stapes is present.

The quadrate ramus of the pterygoid is a deep, thin bar which extends postero-laterally from the posterior part of the pterygoid plate towards the quadrate. The contact between the quadrate ramus and the quadrate is not visible on the right side, where a quadrate is present. On the left side, where the quadrate was lost postmortem, the quadrate ramus lies against the anterior surface of the ventral part of the squamosal flange, extending a short distance into a smooth, shallow, oblong depression that held the quadrate. The right quadrate is in place but only its lateral surface is visible; it appears to conform to the usual dicynodont pattern as described by Watson (1948) and Cluver (1971).

Occiput (Figure 16.5)

The occiput is nearly complete but in localized areas detail was not preserved. The impressions of the bone that are preserved indicate good preservation and considerable detail, but most of the left and ventral parts of the occiput have left only vague impressions on the mold. The squamosals project backwards about 18 mm beyond the occipital plate. Dorsally, beneath the overhanging squamosals, is a large interparietal bone which forms the upper border of the occiput. None of the sutures bordering the tabular bones are visible. No sutures are visible between the supraoccipital, exoccipital, and opisthotic bones, but the preservation of this general area is quite good, suggesting that these bones are tightly fused.

Neither posttemporal fenestra is visible, as these regions left no impression on the mold. The posttemporal fenestra has been thought to be characteristic of all dicynodonts, and was therefore probably present in *Geikia*. However, an undescribed specimen in the Museum of Paleontology, University of California, Berkeley, (UCMP 42401) from Krugerskraal, South Africa (*Daptocephalus* Zone), has a completely preserved fully prepared occiput in which the posttemporal fenestra is absent. This specimen is referable to the Cryptodontidae with close affinities to *Geikia* and, therefore, a remote possibility does exist that the posttemporal fenestra was absent in *Geikia elginensis.*

The foramen magnum is quite large, its greatest vertical dimension being 20 mm and its greatest horizontal dimension 15 mm. The condylar portions of the basioccipital and exoccipital bones are missing and appear to have been broken off prior to fossilization. On either side, ventrolateral to the foramen magnum, is a rather large jugular canal which pierces the opisthotic and communicates with the floor of the braincase through the exoccipital, just inside the foramen magnum. Immediately behind this opening, the hypoglossal foramen pierces the exoccipital and courses ventrolaterally to communicate with the jugular canal. The paraoccipital process of the opisthotic expands laterally where it abuts against the squamosal, and apparently also contracts the quadrate ramus of the pterygoid, and the quadrate.

Mandible (Figure 16.6)

The mandible is not completely preserved, but by combining the right and left sides a lateral view of a complete ramus can be reconstructed, and most of the symphyseal and articular regions can be seen clearly. The mandible is preserved in a position such that the mouth is gaping open at 90 degrees, with the right articular nearly touching the right quadrate and the left articular lying close to the lower edge of the left squamosal, although the quadrate has dropped off on this side.

Anteriorly, the fused dentaries form an upturned symphysis. The tip of the symphysis is broken, but the preserved portion extends 5 mm above the dorsal edges of the dentary ramus, and its contours indicate that it may have extended another 4 mm dorsally (and possibly further). In dorsal view, a 5 mm wide concave dentary table is developed medially in the symphyseal region. The concavity of its surface extends posteriorly along the dorsal edge of the dentary as a narrow groove which almost reaches the dentary-surangular contact. The bone of the anterior and lateral surface of the symphysis has a rough, pitted texture and was probably covered by horn in life. Both the symphysis and the rami appear to be rather lightly built.

Behind the symphysis, the lateral surface of the dentary bears a low, rounded shelf. A large mandibular fenestra is present, behind which lies the large reflected lamina of the angular. Newton (1893) illustrated the reflected lamina as being divided by a horizontal line, giving the appearance of a bipartite structure with the dorsal half merging anteriorly with the body of the angular, and the separate ventral half attaching along its dorsal edge to the bottom of the ramus. This interpretation is incorrect, however, for, as in other dicynodonts, the reflected lamina is a single continuous sheet of bone. It merges anteriorly with the body of the angular, while posteriorly it forms a broad, continuous sheet with free dorsal, ventral, and posterior edges lying lateral to the ramus. The reflected lamina undulates gently, its dorsal half is convex and its ventral half is concave. The surface of the angular behind the reflected lamina is smooth and flat.

The articular is divided into two condyles, a large lateral condyle, and a relatively shallower medial condyle. The lateral condyle is divided into an anterior concave surface (condylar recess of Crompton and Hotton, 1967) and a convex posterior surface which curves back and down nearly 10 mm below the ventral edge of the ramus onto the prominent retroarticular process.

Postcranium (Figure 16.7)

An impression of a left humerus was found lying about 2 mm from the left temporal arch of the skull (Walker, written communication, 1979). The proximal articular surface is missing and most of the distal articular surface left no impression on the mold. Some damage to the entepicondyle and the ectepicondyle ap-

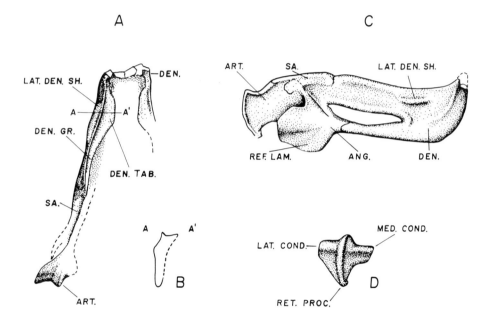

Figure 16.6. *Geikia elginensis*, mandible, A) dorsal view; B) section through A–A' of right ramus; C) composite illustration from right and left sides of lateral view; D) posterior view of left articular. × 0.63.

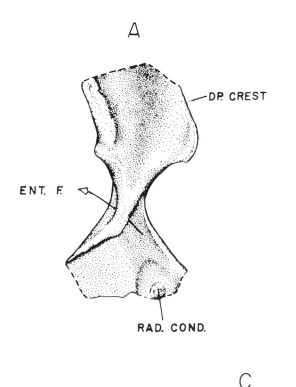

Figure 16.7. *Geikia elginensis,* postcranium, A) left humerus ventral view; B) left humerus dorsal view; C) proximal phalanx or metapodial in (?)ventral view. Dashed lines indicate margins of cast; solid lines are natural borders of bone; cross-hatched area in C is broken surface. × 0.72.

pears to have occurred prior to fossilization. The humerus is twisted such that the deltopectoral crest lies at an angle of about 50 degrees to the widest part of the distal end of the bone. The deltopectoral crest is quite broad and curves antero-ventrally, leaving the proximal ventral surface concave and the dorsal surface convex. The proximal half of the humerus is almost as wide as its estimated length, giving this part of the bone a square appearance in dorsal or ventral view. The humerus is tightly constricted at its midpoint, and expands distally until the distal end is somewhat wider than the proximal end. The radial condyle is partially preserved in ventral view; it does not appear to be strongly developed, but damage to this area leaves some question about its original configuration. A shallow depression on the distal ventral surface surrounds the radial condyle.

In addition to the humerus, the impression of an isolated, partial metapodial or proximal phalanx is also preserved near the left temporal arch (Walker, written communication, 1979). Preservation is poor and only the most general shape of the bone can be discerned. The bone appears to be broken across its shaft and only one end is preserved. This end is mediolaterally expanded, asymmetrical, and somewhat dorsoventrally thickened. The shaft appears to thicken dorsoventrally but this may be an artifact. Even if this thickening reflects its true shape, the bone is rather long and slender, and it resembles the metacarpals of *Diaelurodon* figured by Watson (1960, figure 14). It is therefore likely that the feet of *Geikia* were relatively long with slender digits, and were not the shortened, stout feet characteristic of *Placerias* and many of the larger dicynodonts (Camp, 1956).

Affinities of *Geikia*

The affinities of *Geikia* have been the subject of considerable debate. In the initial description of this genus Newton (1893) noted a marked resemblance between *Geikia* and "*Ptychognathus*" (= *Lystrosaurus*), mentioning similarities in the general structure of the occiput and interorbital region and in the abrupt angular junction between the premaxillary and frontal regions. The only features he found that distinguished the two were the weakly developed caniniform process of the maxilla and absence of a tusk in *Geikia*, and a strongly developed caniniform process and presence of a tusk in *Lystrosaurus*. Newton mentioned that these differences might merely be sexual but did not refer his specimen to *Lystrosaurus*, primarily because no tusked specimens had been found in the Elgin area.

Watson (1909) accepted Newton's opinion regarding the close relationship between *Geikia* and *Lystrosaurus*. Watson and Hickling (1914), however, departed from this view and stated that *Geikia* "is not paralleled by any other known form, but is simply derived from *Dicynodon* by the development of horns [*sic*] on the nasals" (p. 399). They also noted a general tendency among dicynodonts towards the development of nasal bosses.

Von Huene (1940) stated that the short wide skull roof of *Geikia* resembled that of *Cistecephalus*. Any resemblance between the two is, however, superficial at best, since the broadening of the cistecephalid skull roof occurs in the parietals whereas in *Geikia* the frontals have widened while the parietal region remains relatively narrow. Later, von Huene (1948, 1956) erected the monotypic family Geikiidae to accommodate *Geikia*. The apparent basis for establishing this familial distinction was his belief that *Geikia* lacked a preparietal (von Huene, 1956). His reasoning is difficult to understand since Newton (1893, p. 466) stated that none of the skull

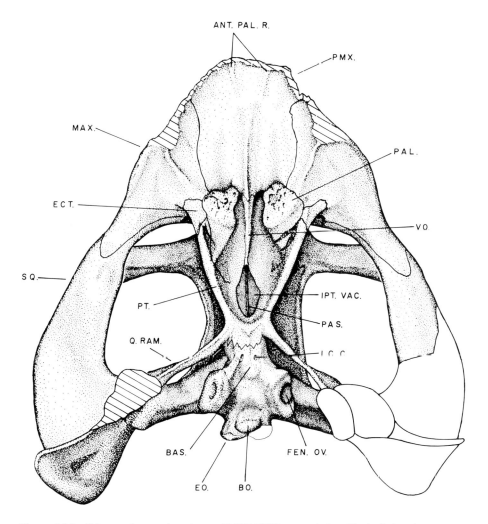

Figure 16.8. *Pelanomodon moschops* (type, AMNH 5325) ventral view. Unshaded regions are restored; cross-hatched regions are broken surfaces. × 0.54.

roof sutures are visible and the condition of the preparietal is therefore indeterminate. Von Huene believed the Geikiidae to be most closely related to the Lystrosauridae and considered the two families to represent a "specialized sidebranch" of the Anomodontia.

Walker (1973) suggested that *Geikia elginensis* was derived from *Dicynodon locusticeps* von Huene (1942), from the Kawinga Formation (Charig, 1963) of Kingori, Tanzania. He stated that *D. locusticeps* demonstrates the initiation of trends further developed in *Geikia elginensis*, including the development of a depressed area between the preorbital protuberances, a broad interorbital region, and an abruptly descending flat anterior premaxillary surface. Walker also stated that the resemblances between *Geikia elginensis* and *Lystrosaurus* are merely superficial, although most of the character states he lists as evidence of close relationship between *G. elginensis* and *D. locusticeps* are found in a number of other dicynodonts and all are present in some species of *Lystrosaurus*.

Much of the confusion surrounding the affinities of *Geikia* appears traceable to the use of only a few characters, most of which seem to be subject to parallelism in phylogenetic analyses. A good example of a problematic character state is the relative broadening of the interorbital region. This has occurred in *Geikia, Platycyclops, Pelanomodon, Aulacocephalodon;* to a lesser extent in *Kingoria,* some species of *Dicynodon* (*sensu* Hotton and Cluver, in preparation) and other Permian forms; it is also seen in *Lystrosaurus* and most large Triassic forms. Similarly, the prefronto-nasal bosses that have fused and expanded to form the pronounced preorbital protuberance in *Geikia* are developed to some degree, although generally unfused, in genera assigned to a number of families. Such characters have little value at high taxonomic levels, but can be useful at a generic or specific level.

The structure of the dicynodont palate, which forms a tightly integrated component of the masticatory apparatus, has become known in detail in a number of genera, and is proving to be quite useful in arriving at hypotheses of relationship and phylogenetic classification at high taxonomic levels. Cluver (1970) demonstrated that the palate of dicynodonts is subject to considerable structural variation, and that the differences may be associated with taxon-specific differences in the masticatory cycle. Toerien (1953) demonstrated the potential usefulness of the structure of the palate in arriving at a phylogenetic classification of dicynodonts, and Cox (1964) used palatal structure to revise the dicynodont genus *Endothiodon* (see also Cluver, 1975). More recently, Hotton and Cluver (in preparation) have used palatal structure as the foundation of a comprehensive revision of South African Permian dicynodonts.

The palate of *Geikia* bears a detailed resemblance to palates of dicynodonts assigned to the family Cryptodontidae (*sensu* Toerien, 1953; Haughton and Brink, 1954). A suite of character states diagnostic of this family includes: palate highly vaulted in most forms, but becoming somewhat shallower in larger forms; large palatines having rugose palatal surface; palatine having wide contact with premaxilla and maxilla; palatal surface of palatine slanting posteroventrally relative to palatal surface of rear part of premaxilla; absence of tusk; absence of postcanine teeth; delicate sharp caniniform process of maxilla, having pronounced lateral ridge; palatal rim strongly developed in front of caniniform process; strongly developed ridge or "keel" on ventral edge of maxilla behind caniniform process; pterygoid separated from maxilla by ectopterygoid and palatine bones; interpterygoidal vacuity lying entirely within roof of interpterygoidal fossa; interpterygoidal vacuity long, approximately equal to or greater than one-half the length of the interpterygoidal fossa.

In *Geikia elginensis* all but one of these character states are known to exist. The single exception involves the separation of the pterygoid from the maxilla by the ectopterygoid. Whether an ectopterygoid is present in *Geikia* is not known, so the presence of this character state can not be confirmed.

Using the above criteria, the following genera can be assigned to the Cryptodontidae (zone definitions and locality data from Kitching, 1977):

Family Cryptodontidae

Geikia Newton (1893)

> Horizon: Upper Permian: Cuttie's Hillock Sandstone, Scotland (*Daptocephalus* Zone); Kawinga Formation, Tanzania (*Cistecephalus* Zone; Haughton, 1932).
> Comments: Here considered to include *"Dicynodon" locusticeps* (von Huene, 1942) as valid species, *Geikia elginensis*.

Neomegacyclops Boonstra (1953)

> Synonyms: *Megacyclops* Broom (1931)
> Horizon: Upper Permian: *Cistecephalus* Zone, Beaufort Series, South Africa; Kawinga Formation, Tanzania (*Cistecephalus* Zone; Haughton, 1932).

Pelorocyclops Van Hoepen (1934)

> Horizon: Upper Permian: Beaufort Series, South Africa.
> Comments: Locality data lost, but probably from *Cistecephalus* Zone.

Eocyclops Broom (1913)

> Horizon: Upper Permian: *Cistecephalus* Zone, Beaufort Series, South Africa.

Rhachiocephalus Seeley (1898)

> Horizon: Upper Permian: *Cistecephalus* Zone and *Daptocephalus* Zone, Beaufort Series, South Africa; Upper Madumabisa Mudstone, Zambia (*Daptocephalus* Zone; Kemp, 1975).

Propelanomodon Toerien (1955)

> Horizon: Upper Permian: *Daptocephalus* Zone, Beaufort Series, South Africa.
> *Pelanomodon* Broom (1938)
> Horizon: Upper Permian: *Daptocephalus* Zone, Beaufort Series, South Africa.

Platycyclops Broom (1932)

> Horizon: Upper Permian: *Cistecephalus* Zone, Beaufort Series, South Africa.

Kitchingia Broom and George (1950)

> Horizon: Upper Permian: *Cistecephalus* Zone, Beaufort Series, South Africa.

Oudenodon Owen (1860)

> Synonyms: *Mastocephalus* Van Hoepen (1934); *Chelyrhynchus* Haughton (1917).
> Horizon: Upper Permian: *Cistecephalus* Zone and *Daptocephalus* Zone, Beaufort Series, South Africa; Upper Madumabisa Mudstone, Zambia (*Daptocephalus* Zone; Kemp, 1975).

Cryptodontidae *incertae sedis*

Haughtoniana Boonstra (1938)

> Horizon: Upper Permian: Upper Madumabisa Mudstone, Zambia (*Daptocephalus* Zone; Kemp, 1975).
> Comments: Palate not preserved.

This list includes, in addition to *Geikia,* only genera recognized by Haughton and Brink (1954). It is not intended to be a revision of the family, which is beyond the scope of this paper. It should be noted, however, that Keyser (1972) suggested that *Rhachiocephalus, Eocyclops, Neomegacyclops, Platycyclops, Pelorocyclops,* and *Kitchingia* may prove to be synonymous. Kitching (1977) believed that *Kitchingia* is a junior synonym of *Rhachiocephalus,* and assigned *Pelanomodon* to the Aulacocephalodon-tidae. Toerien (1953) assigned only *Oudenodon, Kitchingia, Platycyclops,* and *Pelanomodon* to the Cryptodontidae, but dealt with only a few genera and did not attempt a complete revision of the family. Haughton and Brink (1954) included all of the genera listed above, except *Geikia,* but also included *Eosimops* and *Digalodon,* whose palates differ from the other genera.

In addition to the genera listed above, the type specimens of *"Dicynodon" sidneyi* Broom (1940) and *"Dicynodon" tylorhinus* Broom (1913), and an undescribed specimen in the University of California (Berkeley) Museum of Paleontology (UCMP 42401) are referable to the Cryptodontidae. Whether these specimens warrant generic distinction or can be assigned to previously established genera is difficult to assess at present. They are noted here because they may be closely related to *Geikia* and consideration of them in a more detailed analysis of the interrelationships of cryptodontids may require modification of the following discussion.

Below the family level, the palatal structure of cryptodontids becomes less useful in assessing relationships, because differences in character states are largely a result of subtle proportional changes, and few character states are either present or absent. Using such character states is often difficult because nearly all dicynodont specimens have been distorted to a greater or lesser degree, making accurate measurement difficult and often impossible. Furthermore, Hecht (1976) and Hecht and Edwards (1976) pointed out that proportional changes in bony structures are often the result of simple allometric relationships or simple growth patterns, and that this mechanism of morphologic change is highly susceptible to parallelism.

Parallelism has certainly occurred in a number of cranial character states among cryptodontids, but identification of instances of parallel evolution of a given character state at this low taxonomic level is difficult. The polarities of character states can generally be determined by comparison with other taxa believed to be more primitive. In this case the endothiodontinid genera *Brachyprosopus* and *Tropidostoma* were believed to be structurally ancestral to the Cryptodontidae, and the genus *Oudenodon* was considered to be the most primitive cryptodontid (a complete discussion of this will be presented elsewhere; Rowe, in preparation). The major assumption in assessing the affinities of *Geikia,* then, is that the character states found in *Brachyprosopus, Tropidostoma,* and *Oudenodon* are generally primitive for the Cryptodontidae. Even when the polarities of character states are known, however, the identification of parallelism involves a certain element of subjectivity for, in any given group of genera and character states, several phylogenies can be arrived at depending on which shared derived character states are believed to represent true synapomorphies and which have evolved in parallel.

Geikia elginensis was compared to the cryptodontids *Geikia locusticeps, Pelanomodon moschops, Propelanomodon devillersi, Oudenodon marlothi, "Dicynodon" tylorhinus,* and the endothiodontinids *Brachyprosopus broomi* and *Tropidostoma microtrema.* Seven characters were compared. Although it would be desirable to use more characters, the incompleteness of several specimens made this difficult. Table 16.1 summarizes the distribution of these character states.

The character states used in this analysis include the relations of the septomaxilla, the degree of development of the anterior palatal ridges, the width of the interorbital region relative to the intertemporal region, size of the preorbital protuberance or boss, composition of the preorbital protuberance, shape of the premaxilla, and relationship of the premaxilla to the frontals.

In *Tropidostoma* and *Brachyprosopus* and most cryptodontids the septomaxilla lies entirely within the external nares, with no exposure on the lateral surface of the snout. The exposure of the septomaxilla on the side of the snout which occurs in *Geikia* and *Pelanomodon* appears to be a good derived character. In most endothiodontinids and cryptodontids the frontals are separated from the premaxilla by the nasals, which meet along the midline. Contact between the frontals and the premaxilla is, therefore, probably also a derived character state within the Cryptodontidae. However, the distribution of this derived character state conflicts with that of the septomaxilla. Contact between the frontals and the premaxilla occurs in *Propelanomodon* and *Geikia elginensis*, but not in *G. locusticeps*. Since both species of *Geikia* and *Pelanomodon* share other derived character states that are not present in *Propelanomodon*, it seems likely that the frontal-premaxilla contact developed independently in *Geikia* and *Propelanomodon*.

The "square" premaxilla found in both species of *Geikia* and the fusion of the prefrontal and nasal to form a single large protuberance are unusual among dicynodonts, and both appear to represent good derived character states within the Cryptodontidae. It is difficult to be certain of the primitive condition of the preorbital protuberance or bosses. Certainly the presence of a small nasal boss is primitive, but whether a separate small prefrontal boss is primitive or if it has developed independently several times is not clear, since *Oudenodon* has both a prefrontal and nasal boss while *Brachyprosopus, Tropidostoma* (usually, but not always), and *Propelanomodon* have only a nasal boss.

The size of the anterior palatal ridges is subject to a wide range of variation among endothiodontinids and cryptodontids. The large robust ridges in *Oudenodon* are probably not primitive since these ridges are not especially pronounced in *Tropidostoma* and are absent in *Brachyprosopus*. The moderate development of the ridges seen in *Tropidostoma* and *Pelanomodon* may represent the primitive condition, while their enlargement in *Oudenodon* and their reduction and loss in *Geikia* may both be derived states.

Geikia elginensis was found to share the most derived character states with *Geikia* (*Dicynodon*) *locusticeps*, including a "square" premaxilla, single large preorbital protuberance composed of fused nasal and prefrontal bones, and the septomaxilla having exposure on the lateral surface of the snout behind the external nares. Because of this close resemblance, "*Dicynodon*" *locusticeps* is here referred to *Geikia* as a valid species, *Geikia locusticeps*. This assignment may not be altogether satisfactory because considerable differences do exist in the proportions of the interorbital/intertemporal region, the size of the preorbital protuberance, and the relationship between the frontals and the premaxilla. However, a similar range of variation is recognized in the dicynodont genus *Lystrosaurus* (e.g., Cluver, 1971) and, until a comprehensive revision of the Cryptodontidae is available, it seems even more unjustifiable to erect a new genus for this specimen. Walker (1973) believed that

TABLE 16.1

Distribution of character states used in assessing affinities of *Geikia elginensis* to other cryptodontids.

	Geikia elginensis	*Geikia locusticeps*	*Pelanomodon moschops*	*Oudenodon marlothi*	*Propelanomodon devillersi*	*"Dicynodon" tylorhinus*	*Tropidostoma microtrema*	*Brachyprosopus broomi*
Anterior Palatal Ridges (large; reduced; absent)	reduced	absent	reduced	large	absent	reduced	reduced	absent
Interorbital Width (greater than intertemporal width; equal to intertemporal width)	greater	greater	greater	equal	equal	greater	equal	equal
Septomaxilla (exposure on lateral surface of snout; no exposure on snout)	exposure	exposure	exposure	no exposure	no exposure	?	no exposure	no exposure
Preorbital Protuberance(s) Size (large; medium; small)	large	large	medium	small	large	large	small	small
Preorbital Protuberance Composition (nasal only; unfused nasal and prefrontal; fused nasal and prefrontal)	fused prefrontal and nasal	fused prefrontal and nasal	unfused prefrontal and nasal	unfused prefrontal and nasal	nasal only	fused prefrontal and nasal	nasal only	nasal only
Premaxilla Shape (beak tip square; beak tip rounded)	square	square	rounded	rounded	rounded	rounded	rounded	rounded
Frontals-Premaxilla Relations (contact; no contact)	contact	no contact	no contact	no contact	contact	?	no contact	no contact

Dicynodon locusticeps is structurally ancestral to *Geikia elginensis,* and his conclusion is supported by this study.

Among the other cryptodontids, *Geikia* closely resembles *Pelanomodon* in that both genera have the septomaxilla exposed on the lateral side of the snout behind the external nares, the preorbital protuberance involves both enlarged prefrontal and nasal bones although they are unfused in *Pelanomodon,* and the interorbital region is substantially wider than the intertemporal region. In the type specimen of *"Dicynodon" tylorhinus* the prefrontals and nasals fuse to form a large preorbital protuberance and the interorbital region is much wider than the intertemporal region. It therefore may share a more recent common ancestor with *Geikia* than *Pelanomodon* does, but the condition of the septomaxilla is currently unknown, leaving its position uncertain.

Age of Cuttie's Hillock

With the exception of *Geikia elginensis,* all known cryptodontids are from South Africa, Zambia, and Tanzania. They occur in continental sediments of Late Permian age which correlate with the *Cistecephalus* Zone and *Daptocephalus* Zone of the Beaufort Series of South Africa. *Geikia locusticeps* is from a horizon that correlates with the *Daptocephalus* Zone (Haughton, 1932). The stratigraphic range of other cryptodontids favors a Late Permian age rather than an Early Triassic age, and therefore a correlation of the Cuttie's Hillock Sandstone with the *Daptocephalus* Zone. The presence of the pareiasaur *Elginia* in the Cuttie's Hillock fauna tends to support this suggestion inasmuch as pareiasaurs are common members of *Daptocephalus* Zone faunas but are not known to extend into the overlying Early Triassic *Lystrosaurus* Zone. Walker (1973) favored a correlation with the *Lystrosaurus* Zone because the skull of *Elginia* is more spinose than pareiasaurs from the latest known faunas elsewhere. He apparently believed that this extreme specialization required more time to evolve, and consequently *Elginia* is younger than other pareiasaurs. This argument, however, requires tacit assumptions regarding pareiasaur phylogeny and evolutionary rates, for which there seems to be little evidence. The available evidence favors a Late Permian age and *Daptocephalus* Zone correlation for the Cuttie's Hillock fauna.

Summary

The dicynodonts *Geikia* and *Gordonia* are the only known therapsid reptiles from the Late Permian of Europe. Some authors believe that they may be as young as Triassic, but the available evidence favors a Late Permian age, and correlation with the *Daptocephalus* Zone of the Beaufort Series of South Africa. *Geikia elginensis* is assigned to the Cryptodontidae (*sensu* Toerien, 1953; Haughton and Brink, 1954), and is the only member of this family occurring outside of Africa. It is distinctive in the marked enlargement of its preorbital protuberances, its short box-like snout, and the general proportions of its skull. *"Dicynodon" locusticeps* (von Huene, 1942) is assigned as a valid species of *Geikia, G. locusticeps.* Among cryptodontids, *Geikia* is considered to be most closely related to *Pelanomodon.*

Acknowledgments

I gratefully acknowledge Dr. James A. Hopson, University of Chicago, and Dr. Michael A. Cluver, South African Museum, Cape Town, for their careful reading of this manuscript and their many helpful suggestions and corrections. I would like to thank Dr. Alick Walker, University of Newcastle-upon-Tyne, for making available to me his fine casts of *Geikia;* and Dr. Eugene Gaffney, American Museum of Natural History, for making available the type of *Pelanomodon moschops* and the opportunity to study the Broom Collection of Karroo vertebrates in the American Museum. I would also like to thank Dr. Richard Dehm for the opportunity to study the collections of the Institut für Paläontologie und historische Geologie, Munich; Dr. Frank Westphal for the opportunity to examine the collections of the Institut für Geologie und Paläontologie, Tübingen; and Mr. Michael Greenwald for the opportunity to examine the Karroo vertebrates in the University of California Museum of Paleontology. Finally, I would like to thank the Thomas J. Watson Foundation for funding the early stages of research on this project.

I would also like to use this opportunity to express my sincere appreciation and gratitude to Dr. Edwin Colbert for the inspiration his work provided during my years as a student and for the pleasure and education I have received while working with him at the Museum of Northern Arizona.

Literature Cited

Boonstra, L. D. 1938. A report on some Karroo reptiles from the Luangwa Valley, northern Rhodesia. Geol. Soc. London, Quart. J., 94:371–84.
———. 1953. A report on a collection of fossil reptilian bones from Tanganyika Territory. S. Afr. Mus. Ann., 42(1):5–18.
Broili, F., and Schroeder, J. 1936. Ein neuer Anomodontier aus der *Cistecephalus*-Zone. J. Sitzber. Bayer. Akad. Wiss., Math-naturw. Abt., p. 45, figs. 1–5.
Broom, R. 1913. On some new genera and species of dicynodont reptiles, with notes on a few others. Amer. Mus. Natur. Hist., Bull., 32:441–57.
———. 1931. Notices of some new genera and species of Karroo fossil reptiles. Albany Mus., Rec., 4:161–66.
———. 1932. The mammal-like reptiles of South Africa, and the origin of mammals. Witherby, London, 376 pp.
———. 1938. On two new anomodont genera. Transvaal Mus., Ann., 19:274–80.
———. 1940. On some new genera and species from the Karroo Beds of Graaf Reinet. Transvaal Mus., Ann., 20:158–92.
Broom, R., and George, M. 1950. Some new anomodont reptiles in the Bernard Price Collection. S. Afr. J. Sci., 46:275–78.
Camp, C. L. 1956. Triassic dicynodont reptiles. Part II. Triassic dicynodonts compared. Calif., Univ., Mem., 13:305–41.
Charig, A. J. 1963. Stratigraphical nomenclature in the Songea Series of Tanganyika. Geol. Surv. Tanganyika Rec., 10(for 1960):47–53.
Cluver, M. A. 1970. The palate and mandible in some specimens of *Dicynodon testudirostris* Broom and Haughton (Reptilia, Therapsida) S. Afr. Mus., Ann., 56(4):133–53.
———. 1971. The cranial morphology of the dicynodont genus *Lystrosaurus.* S. Afr. Mus., Ann., 56(5)155–274.
———. 1975. A new dicynodont reptile from the *Tapinocephalus* Zone (Karroo System, Beaufort Series) of South Africa, with evidence of the jaw adductor musculature. S. Afr. Mus., Ann., 67(2):7–23.
Cox, C. B. 1959. On the anatomy of a new dicynodont genus with evidence of the position of the tympanum. Zool. Soc. London, Proc., 132:321–67.
———. 1964. On the palate, dentition and classification of the fossil reptile *Endothiodon* and related genera. Amer. Mus. Nov., 2171:1–25.
Crompton, A. W., and Hotton, N. 1967. Functional morphology of the masticatory apparatus of two dicynodonts (Reptilia, Therapsida). Postilla, 109:1–51.
Haughton, S. H. 1917. Investigations in South African fossil reptiles and amphibians. Part 10. Descriptive catalogue of the Anomodontia, with especial reference to the examples in the South African Museum. S. Afr. Mus., Ann., 12(5):127–74.
———. 1932. On a collection of Karroo vertebrates from Tanganyika Territory. Geol. Soc. London, Quart. J., 88:634–68.

Haughton, S. H., and Brink, A. S. 1954. A bibliographical list of Reptilia from the Karroo beds of Africa. Paleontol. Afr., 2:1–187.

Hecht, M. K. 1976. Phylogenetic inference and methodology as applied to the vertebrate record. *In* Evolutionary biology, vol. 9, M. K. Hecht, W. C. Steere, and B. Wallace (eds.), Plenum Press, New York, London, pp. 335–63.

Hecht, M. K., and Edwards, J. L. 1976. The determination of parallel or monophyletic relationships: the proteid salamanders—a test case. Amer. Natur., 110:653–67.

Huene, F. von 1913. Ueber die reptilfuhrenden Sandstein bei Elgin in Schottland. Centralbl. f. Min. etc., 19:617–23.

———. 1940. Die Saurier der Karoo-, Gondwana- und verwandten Ablagerungen in faunistischer, biologischer und phylogenetischer Hinsicht. Neues Jahrb. Min. Geol. Paläontol., 83(B):246–347.

———. 1942. Die Anomodontier des Ruhuhu-Gebietes in der Tübinger Sammlung. Palaeontographica, 94(A):154–84.

———. 1948. Short review of the lower tetrapods. *In* Robert Broom commemorative volume, A. L. DuToit (ed.), Roy. Soc. S. Afr., Spec. Publ., pp. 65–106.

———. 1956. Paläontologie und phylogenie der niedern Tetrapoden. Gustav Fischer Verlag, Jena, 716 pp.

Kemp, T. S. 1975. Vertebrate localities in the Karroo System of the Luangwa Valley, Zambia. Nature, 254:415–16.

Keyser, A. E. 1972. A re-evaluation of the systematics and morphology of certain anomodont Therapsida. Paleont. Afr. 14:15–16.

Kitching, J. W. 1970. A short review of the Beaufort zoning in South Africa. *In* I.U.G.S., 2nd Symposium on Gondwana Stratigraphy and Paleontology, Cape Town and Johannesburg, 1970, S. H. Haughton (ed.), C.S.I.R., Pretoria, pp. 309–12.

———. 1977. The distribution of the Karroo vertebrate fauna. Bernard Price Inst., Mem. 1, 131 pp.

Newton, E. T. 1893. On some new reptiles from the Elgin sandstones. Roy. Soc. London, Phil. Trans., Ser. B, 184:431–503.

Nowack, E. 1937. Zur kenntnis der Karruformation in Ruhuhu-Graben. (D.O.A.). Neues Jahrb. Min. Geol. Paläont., 78(B):380–412.

Owen, R. 1860. On some reptilian fossils from South Africa. Geol. Soc. London, Quart. J., 16:49–54.

Romer, A. S. 1966. Vertebrate paleontology. Univ. of Chicago Press, Chicago, 468 pp.

Seeley, H. G. 1898. *Oudenodon (Aulacocephalus) pithecops* from the Cape Colony. Geol. Mag., 5:107–10.

Smith Woodward, A. 1898. Outlines of vertebrate paleontology for students of zoology. Cambridge Univ. Press, 470 pp.

Smith Woodward, A., and Sherborn, C. D. 1890. A catalogue of British fossil Vertebrata. Dulau and Co., London, 396 pp.

Toerien, M. J. 1953. The evolution of the palate in South African Anomodontia and its classificatory significance. Paleontol. Afr., 1:49–117.

———. 1955. Important new Anomodontia. Paleontol. Afr., 3:65:–75.

Traquair, R. H. 1886. Preliminary note on a new fossil reptile recently discovered at New Spynie, near Elgin. Rep. Brit. Assoc. (for 1885):1024.

Van Hoepen, E. C. N. 1934. Oor die indeling van die Dicynodontidae na aanleiding van nuwe vorme. Palaeontol. Navors. Nas. Mus., Bloemfontein, 11(6):67–101.

Walker, A. D. 1973. The age of the Cuttie's Hillock Sandstone (Permo-Triassic) of the Elgin Area. Scott. J. Geol., 9:177–83.

Watson, D. M. S. 1909. The "Trias" of Moray. Geol. Mag., 46:102–7.

———. 1948. *Dicynodon* and its allies. Zool. Soc. London, Proc., 118:823–77.

———. 1960. The anomodont skeleton. Zool. Soc. London, Trans., 29:131–208.

Watson, D. M. S., and Hickling, G. 1914. On the Triassic and Permian rocks of Moray. Geol. Mag., 51:399–402.

Westoll, T. S. 1951. The vertebrate-bearing strata of Scotland. Int. Geol. Cong. XVIII, Session 11, pp. 5–21.

Abbreviations for Figures

ANG.—angular
ANT. PAL. R.—anterior palatal ridge
ART.—articular
BAS.—basisphenoid
BO.—basioccipital
CAN. PROC.—caniniform process of
    maxilla
DEN.—dentary
DEN. GR.—dentary groove
DEN. TAB.—dentary table
DP. CREST—deltopectoral crest
ECT.—ectopterygoid
ENT. F.—entepicondylar foramen
EO.—exoccipital
EPT.—epipterygoid
EXT. NAR.—external nares
F. M.—foramen magnum
FEN. OV.—fenestra ovalis
FR.—frontal
I.C.C.—internal carotid canal
IP.—interparietal
IPT. VAC.—interpterygoidal vacuity
JUG. FOR.—jugular foramen
LAT. COND.—lateral condyle of articular
LAT. DEN. SH.—lateral dentary shelf
MAX.—maxilla
MED. COND.—medial condyle of
    articular
NAS.—nasal
OP.—opisthotic
ORB.—orbit
PA.—parietal
PAL.—palatine
PAS.—parasphenoid
PMX.—premaxilla
PO.—postorbital
PRF.—prefrontal
PRO.—prootic
PT.—pterygoid
Q.—quadrate
Q. RAM.—quadrate ramus of pterygoid
RAD. COND.—radial condyle
REF. LAM.—reflected lamina of angular
RET. PROC.—retroarticular process
SA.—surangular
SMX.—septomaxilla
SOC.—supraoccipital
SQ.—squamosal
VO.—vomer

COLBERT AS PALEOMAMMALOGIST

by George Gaylord Simpson

EDWIN HARRIS COLBERT, who is now being honored on his seventy-fifth birthday, has become best known as a paleoherpetologist, especially for his research and popular books on dinosaurs, but also on other fossil reptiles and some amphibians. For the most part before he entered that field, he had also made extensive studies on fossil mammals, some of which this note will briefly review.

In 1930 Colbert went to the American Museum of Natural History as a research assistant in the Department of Vertebrate Paleontology, working primarily with Henry Fairfield Osborn, founder and then still *de facto,* although not nominal, chairman of that department. In 1933 Colbert was made an assistant curator there and forthwith started his long career in independent research.

Barnum Brown, although nominally in charge of fossil reptiles and already well known as a great collector of dinosaurs, had also made extensive collections of fossil mammals for the American Museum, especially in Cuba in 1918, in India in 1921 and 1922, in Burma in 1923, and on the island of Samos, Greece, in 1924. William Diller Matthew had made some studies of those collections but had not completed any of them when he left the American Museum for the University of California in 1927 or when he died in 1930. Study of the Cuban collection of ground sloths was eventually completed by the Brazilian paleontologist Carlos de Paula Couto. The collections from the Siwalik Hills of India and from Samos, as well as the less extensive one from Burma, were turned over to Colbert in part as early as 1931 and more extensively when he was put on the curatorial staff in 1933.

From his first publication in 1931 (as co-author with Osborn) well into 1943, almost all of Colbert's research papers were devoted to fossil mammals. The only significant exception was a study of certain phytosaurs written jointly with Robert G. Chaffee, then a student, and published in 1941. Colbert returned to this subject in 1943, and since then most of his research papers have been on reptiles, amphibians, and the Triassic period. About forty-five research papers on mammals were published by him from 1931 to 1943, and a few were published thereafter. I cannot review all of these in the present summary, but some will be noted as of special interest for one reason or another.

Already with some familiarity with the Siwalik faunas, Colbert lost no time in starting publication on them when he joined the curatorial staff. In 1933 he published no fewer than six short (three to fourteen pages) preliminary studies on individual members of those faunas. Other such papers followed, in 1935 including a series of five *Distributional and Phylogenetic Studies on Indian Fossil Mammals.* Later that same year the whole subject was pulled together and expanded in a voluminous memoir, still the most extensive of all of Colbert's research publications: *Siwalik Mammals in the American Museum of Natural History,* running to more than 400 quarto pages with 198 figures and a folded map annexed.

Such a work cannot be summarized in a retrospective note like this, but it can be mentioned that among many other things it included phylogenetic summaries of the Ursidae (bears), Chalicotheriidae (an extinct family of strange-clawed ungulates), Suidae (pigs), Anthracotheriidae (another extinct group, related to the hippopotamus), and Giraffidae, not only of the Siwaliks but of much or all of the world. This magnum opus made Colbert a worthy recipient of the Daniel Giraud Elliot Medal of the National Academy of Sciences.

Much further work has since been done on the Siwalik faunas and is still going on. Inevitably this has led to numerous additions to and some modifications of Colbert's classic work, but that monograph still is the indispensable basis for all later advances. There is an interesting parallel with Matthew's great memoir on the Paleocene faunas of the San Juan Basin, New Mexico, published posthumously in 1937. This, too, has been and is being followed by extensive further study and collecting in the same field, and yet remains indispensable and fundamental for all that later work. Colbert's and Matthew's memoirs here discussed were published in the same format in the Transactions of the American Philosophical Society, and Colbert carried out much of the editorial work on Matthew's manuscript; Matthew had died in 1930. It adds to the human interest of this association that although the two men had not been acquainted, Matthew posthumously became Colbert's father-in-law when Colbert married Margaret Mary Matthew in 1933.

In one of his separate preliminary papers in 1933 Colbert had described and named a peccary from the Siwalik Pliocene. Although now solely North and South American in distribution, peccaries were formerly widespread, although not so common, in Eurasia and Africa, and Colbert suggested that they may have originated in the Old World. Among his other shorter studies are some on fossil North American peccaries, most noteworthy perhaps one of twenty-eight pages and six plates on Pliocene peccaries from the Pacific Coast Region published in 1938 by the Carnegie Institution of Washington.

Having completed his Siwalik studies, Colbert turned to the collections made by Brown in Burma. These came from at least five different faunas, ranging from late Eocene to Pleistocene in age. Most of the specimens are fragmentary and the most unusual but also most tantalizing is part of the lower jaw of a late Eocene primate with only three teeth preserved. This was clearly a new genus, which Colbert named *Amphipithecus,* but he left its affinities in some doubt which still has not been dispelled. There are, however, fairly well-preserved skulls of late Eocene rhinoceroses and anthracotheres in this collection. The whole collection was fully described by Colbert in a 182-page bulletin of the American Museum of Natural History published in 1938.

In 1937 an expedition headed by Helmut de Terra and sponsored jointly by several American institutions (not including the American Museum of Natural History) had collected a few fragments of Pleistocene mammals along the Irrawaddy River in Burma. As noted above, Colbert had previously made an important study of Burmese fossil mammals, and after preparation and curating this new collection was placed in his hands for identification. It added little to previous knowledge, but it led to an extended and updated summary of the nature and relationships of the Burmese Pleistocene mammalian fauna in a publication in 1943 by Colbert in the Transactions of the American Philosophical Society, one of the sponsors of the expedition.

Colbert published several studies on fossil mammals collected in Mongolia by the American Museum's Central Asiatic Expeditions. For example, one on Miocene (Tung Gur formation) carnivores which included several extraordinarily well-preserved skulls, described in 1939 in a bulletin of thirty-five pages.

Brown's large collection from the island of Samos has not been studied as a whole. That fauna is fairly well known, and there are large collections in some other museums. Some individual studies have nevertheless been made on the American Museum's collection and the most noteworthy of these is one by Colbert on a nearly complete skeleton and other specimens, including several skulls, of a fossil aardvark, *Orycteropus gaudrii*. Colbert's description of the whole bony anatomy, published in 1949 in 47 pages of the Bulletin of the American Museum of Natural History, is meticulous. The aardvarks are customarily classified in an order of their own, Tubulidentata, the origin of which had been under considerable question. Here there is again a meeting of the minds of Matthew and Colbert, who had not met in the body. In his Paleocene memoir Matthew had closely compared a skeleton of *Ectoconus*, a primitive ungulate, condylarth, with that of the living aardvark. Colbert now took this up in greater detail, comparing the two by the method of deformed coördinates, which had been devised long before in a different context by D'Arcy Thompson. (Colbert had also used this method in his Siwalik memoir.) The skulls and teeth of aardvarks are very different from those of a condylarth in adaptation to a radical change in food, but from the skull back the aardvark can almost be called a living condylarth. (Condylarths as such have been extinct for millions of years, since the Miocene.)

In 1942 Barnum Brown retired and Colbert was placed in charge of fossil amphibians and reptiles on the American Museum staff. Thereafter, until he in turn retired from the American Museum in 1970, and since 1970 in connection with the Museum of Northern Arizona, Colbert has worked primarily as a paleoherpetologist. Nevertheless, after 1942 he published several short papers and one extensive monograph on fossil mammals. The monograph was in one way another outcome of the Asiatic expeditions mentioned above. While those expeditions were going on during summers, Walter Granger, in charge of paleontology for them, spent the winters of 1921–22, 1922–23, and 1925–26 on his own in Szechwan, China, where he made extraordinary collections of Pleistocene mammals in the vicinity of Yengchingkou. Here on the heights above the right bank of the Yangtze in fissures or natural shafts in ancient (Paleozoic) limestone are entombed remains of a large Pleistocene fauna. The Chinese had long exploited these for sale as "dragon bones" to drug dealers. Largely by judicious purchases from the local diggers Granger amassed a great collection of these fossil bones, ranging from virtually complete individual skeletons to single teeth and other identifiable fragments. Matthew and Granger published a 36-page summary of this collection in 1923 and Osborn and Hooijer separately later published notices of some particular specimens, but the whole fauna had not been adequately studied as such. Colbert and Dirk Hooijer, a Dutch paleontologist who spent some time at the American Museum, performed this task in full detail in a large 1953 monograph, 134 pages (in the new large-page format of the Bulletin of the Museum) with 40 photographic plates and 42 text figures. Although by this date there was nothing new in the collection above the subspecific level, the numerous specimens provided data for studies of variation and for excellent new figures.

That monograph, completed with Hooijer's assistance while Colbert was already heavily committed to research on fossil amphibians and reptiles, may be taken as marking the close of his distinguished career as a paleomammalogist.

Literature Cited

Colbert, E. H. 1933. An upper Tertiary peccary from India. Amer. Mus. Nov., 635:1–9.
————.1935a. Distributional and phylogenetic studies on Indian fossil mammals. I-V. Amer. Mus. Nov., 796:1–20; 797:1–15; 798:1–15; 799:1–24; 800:1–15.
————.1935b. Siwalik mammals in the American Museum of Natural History. Amer. Phil. Soc., Trans., 26:i–x, 1–401.
————.1938a. Pliocene peccaries from the Pacific Coast region of North America. Carnegie Inst. Washington, Publ., 487:241–69.
————.1938b. Fossil mammals from Burma in the American Museum of Natural History. Amer. Mus. Natur. Hist., Bull., 74:255–436.
————.1939. Carnivora of the Tung Gur formation of Mongolia. Amer. Mus. Natur. Hist., Bull., 76:47–81.
————.1941. A study of *Orycteropus gaudryi* from the island of Samos. Amer. Mus. Natur. Hist., Bull., 78:305–51.
————.1943. Pleistocene vertebrates collected in Burma by the American Southeast Asiatic Expedition. Amer. Phil. Soc., Trans., 32:395–429.
Colbert, E. H., and Chaffee, R. G. 1941. The type of *Clepsysaurus pennsylvanicus* and its bearing upon the genus *Rutiodon*. Notulae Naturae, 90:1–19.
Colbert, E. H., and Hooijer, D. A. 1953. Pleistocene mammals from the limestone fissures of Szechwan, China. Amer. Mus. Natur. Hist., Bull., 102:1–134.
Matthew, W. D. 1937. Paleocene faunas of the San Juan Basin, New Mexico. Amer. Phil. Soc., Trans., 30:1–372.
Matthew, W. D., and Granger, W. 1923. New fossil mammals from the Pliocene of Sze-Chuan, China. Amer. Mus. Natur. Hist., Bull., 48(17):563–98.
Osborn, H. F., and Colbert, E. H. 1931. The elephant-enamel method of measuring Pleistocene time. Also stages in the succession of fossil man and Stone Age industries. Amer. Phil. Soc., Proc., 70:187–91. (This was Colbert's first publication.)

A NOTE ON THE SKULL AND MANDIBLE OF A NEW CHOEROLOPHODONT MASTODONT (PROBOSCIDEA, MAMMALIA) FROM THE MIDDLE MIOCENE OF CHIOS (AEGEAN SEA, GREECE)

by Heinz Tobien

It was Paraskevaidis (1940, pp. 417–27) in his monograph on the "Obermiozäne Fauna von Chios," who reported the first proboscidean finds from the Neogene south of Thymianá at the eastern coast of the island of Chios (Figure 18.1). He recognized among other mammalian taxa *"Dinotherium bavaricum* H. v. Meyer var. *Aegäum* n. var.," based on a right P^4, a left M^2 and a fragmentary right M^2 (Plate 14, Figures 1–5), and *"Trilophodon (Mastodon) angustidens* Cuvier" based on the posterior part of a right M_3 with two lophids and a strong, blunt talonid button (Plate 14, Figure 6) (see also Tobien, 1977, p. 489). Valuable additions to the discoveries of mastodonts in the Neogene of Thymianá were added by Besenecker (1973, p. 38).

Further materials were discovered by the Mainz excavations in 1967–68. Among the nine localities investigated in the brick quarry and its environments north of Thymianá (Figure 18.1 and Rothausen, 1977, p. 496), only the localities Thy 1, Thy 2, and Thy 3 produced cranial elements and isolated teeth besides some postcranial skeletal remnants. The presence of proboscideans in Thy 4 and Thy 6 is indicated by some large otherwise indeterminable bone fragments. The bulk of the material came from Thy 1, which is also the site of Besenecker's specimens. (For other vertebrate finds see the preliminary notes of Meletis and Tobien, 1967, 1968; Tobien, 1968, 1969.)

The Thy 1 specimens are imbedded in a medium to coarse, grey green, hard sandstone; intercalated are some thin conglomeratic layers with small pebbles (Rothausen, 1977, p. 511, profile no. 1).

It is this layer which yielded the choerolophodont skull and mandible Thy 1-1 with M1 and M2, briefly described below.

Family: Gomphotheriidae Cabrera, 1929
   Subfamily: Choerolophodontinae Gaziry, 1975
      Genus: *Choerolophodon* Schlesinger, 1917
         Choerolophodon chioticus sp. nov.

Gomphotherium angustidens Tobien, 1973, p. 232, Plate 23, Figure 1.

Diagnosis. A *Choerolophodon* species with a narrow temporal region, and rather long, lower tuskless symphysis. Ptychodonty, choerodonty, and cementodonty weakly developed.

Etymology. After the island of Chios.

Type locality. 2.5 kilometers south of Thymianá, southeast coast of the isle of Chios. Site Thy 1 of the excavations of the Palaeontological Institute, University of

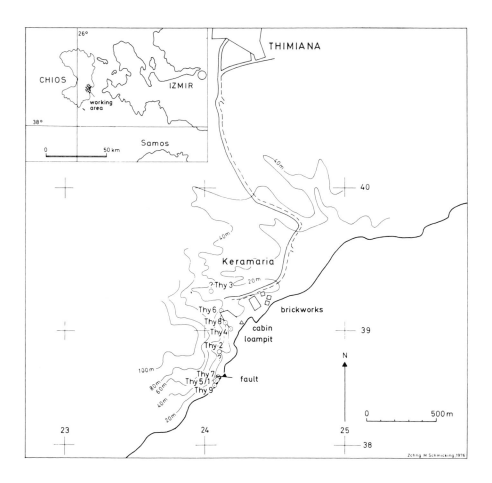

Figure 18.1. Mammal-bearing localities in the middle Miocene beds south of Thymiana, east coast of Chios (after Rothausen, 1977, Figure 1). Thy 1 is the type locality of *Choeroloph-odon chioticus.*

Mainz (Rothausen, 1977, p. 498, Figure 1); Bed 1 of the section Thy 1, 5, 9; lower part of the Keramaria beds (Rothausen, 1977, p. 511, Figure 3).

Age. Middle Miocene : Early Sarmatian (after Besenecker, 1973, p. 16, Figure 2); medial Aragonian in the continental stratigraphic scale.[1]

Holotype. A skull and mandible with deeply worn M1, fully worn M2 and M3 in alveolis, Thy 1-1; Department of Geology and Palaeontology, University of Athens (Greece) (Figures 18.2, 18.3).

Differential diagnosis. *Choerolophodon chioticus* is differentiated from the younger (Vallesian/Turolian) *Ch. pentelicus* (Gaudry and Lartet, 1856) by its still more primitive cheek teeth, particularly by the absence of: (1) the thick cement cover on the molars; (2) the multiplication of accessory mammillae (choerodonty); and (3) the strong corrugation of the enamel surfaces (ptychodonty). Slight choerodont indications are, however, visible on the lingual walls of the M^2. Better-preserved mandibles of *Choerolophodon pentelicus* indicate that the deflection of the mandibular symphysis seems to be stronger than in *Ch. chioticus*.

Ch. chioticus has in common with *Ch. pentelicus* the large upper incisors. They are without an enamel band and are turned upwards, outwards and inwards at the tips.

Measurements. Skull: Alveolar border of I^2 to posterior border of occipital condyles: 925 mm; alveolar border of I^2 to supraoccipital: 1,042 mm; width between jugal arches: 495 mm; maximum external length of I^2: 1,070 mm (left), 1,040 (right).

Length M^1: 83 mm (left), 84 mm (right); M^2: 111 mm (left), 112 mm (right); M^3: 170 mm (right).

Mandible: posterior border of condyles to anterior end of symphysis: less than 1,180 mm; posterior border of condyles to posterior border of symphysis: 690 mm; length of symphysis: less than 512 mm; deflection of symphysis: 38 degrees.

Length: M_1: 78 mm (left), 81.3 mm (right; M_2: 121 mm (left), 121 mm (right).

Description. The skull and mandible which were discovered and excavated in seven blocks during the first Chios campaign (March, 1967) belong to the same individual. Preparation revealed the original position of mandible and cranium, the mandibular condyles articulating with the glenoid surface of the squamosals and the occlusal surfaces of upper and lower dentitions in contact with each other. Both upper and lower cheek teeth are in the same wear stages: M1 deeply worn, M2 in full wear, M3 still in the alveolis (Figure 18.3).

The preparation of the specimen out of the hard, calcified, coarse sandstone— harder than the bone—was a time-consuming task, which was finished in an outstanding manner thanks to the ability and skillfulness of preparator Karl Schuchmann, Palaeontological Institute, University of Mainz. Mr. Schuchmann likewise solved the difficult problem of separating and isolating the mandible from the skull to allow study of the basicranium and occlusal surfaces of the upper and lower dentitions.

As the preparation of the mandible was finished, I originally ascribed the specimen to a bunodont, long-jawed, tuskless, hyperlongirostrine trilophodont mastodont belonging to the *Gomphotherium* group such as *G. angustidens* (Tobien, 1973,

---

1. For details of this biostratigraphic scale, based on sequences of mammalian local faunas in the Tertiary, see Fahlbusch, 1976.

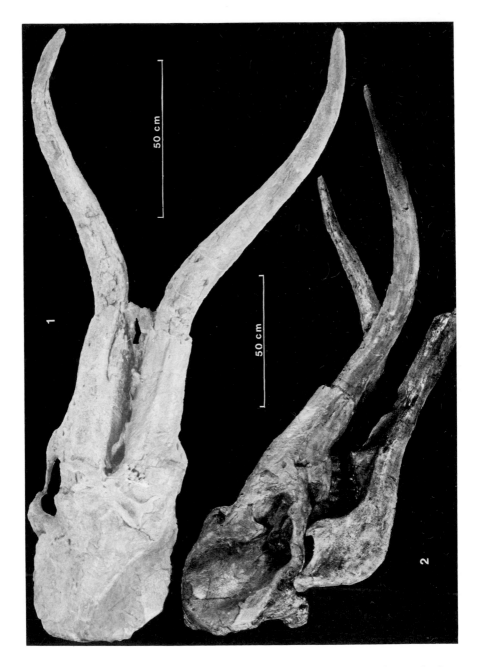

Figure 18.2. *Choerolophodon chioticus* sp. nov., holotype, Department of Geology and Palaeontology, University of Athens, Thy 1-1; 1, cranium, dorsal view, right jugal arch by postmortal deformation somewhat appressed against the cranium; 2, cranium and mandible,
right lateral view.

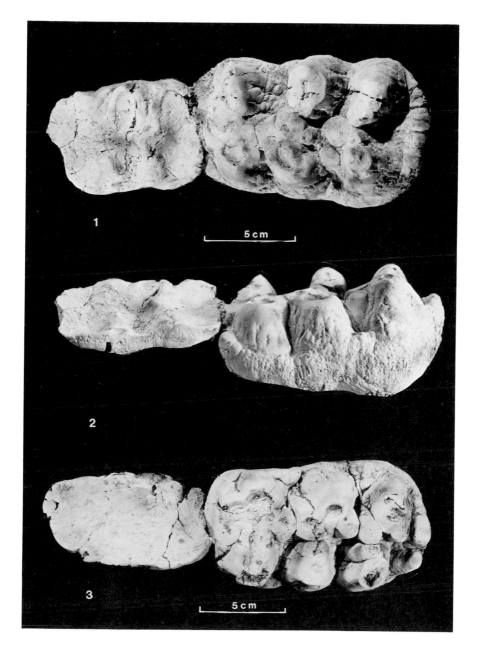

Figure 18.3 *Choerolophodon chioticus* sp. nov., holotype, Department of Geology and Palaeontology, University of Athens, Thy 1-1; 1, left M^1—M^2, occlusal view, × 0.69; 2, left M^1—M^2, lingual view, vertical enamel furrows, indicating incipient ptychodonty, partially worn by tongue wear, × 0.69; 3, right M_1—M_2, occlusal view. Teeth whitened with salmiak.

p. 232, Plate 23, Figure 1). The prepared skull, however, revealed structures quite different from *Gomphotherium.*

The main difference concerns the upper tusks (Figure 18.2). In contrast to *Gomphotherium angustidens,* and to the genus *Gomphotherium* generally, with its incisors downturned and covered with a broad outer enamel band, a diagnostic character for the genus (Schlesinger, 1917, p. 37; Tassy, 1977, p. 1389), the shape and curvature of the Chios tusks are quite different. They have no enamel band and they are turned outward and upward with in incurvation at the tips (Figure 18.2).

This character excludes the allocation of the Chios skull and mandible to the genus *Gomphotherium.* The advanced upper incisors of the Chios specimen, compared with *Gomphotherium* and the Oligocene *Phiomia,* are combined with a rather elongated, narrow, flattened skull and with a hyperlongirostrine tuskless mandible. At the present state of knowledge, only the genus *Choerolophodon* and the closely related genus *Synconolophus* (Tobien, 1973, p. 248; Gaziry, 1976, pp. 104, 105) show this particular combination of characters.

There are bunodont mastodonts with upturned enamelless upper incisors, as the North American trilophodont *Stegomastodon,* but the mandible is brevirostrine. There are, on the other hand, long-jawed, tuskless mandibles with trilophodont molars, in some cases definitely associated with skulls, but the incisors have enamel bands and are downturned, as in the genus *Gomphotherium.* The North American Pliocene offers some examples (Tobien, 1973, p. 232).

There are, however, differences in the Chios specimen compared with *Choerolophodon pentelicus,* which is the Turolian type-species of the genus. *Ch. chioticus,* being definitely older than *Ch. pentelicus,* and associated with a middle Miocene fauna, is more primitive in several respects.

Remarkable characters are the prominent sagittal crests and the narrow skull roof in between. In late Miocene and Pliocene bunodont and zygodont mastodonts, as well as in all elephants, this is quite unusual and obviously an archaic feature. Osborn (1936, Figure 182 A2, Table 1) underlines the relative width of the skull roof in the temporal region in relation to the width of the skull at the postorbital processes and the squamosal processes of the jugal arch respectively. Only the late Oligocene *Phiomia* from the Fayum shows similar sagittal crests and narrow temporal region (Osborn, 1936, Figures 182 A2, 184A).

Other archaic features are the elongated skull proportions and the relatively narrow basicranium. *Gomphotherium* skulls, as e.g. "*Serridentinus productus*" from the Clarendon beds, lower Pliocene of Texas (AMNH 10582, Osborn 1936, Figure 370 A1) in a similar stage of wear, or the slightly younger "*Serridentinus serridens*" from the same beds (AMNH 10673, Osborn, 1936, Figure 366 A1) have a relatively broader cranial base and the jugal arches converge more anteriorly in contrast to the Chios skull (see also the palatal views of different *Trilophodon* species of medial and late Miocene in Osborn, 1936, Figures 253 A1, 267, 270, 271, 278, 279).

The upper incisors are massive, strong, and long (see measurements above). After leaving the sockets they are turned outwards and upwards, and on the terminal part have a slight inward bending (Figure 18.2). The inner walls of both tips have a wear facet of 103 mm length and 30 mm basal width. The position of the incisors is somewhat asymmetrical because of a left-to-right deformation of the whole skull.

The three upper molars of both sides are preserved. The M^1's are deeply worn but the trilophodont morphology is still visible, as it is on the M^2's, which are worn on all three lophs (Figure 18.3). The M^3's are unerupted; their loph formula is the same as *Choerolophodon pentelicus*.

A characteristic feature of the M^2's are the slight vertical furrows on the enamel walls of the posttrites and pretrites, partially removed by tongue and lip wear. This incipient ptychodonty is combined with a thin irregular cement cover (1–2 mm thick) which surrounds the basal parts of the outer and posterior walls of the molar crown. The posterior cingulum is subdivided into seven to nine small enamel knobs. Together with similar knobs on the slopes of the halflophs and in the bottom of the transverse valleys they probably indicate a beginning choerodonty (Figure 18.3).

The most conspicuous detail of the mandible (Figure 18.2) is the long tuskless symphysis. The deflection against the alveolar border of the mandibular body is rather low, about 38 degrees, similar to the hyperlongirostrine mandibles of *"Trilophodon angustidens gaillardi"* (38 degrees), from the late Miocene of Villefranche d'Astarac (Osborn, 1936, Figure 218b), and *"Trilophodon chinjiensis"* (37 degrees) from the lower Chinji of the Siwaliks (Osborn, 1936, Figure 218A).

A comparable deflection of the elongate symphysis shows likewise the mandible of the lower Pliocene from Küçükçekmece near Istanbul, described by Nafiz and Malik (1933, p. 104) under *"Mastodon sp."* The elongate symphysis with the deep gutter and the cement-covered lower M_3 (Nafiz and Malik, 1933, Plate 6) with ptychodont tendencies and only three lophids plus a talonid demonstrates the *Choerolophodon* character of this specimen. In contrast, the deflection of the tuskless symphysis in *"Synconolophus propathanensis"* (circa 60 degrees) from the Dhok Pathan of the Siwaliks (Osborn, 1936, Figure 633) is distinctly stronger.

There are no lower incisors present and there is no indication of postmortal loss of incisors. In order to decide whether lower tusks were present in a juvenile stage, then afterwards lost and the sockets closed by secondary spongy bone [as in *"Trilophodon joraki"* from the early Pliocene of New Mexico, in *Trilophodon (Megabelodon) lulli* from the early Pliocene of Nebraska, or at least partially in the Siwalik *"Trilophodon chinjiensis,"* Tobien, 1973, p. 234], a transverse section was made 40 mm behind the broken anterior extremity of the rostrum. The structure of the spongy bone is homogeneous over the whole surface. Spongy bone secondarily filling empty incisor alveoli would offer a structure different from the surrounding symphyseal bone.

Both M_1's and M_2's are preserved. The two M_3's are totally in alveoli, only the points of some anterior conelets and conules are visible about 40 mm below the wear level of the M_2.

The M_1's are deeply worn, distinctly deeper than the upper first molars (Figure 18.3). There are no enamel remnants on the triturating surface. On the pretrite side, the molars are worn down to the roots and, therefore, the wear surface is labially inclined. The second molars are trilophodont with a talonid of two large mamillae. This is in contrast to the M^2 talon with its more numerous but smaller knobs. All lophids are fully worn, somewhat more than on the upper M^2 (Figure 18.3). The advanced stage of wear of the lower dentition compared with the upper one in associate skulls and mandibles is not unusual in mastodonts (Tobien, 1973, p. 207, footnote 5).

A thin film (1 mm) of cement covers the basal parts of the posterior wall of the

M_2. Although the enamel is polished by lip and tongue wear, there are still some vertical furrows on the labial wall comparable to those on the upper molars, thus indicating a feeble ptychodonty (Figure 18.3).

The structure of the lower second molars is less complicated, the cement deposit less than on their antagonists in the skull.

Discussion

In general *Choerolophodon chioticus*, by its upper upturned and enamelless incisors, and the long, incisorless, downturned mandibular symphysis, demonstrates typical features of the subfamily Choerolophodontinae. The narrow temporal region, the elongated skull base, and the incipient choerodonty, ptychodonty, and cement deposition (cementodonty) indicate a more primitive evolutionary stage for the Chios mastodont compared with the more advanced *Ch. pentelicus* from the upper Miocene (Vallesian/Turolian). This is consistent with the earlier stratigraphic age of *Ch. chioticus* (middle Miocene; Aragonian).

A more detailed description of the skull and mandible of *Ch. chioticus* and of the other proboscidean remains from the Chios sites will be published in the "Annales Géologiques des Pays Helléniques" Athens.

Acknowledgments

The excavations of the Palaeontological Institute of the Johannes-Gutenberg-University Mainz in Chios which produced the choerolophodont skull and mandible, briefly described above, were helpfully supported by the Department of Geology and Palaeontology of the University of Athens (the late Professor Dr. M. K. Mitzopoulos, Professor Dr. J. K. Melentis, now Saloniki, Professor Dr. G. P. Marinos, and Professor Dr. N. K. Symeonidis) and by the Deutsche Forschungsgemeinschaft Bonn. This is thankfully acknowledged. For details see Tobien, 1968, 1969, 1977; Melentis and Tobien, 1967, 1968.

Last but not least, the author profited very much by discussions with and helpfulness from Professor Dr. E. H. Colbert during visits to the American Museum of Natural History, New York.

Literature Cited

Besenecker, H. 1973. Neogen und Quartär der Insel Chios (Ägäis). Dissertation, Johannes-Gutenberg-Universität, Mainz, 196 pp.

Fahlbusch, V. 1976. Report on the International Symposium on mammalian stratigraphy of the European Tertiary. Newsl. Stratigr., 5(2/3):160–67.

Gaziry, A. W. 1976. Jungtertiäre Mastodonten aus Anatolien (Türkei). Geol. Jahrb., B 22:3–143.

Melentis, J. K., and Tobien, H. 1967. Paläontologische Ausgrabungen auf der Insel Chios (eine Vorläufige Mitteilung). Praktika Akad. Athen, 42:147–52.

———. 1968. Paläontologische Ausgrabungen auf der Insel Chios (eine vorläufige Mitteilung). Ann. Géol. Pays Hellén., 19:647–52.

Nafiz, H., and Malik, A. 1933. Vertébrés fossiles de Küçükçekmece. Istanbul, Faculty Sci., Bull., 3/4, 119 pp.

Osborn, H. F. 1936. Proboscidea: A monograph of the discovery, evolution, migration, and extinction of the mastodonts and elephants of the world. Vol. 1, Moeritherioidea, Deinotherioidea, Mastodontoidea. American Museum Press, New York, 802 pp.

Paraskevaidis, J. 1940. Eine obermiozäne Fauna von Chios. Neues Jahrb. Mineral. Geol. Paläontol., 83:363–442.

Rothausen, K. 1977. Die mittelmiozänen Wirbeltierfundstellen südlich Thymiana (Insel Chios, Ägäis, Griechenland). 2. Teil: Geologie: Die Fundstellen und ihre Abfolge. Ann. Geol. Pays Hellén., 28:495–515 (publ. 1979).

Schlesinger, G. 1917. Die Mastodonten des K. K. Naturhistorischen Hofmuseums. Denkschr. Naturhist, Hofmus. I, Geol. Paäontol., 1:1–230.

Tassy, P. 1977. Les Mastodontes miocènes du Bassin aquitain: une mise au point taxonomique. Acad. Sci., C. R., Sér. D., 284:1389–92.

Tobien, H. 1968. Paläontologische Ausgrabungen nach jungetrtiären Wirbeltieren auf der Insel Chios (Griechenland) und bei Maragheh (NW Iran). Jahrb. Vereinigung "Freunde der Univ. Mainz," pp. 51–58.

———. 1969. Wirbeltiergrabungen im Miozän der Insel Chios (Ägäis). 2. vorläufige Mitteilung. Praktika Akad. Athen, 43:151–57.

———. 1973. On the Evolution of Mastodonts (Proboscidea, Mammalia) Part 1: The bunodont trilophodont Groups. Notizbl. hess. L.-Amt Bodenforsch., 101:202–76.

———. 1977. Die mittelmiozänen Wirbeltierfundstellen südlich Thymiana (Insel Chios, Ägäis, Griechenland. 1. Teil: Einleitung. Ann. Géol. Pays Hellén., 28:489–94 (publ. 1979).

CORRELATION OF SIWALIK FAUNAS

by Everett H. Lindsay, Noye M. Johnson,
and Neil D. Opdyke

Introduction

The succession of vertebrate faunas in Siwalik deposits at the forefront of the Himalaya Mountains of southcentral Asia is one of the most impressive vertebrate faunal successions in the world. Fossils from these deposits were made known to the scientific world mainly by Falconer during the interval 1830–60. Notable additions to knowledge of Siwalik mammals were later made by Lydekker, Pilgrim, Matthew, Colbert, Lewis, and others. Collections studied by those named above were made prior to 1933, and are located primarily in four museums: the Indian Museum, Calcutta; the British Museum (Natural History), London; the American Museum of Natural History, New York; and the Yale Peabody Museum, New Haven. In 1935, Colbert published a monograph on vertebrate faunas from the Siwaliks that summarized previous studies and has served as the major reference for vertebrate biochronology and evolution in the Siwaliks for the last 45 years. A complete bibliography of important Siwalik publications prior to 1935 is found in Colbert's monograph.

Fossils collected by Lewis in the 1930s suggested that Siwalik primates might hold the key to the ancestry of man. In 1969 paleontologists from the Yale Peabody Museum, led by Elwyn Simons, traveled to northern India in the quest for more fossils. In 1973, study of the extensive Siwalik deposits in the Potwar Plateau of Pakistan was initiated independently by two groups, each unaware of the other's plans. These groups were the Yale Peabody Museum, collaborating with the Geological Survey of Pakistan, and Dartmouth College, collaborating with Peshawar University of the Northwest Frontier Province of Pakistan, Columbia University, and the University of Arizona.

The Yale–Geological Survey of Pakistan group, led by David Pilbeam, was interested primarily in middle Siwalik fossils and strata where primate remains are most common. Their studies were concentrated in the Soan Synclinorium, southwest of Rawalpindi. The Dartmouth-Peshawar group, led by Noye Johnson, was interested primarily in upper Siwalik fossils and strata, which they planned to correlate with similar deposits in North America by magnetic polarity stratigraphy. Gradually, following 1973, interest of both groups broadened and overlapped, resulting in increased collaboration. By 1980 both groups were effectively combined into one. One result of combined efforts is a composite magnetostratigraphic sequence for Siwalik strata in the Potwar Plateau of Pakistan. This report summarizes development of the magnetostratigraphic framework in the Potwar Plateau, correlation of deposits within that area, and correlation of the Siwalik faunal sequence with North American and European faunal sequences.

310   *Aspects of Vertebrate History*

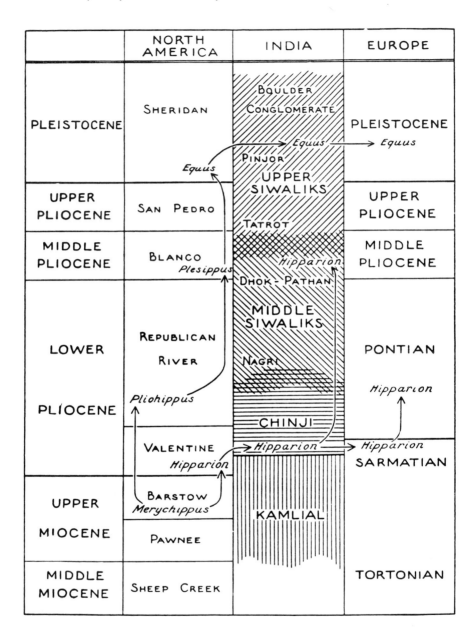

Figure 19.1 Correlation of Siwalik deposits and faunas with North American and European faunal sequences as presented by Colbert in 1935. Reproduced with permission of the American Philosophical Society.

Siwalik Stratigraphy

The term "Siwalik" was applied by Medlicott in 1864 to terrestrial deposits in the Siwalik Hills of present day India. Wynne (1875) extended the concept and term (with some hesitation) to the Potwar Plateau of Pakistan. Subdivision of Siwalik faunas and strata during the early 20th Century was accomplished primarily through the work of Pilgrim, especially his (1913) "Correlation of the Siwaliks with mammal horizons of Europe." During the 1920s there remained a question of whether the Kamlial Formation should be included as the basal unit of the Siwalik Group, or as the uppermost unit of the Murree Group which underlies the Siwalik Group. Subdivision of Siwalik faunas and strata has been relatively stable since 1935 when Colbert summarized the Siwalik faunas (Figure 19.1): Kamlial and Chinji zones were placed in the lower Siwaliks, Nagri and Dhok Pathan zones were placed in the middle Siwaliks, and Tatrot, Pinjor, and Boulder Conglomerate zones were placed in the upper Siwaliks. Lower, middle, and upper Siwalik divisions were conceived and applied as formal biochronologic or chronostratigraphic units by both Pilgrim and Colbert in their studies, but are recognized as informal lithostratigraphic divisions in this study (i.e., as subgroups within the formal Siwalik Group). Actually, biochronologic units have yet to be formalized by stratigraphic codes, although North American Provincial Land Mammal Ages were formalized as quasi-biochronologic units about six years after Colbert's monograph (Wood, et. al., 1941), and chronostratigraphic units were invented about the same time (Schenck and Muller, 1941).

Colbert (1935, pp. 21–26) reviewed the history and rationale for correlation of Siwalik faunas with European and North American faunas. Correlation of Siwalik deposits and subdivisions has been hindered by failure to distinguish lithostratigraphic from biostratigraphic aspects of those deposits. Paleontologists studying Siwalik faunas commonly refer to Siwalik subdivisions as faunal units (e.g., Chinji zone), whereas geologists studying Siwalik sediments commonly refer to those same subdivisions as lithostratigraphic units (e.g., Chinji Formation). This practice persisted for a long time, partly because lower and middle Siwalik sediments are readily divisible into lithostratigraphic units. For example, there is a distinctive lithostratigraphic unit for Kamlial, Chinji, Nagri, and Dhok Pathan formations. However, upper Siwalik deposits lack persistent and characteristic lithologies.

An attempt to establish an internally consistent and structured terminology for Siwalik deposits was made by the Stratigraphic Committee of Pakistan in 1973 (Fatmi, 1973). This committee formalized the well-known Siwalik subdivisions as lithostratigraphic units (formations), and characterized each by a type section, dominant lithologies, distribution, thickness, and predominant vertebrate fossils. Geology and paleontology for the lower and middle Siwaliks was recently summarized by Pilbeam, et al. (1977), and by Moonen, et al. (1978). The principal upper Siwalik lithostratigraphic reference in Pakistan is the Soan Formation, named by Kravtchenko (1964) for strata in the upper part of the Soan Synclinorium south of Rawalpindi. The Soan Formation is poorly fossiliferous in the type area, and our paleomagnetic study indicates the type Soan Formation is *not* equivalent in age to strata that yield upper Siwalik fossils, but is equivalent to strata of the middle Siwalik Dhok Pathan Formation.

Magnetic Polarity Sequence

Six stratigraphic sections in the Potwar Plateau provide a framework for magnetic polarity zonation of Siwalik strata. The location, lithologic properties, and magnetic properties of these six reference sections are presented elsewhere (Johnson, et al., in press). Three of these six reference sections are from the Soan Synclinorium; the Dhok Pathan section is on the north limb of the synclinorium, the Chinji-Nagri and Chakwal-Bhaun sections are on the south limb of the synclinorium. The Dhok Pathan section is that presented by Barndt, et al. (1978), as modified by Tauxe (1979). Several other fossiliferous sections on the north limb of the synclinorium are being compiled by Tauxe and will be presented elsewhere. Three of the reference sections described here (Jalapur, Tatrot-Andar, and Kotal Kund) are from the Bunha River area west of Jhelum. Upper Siwalik magnetostratigraphy and paleontology were summarized by Opdyke, et al. (1979). A bentonite in the Nagri Formation within the Chinji-Nagri section has yielded a fission track date of 9.50 ± 0.63 Ma (Johnson, et al., in press). This, and the fission track dated unit in the upper Siwalik strata of the Kotal Kund section help to "frame" the magnetic polarity sequence in the Potwar Plateau.

The Kotal Kund section is our main reference section for Siwalik strata because it is one of the most complete and lithologically distinctive sections. The Kotal Kund section is approximately 1,800 m thick, of which the upper 800 m are upper Siwalik strata. A volcanic ash high in the Kotal Kund section has yielded a fission track date of 2.35 ± 0.35 Ma. The base of Kotal Kund section is in the Chinji Formation; the stratigraphic section extends lower, into the Kamlial Formation, but those rocks have poor paleomagnetic properties (because of heavy magnetic overprint) and did not yield fossils. The Tatrot-Andar section is a near duplicate of the Kotal Kund section, with the exception that upper Siwalik strata are much thinner in the Tatrot-Andar section. An unconformity between middle Siwalik and upper Siwalik strata at Tatrot has long been recognized (DeTerra and Teilhard de Chardin, 1936). Johnson, et al. (in press), demonstrated that the hiatus represents about 2 million years, i.e., from about 5 Ma in lower Gilbert polarity chron* below the unconformity to about 3 Ma in the Gauss polarity chron above the unconformity.

Comparison of sections from the Soan Synclinorium and the Bunha River area (Figure 19.2) point to the absence, or near absence, of upper Siwalik strata over much of the Soan Synclinorium. Both the Soan Synclinorium and the Bunha River area are covered by flat-lying, poorly indurated sediment called the Potwar silt. The Potwar silt is usually normally magnetized at its base and is therefore considered in the Brunhes polarity chron and less than a million years old.

Development of Siwalik Chronostratigraphy

Correlation of Siwalik faunas has previously been hindered for several reasons. Foremost among these is the absence of good biostratigraphic control for early studies of Siwalik faunas. Siwalik fossils were rarely placed in a stratigraphic section, and when stratigraphic sections were drawn, lithologic units were given faunal

---

*Use of the term "chron" follows the recommendation of the IUGS International Subcommission on Stratigraphic Classification and IUGS/IAGA Subcommision on a Magnetic Polarity Time Scale (1979).

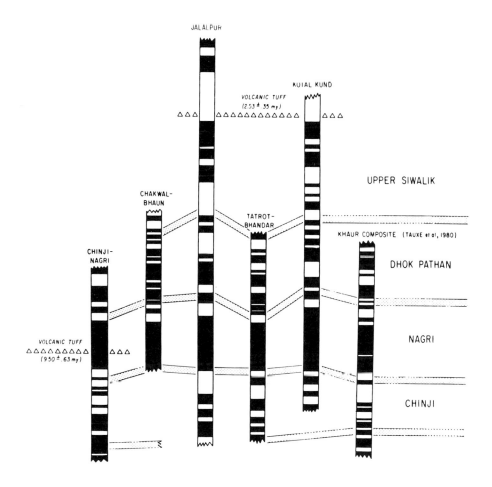

Figure 19.2 Magnetic polarity zonation of Siwalik lithostratigraphic units in six representative sections on the Potwar Plateau (from Johnson, et al., in press).

names. The resulting stratigraphic framework, based on both paleontologic and lithologic criteria, but without defining limits or associations of each has created problems of correlation that must be resolved by detailed biostratigraphic collecting. The addition of magnetostratigraphic zonation to the biostratigraphic framework greatly facilitates correlation, both locally and worldwide.

We recommend restriction of the classical terminology for Siwalik faunas (i.e., Kamlial, Chinji, Nagri, Dhok Pathan, Tatrot, and Pinjor) to lithostratigraphic units in keeping with the formalization of those lithostratigraphic units by the Stratigraphic Committee of Pakistan (Fatmi, 1973). We recommend further that new terminology be developed for Siwalik chronostratigraphic units, and suggest the terms lower Siwalik, middle Siwalik, and upper Siwalik for nonprecise reference to previous faunal divisions. Eventually, the new chronostratigraphic units will supercede the nonprecise lower, middle, and upper Siwalik faunal divisions. The final stratigraphic framework will apply classical terms to rock units and new terms to chronostratigraphic units.

Procedure for developing the new Siwalik chronostratigraphic framework involves four steps: *first*, to relate all fossil localities to stratigraphic sections that have been sampled for magnetic polarity; *second*, to correlate between sections by comparison of similar magnetic polarity sequences; *third*, to group fossil localities from separate but correlated sections into informal faunal units that occur within the same magnetic polarity zone; and *fourth*, to analyze the resulting faunal-magnetostratigraphic sequence for significant faunal changes, usually marked by immigration events. The end result, step four, can be considered a chronozone. Its limits are marked by faunal changes, and its duration is measured by magnetic polarity. We propose these chronostratigraphic units be called "Siwalik mammal zones," e.g., the *Equus-Elephas* Siwalik Mammal Zone. These chronozones will be applicable only to Siwalik strata, but their correlation with terrestrial faunal units on other continents or with marine biozonation is possible by identifying magnetic polarity intervals.

Steps one through three of the above procedure were followed by Opdyke, et al. (1979). They could have named (but did not name because the procedure outlined above had not been proposed) an *Equus-Elephas* Siwalik Mammal Zone for strata in the uppermost Gauss and Matuyama polarity chrons of the Siwaliks. Similarly, Pilbeam, et al. (1979), named Hasal, Dhurnal, and Bora chronozones for faunal divisions placed relative to the magnetic polarity sequence in the Soan Synclinorium of the Potwar Plateau. Those chronozones are bounded by magnetic polarity reversals rather than faunal changes, although some faunal turnover appears near coincident with some of the polarity reversals. The chronozones named by Pilbeam, et al. (1979), represent step three of the procedure outlined above. Note that revision of the polarity sequence by Tauxe (1979) changed the limits (but not the sequence) of the Hasal and Dhurnal magnetic chronozones. Pinpointing fossil localities in the revised and more detailed paleomagnetic framework is now underway. Hence, step three in the chronostratigraphic framework is essentially completed. More rigorous biostratigraphic and taxonomic study is necessary before step four of the procedure can be completed.

Correlation of Siwalik Faunas with Other Continents

As seen in Figures 19.2 and 19.3, the boundary between the Kamlial and Chinji

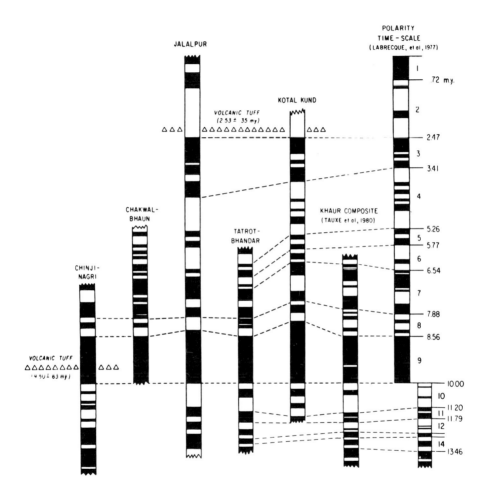

Figure 19.3 Correlation of magnetic polarity sequence in six representative sections on the Potwar Plateau with the magnetic polarity time scale of LaBrecque, et al. (1977), with calibration following Mankinen and Dalrymple (1979) (from Johnson, et al., in press).

formations is approximately in polarity chron 12, although the limits of the Kamlial Formation (and resolution of the magnetic polarity time scale in that interval) are still poorly defined. The Chinji Formation includes part of chron 12, all of chron 11, and most of chron 10. The Nagri Formation includes the top of chron 10, all of chron 9, and most of chron 8. The boundary between the Nagri and Dhok Pathan formations approximates the chron 7–8 boundary, but (as expected) varies in different sections. The Dhok Pathan Formation includes most of chron 7, all of chron 6, and most (if not all) of chron 5. The boundary between the Dhok Pathan and upper Siwalik strata approximates the chron 4–5 boundary. Upper Siwalik strata include most if not all of chron 4, all of chron 3, and most of chron 2.

In terms of the faunas, the lower Siwalik faunas correlate approximately with polarity chron 10–12; middle Siwalik faunas correlate approximately with polarity chrons 5–9; and upper Siwalik faunas correlate approximately with polarity chrons 2–4. Based on calibration of the magnetic polarity time scale of LaBrecque, et al. (1977), as modified by Mankinen and Dalrymple (1979), the Siwalik strata and faunas can be placed in a calibrated time frame (Figure 19.3). As noted above, limits of these Siwalik faunal units are not precise; better accuracy and resolution is expected upon completion of step four in the chronostratigraphic procedure.

Correlation of these poorly limited Siwalik faunal units with the calibrated magnetic polarity time scale allows broad correlation with the North American and European faunal sequences seen in Figure 19.4. The lower Siwalik faunas occur approximately between 10 and 12.5 Ma, the middle Siwalik faunas occur approximately between 5 and 10 Ma, and the upper Siwalik faunas occur approximately between 1 and 5 Ma. This places the lower Siwalik faunas broadly equivalent to the late Barstovian and early Clarendonian Land Mammal Ages of North America and the Vallesian Land Mammal Age of Europe. Middle Siwalik faunas are broadly equivalent to late Clarendonian and most of Hemphillian Land Mammal Ages in North America, and the Turolian and beginning of Ruscinian Land Mammal Ages in Europe. Upper Siwalik faunas are broadly equivalent to Blancan and Irvingtonian Land Mammal Ages of North America, and Ruscinian and Villafranchian (or Villanyian) Land Mammal Ages of Europe.

The above correlations, based on magnetic polarity zonation of Siwalik strata, are not significantly different from those given by Pilgrim and Colbert. Many of the terms and epoch boundaries have been changed, but an overall similarity is notable. Greater resolution and accuracy of Siwalik correlations is promised by the chronostratigraphic framework now being developed in the Potwar Plateau.

An interesting result from this Siwalik magnetostratigraphic framework is realization that the *Hipparion* datum of the Siwaliks is significantly younger than the *Hipparion* datum of Europe. Pilbeam, et al. (1979), noted the appearance of *Hipparion* in the Siwaliks at the base of magnetic chron 9, or the base of the Dhurnal chronozone. Our calibration of that datum is 10.0 Ma. We know of no earlier

---

Figure 19.4 Magnetic polarity sequence of Siwalik deposits in the Potwar Plateau and correlation of the Siwalik vertebrate sequence with those of North America and Europe. Slanted vertical lines represent overlap of Siwalik lithostratigraphic units (in the Potwar Plateau sequence), and uncertainty in the limits of vertebrate land mammal ages relative to the magnetic polarity time scale. Age limits of magnetic chrons are interpolated from position of those chrons given by LaBrecque, et al. (1977), as calibrated by Mankinen and Dalrymple (1979).

NORTH AMERICA   POTWAR PLATEAU   EUROPE

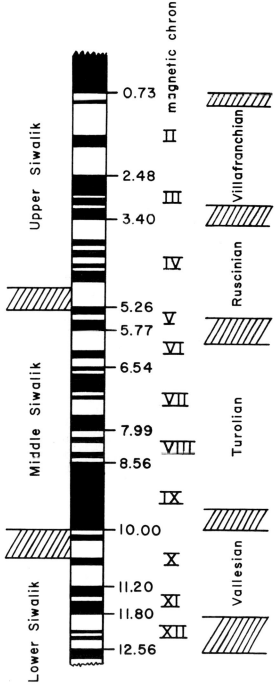

records from the sections in the Bunha River area; to our knowledge, the date of 10.0 Ma for the Siwalik *Hipparion* datum is reliable. The *Hipparion* datum in Europe "converges" on a date of 12.5 Ma (Berggren and Van Couvering, 1978; and Bernor, et al., in press), approximately 2 million years earlier than the Siwalik *Hipparion* datum. L. L. Jacobs (personal communication, 1980) records the appearance of the murid genus *Progonomys* from Yale–Geological Survey of Pakistan locality 259, placed below the Siwalik *Hipparion* datum by Pilbeam, et al. (1979, Figure 14). *Progonomys* commonly follows (is later than) the appearance of *Hipparion* in Europe. The record of *Progonomys* in the Siwaliks supports a later appearance of *Hipparion* in the Siwaliks than in Europe. Thus, the first occurrence of *Hipparion* in the Siwaliks appears closer temporally to the Turolian faunas of Maragheh (Iran) and Mt. Luberon (France) than to the Vallesian faunas of Höwenegg (Germany) and Bou Hanifia (North Africa).

Acknowledgments

We are deeply indebted to our friends and colleagues who participated in the Yale–Geological Survey of Pakistan and Dartmouth–Peshawar field studies, and to their enthusiastic, diligent, and good-natured support. This work has been supported by research grants from the National Science Foundation (EAR 74-13860), and the Smithsonian Foreign Currency Program (SFCP FC80254100).

Literature Cited

Barndt, J.; Johnson, N. M; Johnson, G. D.; Opdyke, N. D.; Lindsay, E. H.; Pilbeam, D.; and Tahirkheli, R. A. K. 1978. The magnetic polarity stratigraphy and age of the Siwaliks near Dhok Pathan Village, Potwar Plateau, Pakistan. Earth Planet. Sci. Lett., 41:355–64.

Berggren, W. A., and Van Couvering, J. A. 1978. Biochronology. Amer. Ass. Petrol. Geol., Studies in Geology, 6:39–55.

Bernor, R. L.; Woodburne, M. O.; and Van Couvering, J. A. In press. Contribution to the chronology of some Old World Miocene faunas based on hipparionine horses. Palaeogeogr. Palaeoclimatol. Palaeoecol.

Colbert, E. H. 1935. Siwalik Mammals in the American Museum of Natural History. Amer. Phil. Soc., Trans., 26:1–401.

De Terra, H., and Teilhard de Chardin, P. 1936. Observations on the upper Siwalik Formation and later Pleistocene deposits in India. Amer. Phil. Soc., Proc., 76:791–822.

Fatmi, A. N. (ed.). 1973. Lithostratigraphic units of the Kohat-Potwar Province, Indus Basin, Pakistan. Geol. Surv. Pakistan, Mem., 10:1–80.

IUGS International Subcommission on Stratigraphic Classification and IUGS/IAGA Subcommission on a Magnetic Polarity Time Scale 1979. Magnetostratigraphic polarity units—A supplementary chapter of the ISSC International Stratigraphic Guide. Geology, 7:578–83.

Johnson, N. M.; Opdyke, N. D.; Johnson, G. D.; Lindsay, E. H.; and Tahirkheli, R. A. K. In press. Magnetic polarity stratigraphy and ages of the Chinji, Nagri and Dhok Pathan formations of the Potwar Plateau, Pakistan. Palaeogeogr. Palaeoclimatol. Palaeoecol.

Kravtchenko, K. N. 1964. Soan Formation—upper unit of Siwalik Group in Potwar. Science and Industry (Pakistan), 2:230–33.

LaBrecque, J. L.; Kent, D. V.; and Cande, S. C. 1977. Revised magnetic polarity time scale for the Late Cretaceous and Cenozoic. Geology, 5:330–35.

Mankinen, E. A., and Dalrymple, G. B. 1979. Revised geomagnetic polarity time scale for the interval 0–5 m.y.b.p. J. Geophysical Res., 84:615–26.

Medlicott, H. B. 1864. On the geological structure and relations of the southern portions of the Himalayan ranges between the rivers Ganges and Ravee. Geol. Surv. India, Mem., 3:1–206.

Moonen, J. J. M.; Sondaar, P. Y.; and Hussain, S. T. 1978. A comparison of larger mammals in the stratotypes of the Chinji, Nagri, and Dhok Pathan formations (Punjab, Pakistan). Kon. Nederl. Akad. Wetenschappen, Proc. ser. B, 81:425–36.

Opdyke, N. D.; Lindsay, E. H.; Johnson, N. M.; Tahirkheli, R. A.; and Mirza, M. A. 1979. Magnetic polarity stratigraphy and vertebrate paleontology of the upper Siwalik subgroup of northern Pakistan. Palaeogeogr, Palaeoclimatol, Palaeoecol., 27:1–34.

Pilbeam, D.; Barry, J.; Meyer, G. C.; Shah, S. M. I.; Pickford, M. H. L.; Bishop, W. W.; Thomas, H.; and Jacobs, L. L. 1977. Geology and paleontology of Neogene strata of Pakistan. Nature, 270:684–89.

Pilbeam, D. R.; Behrensmeyer, A. K.; Barry, J. C.; and Shah, S. M. I. (eds.). 1979. Miocene sediments and faunas of Pakistan. Postilla, 179:1–45.

Pilgrim, G. E. 1913. Correlation of the Siwaliks with mammal horizons of Europe. Geol. Surv. India, Rec., 43:264–326.

Schenck, H. G., and Muller, S. W. 1941. Stratigraphic terminology. Geol. Soc. Amer., Bull., 52:1419–26.

Tauxe, Lisa. 1979. A new date for *Ramapithecus*. Nature, 282:399–401.

Wood, H. E. II; Chaney, R. W.; Clark, J.; Colbert, E. H.; Jepsen, G. L.; Reeside, J. B., Jr.; and Stock, C. 1941. Nomenclature and correlation of the North American continental Tertiary. Geol. Soc. Amer., Bull., 52:1–48.

Wynne, A. B. 1875. Geological notes on the Khareean Hills in the upper Punjab. Geol. Surv. India, Rec., 8:46–49.

LATE CRETACEOUS AND EARLY TERTIARY VERTEBRATE PALEONTOLOGICAL RECONNAISSANCE, TOGWOTEE PASS AREA, NORTHWESTERN WYOMING

by Malcolm C. McKenna

Introduction

This paper deals primarily with the vertebrate paleontology of an area east of Jackson Hole, Wyoming, along the border of Teton County with Fremont County (Figure 20.1). The summit of Togwotee Pass, one of the principal routes into Jackson Hole, lies about a mile west of the county line in Teton County. Little has been published about the many vertebrate faunas represented in the various rock units of eastern Teton County, but it is noteworthy, especially in this volume, that one of the first paleontological studies of a vertebrate fauna in the area was written by E. H. Colbert (1943). Colbert's study dealt with a Miocene fauna on Pilgrim Creek, about 25 miles west of the summit of Togwotee Pass. At the time his paper was published, nothing was known of the Cretaceous, Paleocene, Eocene, or Oligocene faunas discussed here.

As one travels westward toward the thrust fault belt in Wyoming the stratigraphic section becomes generally thicker, lithologies become more diverse, and the structural history more complex. As a result, synthesis of the geological evolution of western Wyoming has required much detailed work in many scientific disciplines and has necessitated many localized studies of key areas. Nowhere is this more apparent than in the (composite) nearly 40,000 feet of Tertiary and Late Cretaceous rocks of Teton County and adjacent areas to the east (Keefer, 1965; Love, 1956a, 1956b, 1973; Love, et al., 1978). Nevertheless, little has been published on the vertebrate paleontology of the area.

The present summary of vertebrate paleontological reconnaissance in the vicinity of Togwotee Pass is based on results of field work conducted during parts of twenty-three seasons. Prospecting has been carried out by various parties under my direction, at first under the auspices of the University of California and later, since 1960, as part of a continuing program of the American Museum of Natural History and Columbia University. Numerous students and volunteers have participated in the paleontological exploration of the region since 1956. Most of the work has been done in collaboration with Dr. J. D. Love, whose U.S.G.S. Oil and Gas Investigations Preliminary Chart 27 (1947) led to our initial interest in the area. Recently the tempo of purely geological work has increased as detailed 7½-minute topographic quadrangle maps have become available. Lindsey (1972) and Love (1973) have greatly increased our knowledge of the Harebell Formation and Pinyon Conglomerate of the region. W. L. Rohrer's geologic mapping in the Kisinger Lakes (1966), Fish Lake (1968), and Sheridan Pass (unpublished, available on open file, U.S.G.S., Denver) 7½-minute quadrangles has contributed enormously to knowledge of an area mainly south of Togwotee Pass. Smedes and Prostka (1972)

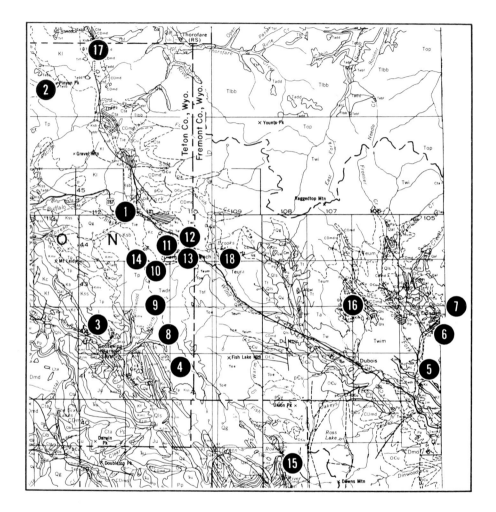

Figure 20.1. Map showing places cited in the text: 1. Harebell vertebrate locality, road-cut, south side of Highway 26/287; 2. Pinyon Peak; 3. Coal-bearing Paleocene underneath referred Pinyon Conglomerate near Goosewing Ranger Station; 4. Love Quarry; 5. Type Indian Meadows Formation; 6. Type locality, Aycross Formation; 7. Type Tepee Trail Formation; 8. Purdy Basin (Buckskin Ridge is at SE end); 9. Hardscrabble Creek; 10. Big Bend of North Fork of Fish Creek; 11. Locality L-41 in Aycross Formation; 12. Bridgerian locality at the summit of Togwotee Pass; 13. Moccasin Basin; 14. Tripod Fault; 15. White Rock Fault; 16. Keefer's Twdr$_4$ and Twdr$_5$; 17. Mink Creek/Fox Creek White River/Arikaree; 18. Pinnacle Buttes. Map based on geologic map of Wyoming (Love, et al., 1955).

and Love, et al. (1978), obtained very important new information, including many K-Ar dates, on the volcanic and volcaniclastic rocks of the Absaroka Volcanic Supergroup that crop out at and near Togwotee Pass. Chadronian and Arikareean faunas and sediments north of the pass along the Buffalo Fork Thrust trace have been described by Love, et al. (1976). Radioisotopic dating of various key Tertiary horizons in the region has led to intensified prospecting for fossil vertebrates in the rocks in which these horizons appear. Radioisotopic calibration of the biostratigraphic record has been a major goal of research in the region (McKenna, et al., 1973; Berggren, et al., 1978; Hardenbol and Berggren, 1978). Indeed, radioisotopic calibration of the nonmarine Eocene biostratigraphic record of North America (Figure 20.2) is greatly facilitated by the joint occurrence of datable rocks and biostratigraphically useful fossils in the Wind River Basin and Togwotee Pass area. In the research reported here, an attempt has been made to find and make known the fossil assemblages that might occur interbedded in datable rocks near their volcanic source rather than to search for undiscovered datable rocks in areas away from such sources in sediments where fossils were already well known. Paleomagnetic stratigraphy of the same rocks is currently under active study (J. J. Flynn, in preparation).

Detailed descriptions of the localities mentioned in this paper are available to qualified investigators. This paper is a modified and updated version of a manuscript privately printed and distributed in 1972.

Potassium-Argon ages in this paper have not been corrected by the factors suggested by Dalrymple (1979) but are given as originally cited.

Synopsis of Late Cretaceous and Early Tertiary Geological History and Vertebrate Faunas of the Togwotee Pass Area

Upper Cretaceous Rocks

Toward the end of the Cretaceous, northwestern Wyoming became involved in compressional folding and thrust faulting began to the west. The first of many wedges of predominantly quartzite roundstone conglomerate spread eastward, within the Harebell Formation (Lindsey, 1972; Love, 1973). The Harebell also contains ashy beds and a brackish-water marine tongue that must represent an ancient connection with the Mid-Continent Sea; the link was probably via the north, across the area of present-day Yellowstone Park (Love, 1956a, 1956b, 1973). A locality in the Harebell along Highway 26/287, west of the summit of Togwotee Pass, has yielded the following vertebrates (preliminary identifications):

Amia (including *Kindleia*) sp. (fish)
Prodesmodon sp. (salamander)
crocodilian
ceratopsian dinosaur tooth fragments
?camptosaurian dinosaur tooth fragments
?nodosaurian dinosaur tooth fragments

Several other dinosaur-yielding localities in the Harebell were reported by Love (1973). The age of the Harebell is probably Maastrichtian. It is locally about 5,000 feet thick but varies widely in preserved thickness. Following the deposition of the Harebell, the first uplift of the Washakie Range (Love, 1939), an important

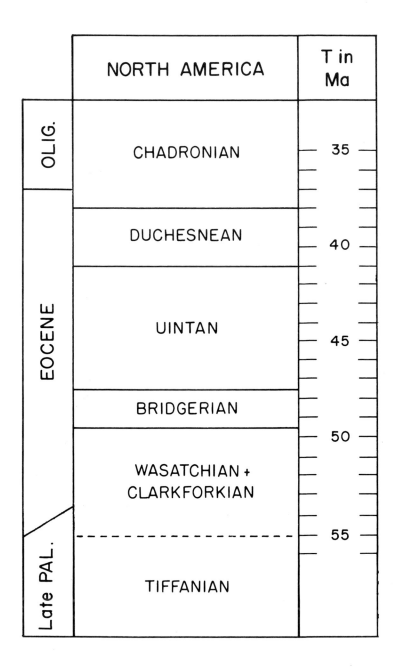

Figure 20.2. Radioisotopic calibration of the North American continental Eocene (taken from Berggren, et al., 1978, with slight modification). The K-Ar ages should be increased by 2.6 percent to be in line with others calculated on the basis of decay and abundance constants recommended by the Subcommission on Geochronology of the IUGS (Dalrymple, 1979).

southeast-northwest trending range whose Precambrian igneous core and flanking Paleozoics locally peek out from beneath the early Tertiary Absaroka Volcanic Supergroup, caused a local 90-degree unconformity beneath the next unit to be deposited, the Pinyon Conglomerate. The age of one Pinyon locality near the base of the principal reference section at the type locality (east side of Pinyon Peak, about 25 miles northwest of the summit of Togwotee Pass), is latest Maastrichtian on the basis of a find of *Leptoceratops* (see McKenna and Love, 1970). The Pinyon Conglomerate is more than 3,700 feet thick at its type locality (Love, 1973) and was rapidly deposited. It is not known whether all of the principal reference section of the Pinyon is Maastrichtian in age. The Pinyon is noteworthy for its enormous mass of pressure-marked quartzite roundstones derived from the Precambrian Belt Supergroup and other quartzite sources. At its type locality on Pinyon Peak and in contiguous exposures north of the Buffalo Fork River the Pinyon is cartographically discontinuous with and lithologically somewhat distinct from referred Pinyon Conglomerate south of the Buffalo Fork River and Highway 26/287. In the southern deposits relatively abundant mafic volcanic and ignimbrite roundstones as well as pressure-marked quartzite roundstones occur (Lindsey, 1970, 1972). At its type locality on Pinyon Peak the Pinyon Conglomerate is overlain unconformably by the Langford Formation and early mafic volcanics of the Absaroka Volcanic Supergroup from which a whole-rock K-Ar date of 49.4 ± 1 Ma has been obtained from a basalt sample collected from the summit of Pinyon Peak (McKenna, 1972; Love, 1973, Figure 18; Love, et al., 1978).

Paleocene Rocks

For a record of Paleocene events other than erosion, one must move either westward to Pilgrim Creek (Love, 1973) or south of the Buffalo Fork River, where abundant sediments of Paleocene age are exposed, especially in the Gros Ventre River drainage. Following the folding of the Spread Creek, Sohare, and other anticlines in the late Maastrichtian, deposition of low energy Paleocene coal-bearing sediments took place, at least locally near the Goosewing Ranger Station on the Gros Ventre River (Love, 1956b, 1973). Invertebrates and pollen are the basis of the date for these sediments (Love, 1973). These low energy deposits were followed by a pulse of conglomerate, as much as 1,400 feet thick, mostly composed of quartzite roundstones but with a significant but small percentage of volcanic roundstones as well, including ignimbrite roundstones. This conglomerate is interbedded with and replaced stratigraphically upward by a 900-foot-thick greenish gray and brown sandstone and shale sequence (Love, 1947, 1973). The conglomerate has been identified as Pinyon Conglomerate by most workers in the region, but the other two Paleocene units are still unnamed.

Love's original report on these rocks (Love, 1947) listed fragmentary specimens of a ptilodontine multituberculate, a taeniodont, and a turtle from his locality 14 within the uppermost sequence. This site has now been developed as Love Quarry. It is in the NW¼, SW¼, Sect. 9, T 41 N, R 110 W, Sheridan Pass 7½-minute Quadrangle, Teton County, Wyoming (Figure 20.3). The fossils occur in a clay-ball conglomerate at the base of a sandstone channel 280 feet above the last quartzite roundstone conglomerate and 800 feet above the main mass of referred Pinyon Conglomerate. Acetic and formic acid etching of blocks of clay-ball conglomerate from Love Quarry has yielded numerous specimens of the fossil vertebrates listed

Figure 20.3. Air oblique view northeast across Fish Creek (in foreground) showing Tertiary mappable units described by Love (1947, 1973). Indicated are Devils Basin Creek (DB); Mid-Tiffanian Love Quarry (LQ); Pinyon Conglomerate (TKp); greenish-gray and brown sandstone and shale sequence (Ts), site of Love, 1973, fig. 32 (X); Tiffanian leaf and mollusc locality L-66-87 of Love, 1973 (L); lower variegated sequence containing Clarkforkian and Wasatchian vertebrate fossils (Tlv); coal-bearing unit (C); upper variegated sequence containing Wasatchian vertebrate fossils (Tuv); Aycross Formation (Ta); and Wiggins Formation in Absaroka Range (Twi). Photograph by R. C. Casebeer, reprinted with modification of the legend from Figure 31 of Love, 1973, with the author's permission.

below. Invertebrates also occur abundantly in Love Quarry and above the quarry at various stratigraphic levels in the greenish gray and brown sandstone and shale sequence.

fish
amphibian
crocodilian
turtle
glyptosaurine anguid lizard, cf. *Odaxosaurus* or *Proxestops jepseni*
Multituberculata:
  Neoplagiaulax fractus
  Ectypodus powelli
  cf. *Prochetodon cavus* or *Ptilodus* sp.
small eutherians:
  Mckennatherium cf. *M. ladae*
  Leptacodon sp.
  cf. *Litolestes notissimus*
  cf. *Elpidophorus elegans*
  ?Mixodectes sp.
  ?Prodiacodon sp.
  Propalaeosinopa thomsoni
Primates:
  Zanycteris paleocenus
  ?Paromomys depressidens
  Ignacius frugivorus
  Carpodaptes hobackensis
  Plesiadapis rex
Carnivora:
  Didymictis sp.
Condylarthra:
  Thryptacodon antiquus
  Mimotricentes subtrigonus
  Phenacodus grangeri
  Phenacodus cf. *P. brachypternus*
  Ectocion (including *Gidleyina*) sp.
  Protoselene cf. *P. opisthacus*
  Promioclaenus cf. *P. aquilonius*
  Litomylus sp.

The age of the assemblage from Love Quarry is mid-Tiffanian (Gingerich, 1976; Rose, 1977). Love Quarry is approximately the same age as Cedar Point Quarry in the Fort Union Formation of the Bighorn Basin (Gingerich, 1976, p. 54) and, nearer at hand, the Battle Mountain local fauna in the Hoback Basin, about 30 miles to the southwest across the Gros Ventre Range (Dorr, 1952, 1958, 1978).

In view of the high energy of deposition and the stratigraphic continuity of the greenish gray and brown sandstone and shale sequence with the underlying referred Pinyon Conglomerate, it is probable that the referred Pinyon of the Sheridan Pass Quadrangle and perhaps all of the referred Pinyon south of Highway 26/287 and the Buffalo Fork River is about 8 to 10 million years younger than the lower part (and perhaps all) of the Pinyon at its type locality at Pinyon Peak north of the highway.

Love (1947) regarded the greenish gray and brown sandstone and shale sequence to be of Paleocene or earliest Eocene age. In view of the pre-Wasatchian and even possibly pre-Clarkforkian age of the lower part of the overlying lower variegated sequence of Love (1947), all of the greenish gray and brown sandstone and shale sequence of Love (1947) and probably the southern referred Pinyon Conglomerate as well can be regarded now as Tiffanian (late Paleocene) in age.

The Lower Variegated Sequence

Toward the end of what might be informally called pre-Wasatchian Tertiary time in order to avoid complications with regard to the definition of the Paleocene and its relationships to terms like Sparnacian, Clarkforkian, and Wasatchian (Schorn, 1971; Gingerich, 1975; Gingerich and Rose, 1977; Savage and Russell, 1977; Dashzeveg and McKenna, 1977), structural and depositional events became quite complex in the Togwotee Pass area. Detailed mapping and reconnaissance primarily in the Fish Lake, Sheridan Pass, Lava Mountain, and Tripod Peak 7½-minute quadrangles south of Highway 26/287 indicate uplift of the Wind River Mountains in the Union Pass area to the southeast, more uplift of the west end of the Washakie Range to the north (South Fork Trapdoor of Buffalo Fork Uplift of Bengtson, 1956), development of the Tripod Fault cutting referred Pinyon Conglomerate, formation of the Box Creek Syncline (Bengtson, 1956), and the initial folding and beginning of stripping of the Blackrock Anticline. Similar structural activity occurred at many places in the Rocky Mountains at about the same time.

Following these events, although in part contemporaneously with them, deposition of variegated (but generally nontuffaceous) sediments began in latest Tiffanian or Clarkforkian time and continued through the Clarkforkian into early Wasatchian time. These rocks are now exposed in the Sheridan Pass, Lava Mountain, and eastern Tripod Peak 7½-minute quadrangles and were dubbed the lower variegated sequence by Love (1947) in a section of his chart concerned with what he designated as lower Eocene rocks. The maximum thickness of this unit is about 900 feet, but the upper part may have been removed by erosion in most of the Sheridan Pass Quadrangle. Wasatchian mammals have not been recovered from the upper part of the lower variegated sequence of that area. The lower variegated sequence can be divided into three main facies, which intergrade: (1) variegated, generally nontuffaceous beds; (2) reduced and coal-bearing, greenish gray and brown sediments in the lower part of the section in the southeast, resembling rocks of the underlying stratigraphic unit; and (3) arkosic sands higher in the section in the southeast. In the northwest, in the Lava Mountain and Tripod Peak quadrangles, debris of pre-Tertiary rocks (but not Precambrian quartzite roundstones) is present in the Red Creek/Hardscrabble Creek area, derived mainly from the northeast side of the nearby Tripod Fault. The size of the clasts increases stratigraphically upward and geographically northeastward.

The earliest known early Tertiary record of vulcanism in the area is present in Clarkforkian sediments near the base of the lower variegated sequence. It consists of a very thin bentonitic bed less than an inch thick in a coal-bearing and kerogenic shale in the SW¼, SE¼, Sect. 13, T 42 N, R 111 W, Sheridan Pass 7½-minute Quadrangle.

The vertebrate fossils of Love's (1947) lower variegated sequence are here allocated to two and possibly even three biostratigraphic assemblages. At the very base of the unit in the Sheridan Pass Quadrangle and probably also in the Tripod Peak Quadrangle it is possible that a latest Tiffanian or earliest Clarkforkian level is separable on the basis of the occurrence of *Apheliscus sabulosus* and an undescribed new pantodont (K. D. Rose and D. W. Krause, in preparation). These lowest beds are not very fossiliferous, but concentrations have been found in the SE¼, SW¼, SE¼, Sect. 13, T 42 N, R 111 W ("Low Locality") and in the NE¼, NE¼, SE¼, Sect. 31, T 42 N, R 110 W ("Rohrer Locality") of the Sheridan Pass Quadrangle and along the lower slopes of the bluff facing Red Creek in the Tripod Peak Quadrangle. Above these lowest beds the lower variegated sequence produces middle Clarkforkian fossils throughout the area and early Wasatchian fossils along the slope northeast of Hardscrabble Creek in the Tripod Peak and Lava Mountain quadrangles. The faunal lists presented below are based upon previously unpublished studies by a number of workers, especially E. Manning, K. D. Rose, and myself.

?Late Tiffanian or Clarkforkian (especially the "Low Locality" and the "Rohrer Locality")
    small eutherians:
        Apheliscus (including *Phenacodaptes*) *sabulosus* (Tiffanian elsewhere)
    Pantodonta:
        New genus and species, being allocated to a new family by Rose and Krause, in preparation. This animal is known elsewhere only in Tiffanian sediments (*Plesiadapis churchilli* and *Plesiadapis simonsi* zones) of the Bighorn Basin, northwestern Wyoming.
    Primates:
        Plesiadapis dubius
        Phenacolemur sp. or *Ignacius* sp.
    Tillodontia:
        Esthonyx sp.
    Carnivora:
        Didymictis protenus proteus
    Condylarthra:
        Phenacodus vortmani
        Phenacodus primaevus
Clarkforkian (all other localities in the Sheridan Pass Quadrangle; Red Creek localities in the Tripod Peak Quadrangle)
    turtle
    crocodilian
    lizard
    bird
    small eutherians:
        Apheliscus insidiosus
        Leipsanolestes siegfriedti
        cf. *Planetetherium mirabile*
        Palaeosinopa didelphoides
        Palaeosinopa cf. *P. didelphoides*

Pantodonta:
  Coryphodon sp.
Taeniodonta:
  cf. *Ectoganus* sp.
Primates:
  Phenacolemur pagei
  Carpolestes cf. *C. nigridens*
  Plesiadapis dubius
  Plesiadapis cookei
Rodentia:
  Paramys cf. *P. atavus*
  Paramys cf. *P. excavatus* (Wasatchian elsewhere)
  ischyromyid, indet.
carnivorous mammals:
  Dissacus praenuntius
  Anacodon(?) *nexus*
  ?*Chriacus* sp.
  Oxyaena transiens
  Oxyaena platypus
  cf. *Prolimnocyon atavus*
  Dipsalodon, new species (K. D. Rose, in press)
  ?*Dipsalodon matthewi*
  Didymictis protenus proteus
  Viverravus politus
  Viverravus cf. *V. acutus*
Tillodontia:
  Esthonyx ancylion
Notoungulata:
  Arctostylops steini
Condylarthra:
  Ectocion parvus
  Ectocion osbornianus
  Phenacodus vortmani
  Phenacodus primaevus
  Aletodon gunnelli
  Haplomylus, new species (K. D. Rose, in press)
Early Wasatchian (northeast bank of Hardscrabble Creek, Lava Mountain and
  Tripod Peak quadrangles, and north bank of Papoose Creek, Lava Mountain
  Quadrangle). These identifications are preliminary:
turtle
crocodilian
glyptosaurine anguid lizard
Primates:
  Phenacolemur sp.
  Pelycodus sp.
Rodentia:
  Paramys sp.
carnivorous mammals:

Oxyaena gulo
Tillodontia:
   Esthonyx spatularius
Condylarthra:
   Ectocion osbornianus (Papoose Creek only)
   Haplomylus sp.
   Hyopsodus sp.
Pantodonta:
   Coryphodon sp.
Perissodactyla:
   Hyracotherium angustidens
Artiodactyla:
   Diacodexis olsoni
   Diacodexis sp.
   artiodactyl, genus and species indet.

As with most fossil vertebrate collections made from outcrops in variegated sediments, these fossils from the various levels within Love's (1947) lower variegated sequence are fragmentary and therefore are difficult to identify. Most of the specimens were obtained over a period of many years by meticulous and oft-repeated surface prospecting of many small patches of badlands and various isolated outcrops. Although not common anywhere in the area, fossils were most often encountered on the Clarkforkian outcrops in Purdy Basin in the Sheridan Pass Quadrangle and in Wasatchian outcrops along the bluffs northeast of Hardscrabble Creek at the boundary between the Tripod Peak and Lava Mountain quadrangles. No quarries were developed and no underwater screening was attempted in this part of the stratigraphic section. The faunal samples are no doubt biased in favor of large species, identifiable remains of which are more easily seen in the field than those of tiny forms.

Coal-Bearing Unit

After deposition of the lower variegated sequence a coal swamp and its associated sediments were present in an outcrop distance of at least 7 or 8 miles in a northwest-southeast direction in part of the Sheridan Pass Quadrangle, although these sediments grade southeastward into arkosic debris coming from the Wind River Mountains in the Union Pass area. The thickest coal bed, which is as much as 63 feet thick but contains shale partings, is at the southeast end of Purdy Basin (SW¼, NW¼, Sect. 20, T 42 N, R 110 W), where it was once briefly mined. Quiet, shallow, subaqueous deposition interspersed with pulses of sand and silt deposition from several directions and sources resulted in up to about 300 feet of chemically reduced sediments containing abundant molluscs, ostracods, coal, bentonitic claystones, and limestones, but no known mammals. The bentonitic content of the coal of this unit, presumably noticeable especially because of the low energy and relatively sluggish deposition of those times, records early but not very strong episodes of vulcanism somewhere upwind or upstream to the north. High-resolution dating of this 300-foot-thick coal-bearing unit which lies above the lower variegated sequence is not yet possible, but it is apparently conformable with the overlying upper variegated sequence. It is apparently not conformable, however, with the underlying lower variegated sequence. Wasatchian sediments of the latter have not been encountered in the Sheridan Pass Quadrangle, but Clarkforkian fossils occur a short distance below the contact. That the coal-bearing unit bevels across the underlying lower variegated sequence is a strong but geologically unsubstantiated possibility.

Upper Variegated Sequence

Following the deposition of the chemically reduced coal-bearing unit, a return to drier conditions took place and variegated sedimentation resumed for a thickness of up to 800 feet. These rocks are only part of the upper variegated sequence of Love (1947), which, in addition, included rocks identified here as Aycross Formation. This second variegated unit, above the coal-bearing middle unit, is slightly tuffaceous but apparently grades southeastward into the same lithosome of arkosic debris that characterizes the southeastward facies of the two units beneath it. To the northwest, on Tripod Peak, it contains abundant quartzite roundstones as well as boulders of Paleozoic (?mainly Tensleep Formation) sediments. Near the north side of Beauty Park and the big bend of the North Fork of Fish Creek (SE¼, NW¼, Sect. 18, T 43 N, R 110 W) in the Lava Mountain Quadrangle, cobbles and large boulders up to five feet in diameter are common. The abundant Tensleep and Madison Formation boulders that occur in the second variegated sequence in the Lava Mountain and Tripod Peak quadrangles indicate renewed uplift northeast of the Tripod Fault.

The age of the second variegated sequence is clearly still Wasatchian. A few fossils have been obtained on Buckskin Ridge (SE¼, SW¼, Sect. 29, T 42 N, R 110 W) from a few feet beneath the unconformity with the overlying Aycross Formation in the Sheridan Pass Quadrangle (Rohrer's Twcc/Twpu contact, unpublished). These include (preliminary identifications):

> crocodilian
> turtle
> ?*Coryphodon* sp.
> ?*Phenacodus* sp.
> *Hyracotherium* sp.
> ?*Diacodexis* sp.

The rock-stratigraphic names for the Tertiary sedimentary units beneath the Aycross Formation south of Togwotee Pass could be argued about endlessly on the basis of present knowledge. Probably, new names should be applied. These rocks should not be referred to the Hoback Formation (Dorr, 1952, 1978; Dorr, et al., 1977), because that would require severe stretching of original definitions and winking at paleogeographic and cartographic discontinuity. The structural evolution of the Gros Ventre Range is intimately interlinked with this problem (Love, 1977). Extension of such terms as Fort Union or Wasatch to the area would be unjustifiable as well.

Extension to the Togwotee Pass area of the Indian Meadows Formation is also a problem. Although the uppermost beds (early Wasatchian) of the lower variegated sequence in the Lava Mountain and Tripod Peak quadrangles are approximately the same age as the lowermost beds of the type Indian Meadows Formation (Love, 1939; Keefer, 1970, plate 1; M. C. McKenna, unpublished data), that unit has not been traced westward into the Togwotee Pass area from the type locality (Blue Holes 7½-minute Quadrangle southeast of Dubois in the Wind River Basin). A structural high running from the northern Wind River Mountains to the Washakie Range, northeast of a line from the White Rock Thrust (Richmond, 1945) of the Green River Lakes area to the Tripod Fault of the Togwotee Pass area, prevented depositional continuity. Rock-stratigraphic names for the first and second sets of Love's (1947) variegated beds and the coal-bearing sequence between them in the area south of Togwotee Pass are desirable but have not been published.

Aycross Formation

Deposition of the second variegated sequence was terminated by uplift of Pinyon Conglomerate or other quartzite roundstone-producing units to the west, with consequent deposition of pressure-marked quartzite roundstones and younger volcanic roundstones into the basal conglomerate of the unconformably overlying lower part of the Aycross Formation (Rohrer's Twcc, unpublished, as here interpreted, see also Love, et al., 1978). Volcanic activity to the north increased substantially and about 700 to 1,500 feet of volcaniclastics, interspersed with cobble conglomerates from a more westerly source, were laid down. These predominantly volcaniclastic deposits contain datable volcanic debris about 500 feet above the base (Rohrer, unpublished), the Coyote Creek flora (MacGinitie, 1974), fossil land molluscs, and a few mammalian and other vertebrate remains near the base of the unit on Buckskin Ridge in the Sheridan Pass Quadrangle (preliminary identifications):

Landslide Locality, (boundary of NE¼ and NW¼, SE¼, Sect. 29, T 42 N, R 110 W), base of Rohrer's Twc, in volcaniclastic sandstone of the lower part of the Aycross Formation. Late Wasatchian or early Bridgerian.
lizard
tiny insectivore, cf. *Scenopagus* (probably new genus and species)
Viverravus cf. *V. lutosus*
Hyopsodus sp.
Lophiparamys sp.
small ischyromyid rodent, genus and species indet.
Tapiroid Tooth Locality, (NE¼, NW¼, Sect. 17, T 42 N, R 110 W), lower part of the Aycross Formation, about 500 feet above base of Rohrer's Twc, in association with K-Ar sample whose date is not yet released. Late Wasatchian or early Bridgerian.
tapiroid perissodactyl, either *Heptodon posticus* (Lysite and Lost Cabin "faunal zones" elsewhere) or *Helaletes intermedius* (late Bridgerian elsewhere).

Subsequent to deposition of the lower part of the Aycross Formation, uplift and folding together with erosion of uplifted areas resulted in a local disconformity beneath the upper part of the Aycross, a highly tuffaceous third variegated unit that Love (1947) regarded as part of his upper variegated sequence and that Rohrer (1968) regarded as still part of the Wind River Formation (Rohrer's Twlc and Twl of the Sheridan Pass Quadrangle and Twrc and Twr of the Fish Lake Quadrangle). It may be correlative with the $Twdr_4$ and $Twdr_5$ of Keefer (1956) north of Dubois, and with the Cathedral Bluffs/Wilkins Peak complex of the Green River Basin. Deposition of this highly tuffaceous suite of sediments occurred in a manner that cut across various units of the lower Aycross in the Sheridan Pass and Lava Mountain quadrangles (Figure 20.4); at Tripod Peak west of the Lava Mountain Quadrangle the unit lapped onto earlier Wasatchian rocks which remained a local source

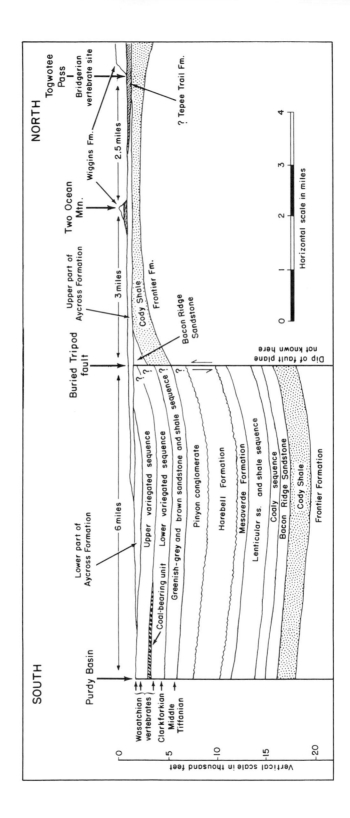

of reworked quartzite and other roundstones. These roundstones and perhaps others from a more northwestern source provide puddingstones, especially in Moccasin Basin (Sect.9, T 43 N, R 110 W) in the Lava Mountain Quadrangle. Closely similar quartzite roundstones occur in the Aycross on the north flank of Coulee Mesa, about four miles south of the type locality of the Aycross Formation east of Dubois, Wyoming (Love, 1939, p. 68). A frequently encountered feature of the referred upper Aycross of the Togwotee Pass area is the presence of abundant doubly terminated quartz crystals derived from nearby vents to the north. Petrified wood is also locally abundant. At the northern border of the Lava Mountain and Tripod Peak quadrangles the sediments identified here as upper Aycross lie unconformably upon Cretaceous Cody Shale with about a hundred feet of relief in the Cody. In that area approximately 15,000 feet of Late Cretaceous, Paleocene, and Eocene rocks had thus been stripped away along the Blackrock Anticline before upper Aycross deposition began. At least part of this stripping occurred immediately before the onset of local upper Aycross deposition, for otherwise there would be no relief in the Cody.

A fossil vertebrate locality (L-41) at the south end of Blackrock Meadows (UTM grid system, zone 12: 4,841,795 meters N; 569,370 meters E; T 44 N, R 111 W, Tripod Peak 7½-minute Quadrangle, Teton County, Wyoming) was found in the Aycross by Love (1947; Love, et al., 1978, plate 5). On his chart Love (1947) listed *Paramys* sp., *Didymictis protenus*, and *Hyopsodus* cf. *H. miticulus* in measured section no. 6. These fossils cannot now be found in the U.S.G.S. collections. Love (1947) referred this third set of variegated beds to his upper variegated sequence and this was compatible with the fossil identifications made at that time. Rohrer's and my own independent mapping and reconnaissance, however, indicate that the situation is different and more complex than Love believed to be the case in 1947. More complete fossil evidence from Love's locality L-41 confirms that the unit here identified as an extension of the Aycross Formation was indeed deposited significantly later than the rest of Love's (1947) upper variegated sequence. The latter is not so tuffaceous, and it contains fairly abundant Paleozoic debris derived from nearby contemporaneous tectonic activity and related erosion. Love, et al. (1978), concurred with these conclusions.

The fauna of Love's locality L-41 has been sampled by the screening process (McKenna, 1962) and has yielded a small number of specimens, most of which consist of isolated teeth and tooth fragments. Excessive bentonitic content of the quarried matrix has prevented large-scale sampling because of the required time-consuming multiple washing and drying of the matrix. In addition to the sample from locality L-41, a few isolated teeth and fragments have been recovered from scattered Aycross outcrops nearby. The known fauna is as follows:

crocodilian
marsupial, cf. *Peratherium* or *Herpetotherium* sp.

Figure 20.4. Simplified partly diagrammatic section from Purdy Basin to Togwotee Pass, showing general relations of major Upper Cretaceous and Tertiary sequences on opposite sides of Tripod Fault; Tripod Peak, Lava Mountain, and Togwotee Pass 7½-minute quadrangles, Teton County, Wyoming. Dark stipple: ? Tepee Trail Formation. Light stipple: Cody Shale, emphasized to show effect of fault in Eocene time. Section based upon unpublished field work and personal communications from J. D. Love.

small erinaceoid insectivore, genus and species indet.
Microsyops cf. *M. elegans*
Microsyops cf. *M. alfi* (early Wasatchian elsewhere)
Notharctus tenebrosus
omomyine or anaptomorphine primate, genus and species indet.
Sciuravus sp.
ischyromyid rodent, cf. *Paramys* sp.
ischyromyid rodent, cf. *Reithroparamys* sp.
?carnivore, genus and species indet.
Hyopsodus sp.
?*Hyracotherium* sp.
several additional mammalian taxa represented by tooth fragments

This assemblage is a poor one, but it suggests late Wasatchian or early Bridgerian age. This determination is not in conflict with the range of ages known from the Aycross at its type locality northeast of Dubois, Wyoming. The lowest of three vertebrate assemblages identified by Wood, et al. (1936), in the rocks exposed in what was later to be designated the type locality of the Aycross Formation (Love, 1939) apparently has a Wasatchian age, but the upper part of the Aycross at its type locality is Bridgerian in age and most of the lower part is either early Bridgerian, latest Wasatchian, or both. A K-Ar date of 49.2 ± 0.5 Ma from a grey biotite tuff has been obtained from the Aycross in its type area (Love, et al., 1976; 1978, plate 5). Bridgerian mammals have been reported to occur both above and below the dated tuff (Wood, et al., 1936; Love, 1939; Love, et al., 1976, p. 17; Love, et al., 1978, plate 5).

Lahar Deposits

The "black rock" of Blackrock Meadows (southwest corner of Togwotee Pass 7½-minute Quadrangle), west of the summit of Togwotee Pass, was apparently the next stratigraphic unit to be deposited in the Togwotee Pass area. Lahar deposits from a northern source are well-exposed in a roadcut on the north side of Highway 26/287 at the east end of Blackrock Meadows (UTM grid system, zone 12: 4,846,126 meters N; 574,134 meters E), a little more than a mile northwest of the summit of the pass. Chunks of petrified wood occur in this chaotically deposited unit. These rocks have been correlated with the lower part of the Thoroughfare Creek Group of the Absaroka Volcanic Supergroup by Smedes and Prostka (1972, figure 8: Langford and/or Two Ocean formations).

?Tepee Trail Formation

Lapping up onto the east side of the lahar deposits and covering them as well is a greenish and brownish volcaniclastic unit that resembles the Tepee Trail Formation at and near its type locality in East Fork Basin some 70 miles to the east. If correctly identified, the occurrence at Blackrock Meadows is the westernmost known extension of the Tepee Trail Formation (Smedes and Prostka, 1972), but Love, et al. (1978), prefer to include these rocks in the Aycross Formation. Locally, near Barbers Point on the Brooks Lake exit road in the northeastern Lava Mountain Quadrangle, this unit contains quartzite roundstones. The unit is especially well-exposed in badlands at the summit of Togwotee Pass, where several hundred feet above its local base it has yielded the following vertebrate assemblage:

fish
crocodilian
tortoise
Glyptosaurus sp.
another lizard, genus and species indet.
small insectivore, genus and species indet.
?*Scenopagus* sp.
Microsyops cf. *M. elegans*
Washakius sp.
Tillomys sp.
Sciuravus sp.
small ischyromyid rodent
a second small ischyromyid rodent
cf. *Reithroparamys* sp.
?carnivore, genus and species indet.
a second carnivore, genus and species indet.
Hyopsodus sp.
cf. *Trogosus* sp.
Hyrachyus eximius
Palaeosyops sp.
hyracotheriine horse, cf. *Orohippus* sp.
artiodactyl, genus and species indet.
several additional taxa represented by tooth fragments

These identifications suggest that the age of this assemblage from the badlands at the summit of Togwotee Pass is Bridgerian, approximately equivalent faunally to assemblages from the Bridger "A-B" of the Green River Basin but also similar to the fauna of the Bridgerian part of the Aycross Formation at its type locality northeast of Dubois, Wyoming. K-Ar dates of 47.8 ± 1.3 Ma (sanidine) and 47.9 ± 1.3 Ma (biotite) have been obtained from a sample collected 5,800 feet north-northwest of Togwotee Pass (Smedes and Prostka, 1972, p. 30 and figure 8, loc. h, relocated) beneath the fossil vertebrate bearing badlands at the summit of the pass (Love, et al., 1978, plate 5). These dates and the biostratigraphic data are compatible with the suggestion by Love, et al. (1978), that the rock unit in which they occur is an extension of the Aycross Formation rather than the Tepee Trail Formation. That question, however, should be settled on the basis of lithogenetic and cartographic data alone.

At the type section of the Tepee Trail Formation an early Uintan vertebrate assemblage has been obtained from unit 24 (bone bed A) in East Fork Basin northeast of Dubois, Wyoming (NW¼, NE¼, NE¼, Sect. 4, T 43 N, R 104 W, East Fork Basin 7½-minute Quadrangle). This assemblage is derived from a level at least 500 feet above the local base of the Tepee Trail Formation, which in its type area and to the southeast lies on a surface of considerable relief cut in rocks referred to the Aycross Formation. A few fossils have been collected from other sites within the Tepee Trail Formation in the East Fork Basin area. The assemblage from bone bed A (unit 24) in the type section contains the following vertebrates:

salamander
lizards:
    Scincidae, new genus and species, cf. *Contogenys* and *Paracontogenys* (J. A. Gauthier, in preparation)
    Iguanidae, new genus and species (J. A. Gauthier, in preparation)
Multituberculata:
    neoplagiaulacid, genus and species indet.
Marsupialia:
    Herpetotherium sp.
    Peradectes sp.
small insectivorous eutherians:
    Palaeictops, new species (M. J. Novacek, in preparation)
    erinaceid insectivore, cf. *Ocajila* sp.
    Oligoryctes sp.
    microchiropteran bat, genus and species indet.
    epoicothere, cf. *Tetrapassalus* sp. (See Rose, 1978)
    Apatemys sp.
    Uintasorex parvulus
    a second uintasoricine species
    Nyctitherium sp.
worlandiine dermopteran, new genus and species
Primates:
    cf. *Trogolemur myodes*
    Phenacolemur sp.
    omomyid, genus and species indet.
Rodentia:
    ?*Reithroparamys* sp.
    Microparamys cf. *M. dubius*
    Sciuravus, new species
    ?*Protadjidaumo typus*
    small eomyid
    rodent, family, genus, and species indet.
    a second rodent, family, genus, and species indet.
Condylarthra:
    Hyopsodus cf. *H. uintensis*
Carnivora:
    Protictis, new species (J. J. Flynn, in preparation)
Perissodactyla (See MacFadden, in press):
    Epihippus uintensis
    Dilophodon minusculus
    Amynodon sp.
    titanothere, genus and species indet.
Artiodactyla:
    Achaenodon sp.
    dichobunid sp.
    ?homacodont sp.
    low-crowned selenodont artiodactyl, genus and species indet.

In addition to these taxa, G. E. Lewis (1973) reported an artiodactyl jaw that he had collected in 1938 from a few feet above bone bed A in the type section. While the specimen was still in the field, Lewis correctly identified it as *Achaenodon*, but then it was misplaced. Thirty-five years later Lewis (1973) reidentified the supposedly lost specimen as *Parahyus vagus* on the basis of an incorrectly scaled photograph. I have found the missing jaw at Yale University, where it was masquerading as a Siwalik *Tetraconodon*, and I confirm Lewis's original field identification (McKenna, in preparation). The jaw actually belongs to a species of *Achaenodon*, but also resembles *Parahyus vagus* in many respects except its much larger size. M_1-M_3 length is 105 mm, not approximately 65 mm as suggested by Lewis's photograph (Lewis, 1973, Figure 1). Presumably the species is the same as that reported here from bone bed A, but precise identification is postponed pending detailed studies of achaenodont systematics.

The assemblage from unit 24 (bone bed A) of the type Tepee Trail Formation is apparently later in age than any known Bridgerian assemblage. It appears more primitive than the "Uinta B" fauna, however. Possibly it is equivalent in age to the almost unknown "Uinta A" fauna, but to elaborate on that suggestion would first require more work in the Uinta Basin of Utah. Complicating the problem is the fact that the known type Tepee Trail assemblage from unit 24 is ecologically rather distinctive and comes from sediments deposited at a higher paleoaltitude, near volcanic vents, than is the usual case with Bridgerian or Uintan localities. The commonest species present is a normally rare hedgehog and many of the other taxa in the assemblage appear to be new, perhaps partly because an unusual community is being sampled for the first time. Caution is called for in the interpretation of the assemblage, yet there is already sufficient information to indicate that the age of the assemblage is intermediate between known Bridgerian ones and known Uintan assemblages above the "Uinta A." For the present I regard the locality in unit 24 as early Uintan in age on the vertebrate scale used in North America and thus I interpret part of the type section of the Tepee Trail Formation, at least 500 feet above its local base, to be Uintan. Another 900 feet of Tepee Trail above that level in the type section is for this reason Uintan or later in age on the same scale.

Referred Tepee Trail sediments at the summit of Togwotee Pass (Smedes and Prostka, 1972) are older than the assemblage from unit 24 (bone bed A) of the type section of the Tepee Trail Formation. Although it is possible that the Tepee Trail transgresses time from earlier in the west to younger in the east as claimed by Rohrer and Obradovich (1969), I regard this as unlikely. Love, et al. (1978), regard the unit at the summit of Togwotee Pass as part of the Aycross Formation, which is more in keeping with the age of the unit. It will be important to search for younger faunas higher in the referred Tepee Trail (or Aycross) at Togwotee Pass and older ones lower in the type Tepee Trail 70 miles to the east.

Inasmuch as the so-called "Tepee Trail fauna" of the Badwater area of the northeastern Wind River Basin represents a spectrum of faunas, including Bridgerian assemblages from localities 17 and 18 in the "Green and Brown Member" (C. C. Black, personal communication), much further paleontological work needs to be done before detailed biostratigraphic correlation becomes a reality. In addition, much petrographic, geochemical, and sedimentological study remains to be accomplished before the relationships of these faunas and rock-stratigraphic units will be much better understood.

Wiggins Formation

Lying above the Tepee Trail Formation (?Aycross) at Togwotee Pass is the spectacular, cliff-forming Wiggins Formation (Love, 1939; Smedes and Prostka, 1972). The thickness of the Wiggins locally reaches 4,000 feet. The formation consists of near-source volcaniclastics, lavas, mudflows, lahar deposits, vent facies, and air-fall tuffs. The Wiggins has been claimed to intertongue with the Tepee Trail Formation by Rohrer and Obradovich (1969). In my opinion, although this relationship is possible in some areas, it has not yet been demonstrated.

The Wiggins had been regarded as Chadronian in age by most workers until 1972. However, the only identifiable "Wiggins" vertebrates, from a site on Mink Creek on the Buffalo Fork Thrust trace three miles south of Yellowstone Park (Love, 1952, 1956a), have proven to occur in sediments of the White River Group, channeled into the Wiggins (Love, et al., 1976). These Chadronian sediments are in turn overlain by Arikareean sediments. Fossil vertebrates do occur in genuine Wiggins, however, in the Monument Peak Quadrangle of the southeastern Absarokas. Those remains are unfortunately indeterminate titanothere bones and tooth fragments (Love, 1939) which are not identifiable to taxonomic levels useful in correlation. Other identifiable vertebrates from genuine Wiggins outcrops have been recovered recently by parties from The University of Wyoming (J. G. Eaton, in preparation).

Rohrer and Obradovich (1969) published two K-Ar dates (46.2 ± 1.8 and 46.5 ± 2.3 Ma) for volcanics at what they interpreted to be the base of the Wiggins in the Pinnacle Buttes area on the east side of Togwotee Pass in the Kisinger Lakes Quadrangle. Probably all or most of the "true" Wiggins is Uintan in age on the vertebrate scale determined by Evernden, et al. (1964).

No post-Wiggins Tertiary strata are known to be present in the Togwotee Pass area.

Postscript

When Love (1947) published the first geological description of the Togwotee Pass area the paleontological interest of the thick Upper Cretaceous and lower Tertiary rocks of the area became obvious. In cooperation with Love and other geologists a long campaign of paleontological reconnaissance began and is still continuing. A number of biostratigraphically distinguishable assemblages of fossil vertebrates either found originally by Love or discovered since 1947 have been sampled repeatedly over the years and these are now well enough known to be useful in correlation. Some of the assemblages are now adequate to serve as the basis of detailed scientific studies as reconnaissance gives way to more sharply focused studies of the individual geologic and paleontologic problems that abound in the area.

An important aspect of the Togwotee Pass area that was not appreciated when the present work began but which has become highly significant as the work progressed is the potential for radioisotopic dating that is provided by volcanic rocks and volcaniclastic sediments that are intimately associated with the fossil assemblages (Smedes and Prostka, 1972; Love, et al., 1976, 1978; Berggren, et al., 1978). Few areas in the world provide such a potentially useful combination of faunal and radioisotopic information, permitting calibration of the geological time scale and providing information useful far beyond the confines of northwestern Wyoming. Paleomagnetic stratigraphy can also be expected to be integrated with this work.

The biostratigraphic and temporal framework of the Togwotee Pass area will continue to improve as work continues, but it is important now that purely geological research on the area be accelerated. As for instance in the Hoback Basin to the south, the stratigraphic and structural history of the Togwotee Pass area are intimately interlinked. Interdisciplinary study of the structural and stratigraphic evolution of this part of Wyoming will have important significance in the search for energy resources as well as in shaping ideas of a more general nature.

Acknowledgments

I thank C. C. Black, M. R. Dawson, J. G. Eaton, R. D. Estes, J. J. Flynn, E. S. Gaffney, H. Galiano, J. A. Gauthier, P. D. Gingerich, R. M. Hunt, L. Krishtalka, J. D. Love, B. J. MacFadden, E. Manning, M. J. Novacek, K. D. Rose, D. A. Russell, D. E. Russell, R. E. Sloan, M. S. Stevens, L. Van Valen, and many others who have helped with field work and identifications. J. A. Lillegraven and J. D. Love provided helpful criticism and advice. I am especially indebted, however, to O. Simonis, whose patience and skill in the laboratory made possible the safe removal of delicate specimens from matrix that was in many instances nearly intractable.

Literature Cited

Bengtson, C. A. 1956. Structural geology of the Buffalo Fork area, northwestern Wyoming, and its relation to the regional tectonic setting. Wyo. Geol. Assoc., Guidebk., 11th Ann. Field Confer., Jackson Hole, 1956:158–68.

Berggren, W. A.; McKenna, M. C.; Hardenbol, J.; and Obradovich, J. D. 1978. Revised Paleogene Polarity Time Scale. J. Geol., 86:67–81.

Colbert, E. H. 1943. A Miocene oreodont from Jackson Hole, Wyoming. J. Paleontol., 17:298–305.

Dalrymple, G. B. 1979. Critical tables for conversion of K-Ar ages from old to new constants. Geology, 7:558–60.

Dashzeveg, D., and McKenna, M. C. 1977. Tarsioid primate from the early Tertiary of the Mongolian People's Republic. Acta Palaeontol. Pol., 22(2):119–37.

Dorr, J. A. 1952. Early Cenozoic stratigraphy and vertebrate paleontology of the Hoback Basin, Wyoming. Geol. Soc. Amer., Bull., 63:59–94.

———. 1958. Early Cenozoic vertebrate paleontology, sedimentation, and orogeny in central western Wyoming. Geol. Soc. Amer., Bull., 69:1217–44.

———. 1978. Revised and amended fossil vertebrate faunal lists, early Tertiary, Hoback Basin, Wyoming. Wyo., Univ., Contrib. Geol., 16(2):79–84.

Dorr, J. A.; Spearing, D. R.; and Steidtmann, J. R. 1977. The tectonic and synorogenic depositional history of the Hoback Basin and adjacent areas. Wyo. Geol. Assoc., Guidebk., 29th Ann. Field Confer., 1977:549–62.

Evernden, J. F.; Savage, D. E.; Curtis, G. H.; and James, G. T. 1964. Potassium-argon dates and the Cenozoic mammalian chronology of North America. Amer. J. Sci., 262:145–98.

Gingerich, P. D. 1975. Discussion. "What is type Paleocene?" Amer. J. Sci., 275(8):984–85.

———. 1976. Cranial anatomy and evolution of early Tertiary Plesiadapidae (Mammalia, Primates). Mich., Univ., Mus. Paleontol., Papers on Paleontol., 15:1–141.

Gingerich, P. D., and Rose, K. D. 1977. Preliminary report on the American Clark Fork Mammal fauna, and its correlation with similar faunas in Europe and Asia. Géobios, Mém. spécial, 1:39–45.

Hardenbol, J., and Berggren, W. A. 1978. A new Paleogene numerical time scale. Amer. Assn. Petrol. Geol., Studies in Geol., 6:213–34.

Keefer, W. R. 1956. Tertiary rocks in the northwestern part of the Wind River Basin, Wyoming. Wyo. Geol. Assn., Guidebk., 11th Ann. Field Confer., Jackson Hole, 1956:109–16.

———. 1965. Stratigraphy and geologic history of the uppermost Cretaceous, Paleocene, and lower Eocene rocks in the Wind River Basin, Wyoming. U. S. Geol. Surv., Prof. Paper, 495-A:1–77.

————. 1970. Structural geology of the Wind River Basin, Wyoming. U. S. Geol. Surv., Prof. Paper, 495-D:i–iv, 1–35.

Lewis, G. E. 1973. A second specimen of *Parahyus vagus* Marsh, 1876. U. S. Geol. Surv., J. Res., 1 (2):147–49.

Lindsey, D. A. 1970. Facies and paleocurrents in conglomerates of the Harebell Formation and Pinyon Conglomerate, northwestern Wyoming. Geol. Soc. Amer., Abstr. 2(5):341.

————. 1972. Sedimentary petrology and paleocurrents of the Harebell Formation, Pinyon Conglomerate, and associated coarse clastic deposits, northwestern Wyoming. U. S. Geol. Surv., Prof. Paper, 734-B:i–vi, 1–68.

Love, J. D. 1939. Geology along the southern margin of the Absaroka Range, Wyoming. Geol. Soc. Amer., Spec. Pap. 20:1–134.

————. 1947. Tertiary stratigraphy of the Jackson Hole area, northwestern Wyoming. U. S. Geol. Surv., Oil and Gas Invest. Prelim. Chart 27.

————.1952. Preliminary report on uranium deposits in the Pumpkin Buttes area, Powder River Basin, Wyoming. U. S. Geol. Surv., Cir. 176:i–vi, 1–37.

————. 1956a. Cretaceous and Tertiary stratigraphy of the Jackson Hole area, northwestern Wyoming. Wyo. Geol. Assoc., Guidebk., 11th Ann. Field Confer., Jackson Hole, 1956: 76–94.

————. 1956b. Summary of geologic history of Teton County, Wyoming, during Late Cretaceous, Tertiary, and Quaternary times. Wyo. Geol. Assoc., Guidebk., 11th Ann. Field Confer., Jackson Hole, 1956:140–50.

————. 1973. Harebell Formation (Upper Cretaceous) and Pinyon Conglomerate (Uppermost Cretaceous and Paleocene), northwestern Wyoming. U. S. Geol. Surv., Prof. Paper, 734-A:i–iv, 1–54.

————. 1977. Summary of Upper Cretaceous and Cenozoic stratigraphy, and of tectonic and glacial events in Jackson Hole, northwestern Wyoming. Wyo. Geol. Assoc., Guidebk., 29th Ann. Field Confer., 1977:585–93.

Love, J. D.; McKenna, M. C.; and Dawson, M. R. 1976. Eocene, Oligocene, and Miocene rocks and vertebrate fossils at the Emerald Lake Locality, 3 miles south of Yellowstone National Park, Wyoming. U. S. Geol. Surv., Prof. Paper, 932-A:i–iv, 1–28.

Love, J. D.; Leopold. E. B.; and Love, D. W. 1978. Eocene rocks, fossils, and geologic history, Teton Range, northwestern Wyoming. U. S. Geol. Surv., Prof. Paper, 932-B:i–iv, 1–40.

Love, J. D.; Weitz, J. L.; and Hose, R. K. 1955. Geologic Map of Wyoming. U. S. Geol. Surv. Scale 1:500,000.

MacFadden, B. J. (In press). Eocene perissodactyls from the type section of the Tepee Trail Formation of northwestern Wyoming. Wyo., Univ., Contrib. Geol.

MacGinitie, H. D. 1974. An early middle Eocene flora from the Yellowstone-Absaroka Volcanic Province, northwestern Wind River Basin, Wyoming. Calif., Univ., Publ. Geol. Sci., 108:i–vi, 1–103.

McKenna, M. C. 1962. Collecting small fossils by washing and screening. Curator, 5(3): 221–35.

————. 1972. Vertebrate paleontology of the Togwotee Pass area, northwestern Wyoming. *In* West, R. M., Coordinator, Guidebook, Field Conference on Tertiary Biostratigraphy of southern and western Wyoming, Aug. 5–10, 1972:80—101. Privately printed and distributed.

McKenna, M. C., and Love, J. D. 1970. Local stratigraphic and tectonic significance of *Leptoceratops*, a Cretaceous dinosaur in the Pinyon Conglomerate, northwestern Wyoming. U. S. Geol. Surv., Prof. Paper, 700-D:55–61.

McKenna, M. C.; Russell, D. E.; West, R. M.; Black, C. C.; Turnbull, W. D.; Dawson, M. R.; and Lillegraven, J. A. 1973. K-Ar recalibration of Eocene North American Land-Mammal "Ages" and European Ages. Geol. Soc. Amer., Abstr., 5(7):733.

Richmond, G. M. 1945. Geology of northwest end of the Wind River Mountains, Sublette County, Wyoming. U. S. Geol. Surv., Oil and Gas Invest., Prelim. Map 31.

Rohrer, W. L. 1966. Geologic map of the Kisinger Lakes Quadrangle, Fremont County, Wyoming. U. S. Geol. Surv. Geol., Quad. Map, GQ-527.

————. 1968. Geologic map of the Fish Lake Quadrangle, Fremont County, Wyoming. U. S.

Geol. Surv., Geol. Quad. Map, GQ-724.

———. (MS). Preliminary geologic map of the Sheridan Pass Quadrangle, Fremont and Teton counties, Wyoming. U. S. Geol. Surv. Open File.

Rohrer, W. L., and Obradovich, J. D. 1969. Age and stratigraphic relations of the Tepee Trail and Wiggins Formations, northwestern Wyoming. U. S. Geol. Surv., Prof. paper, 650-B: 57–62.

Rose, K. D. 1977. Evolution of carpolestid primates and chronology of the North American middle and late Paleocene. J. Paleontol., 51(3):536–42.

———. 1978. A new Paleocene epoicotheriid (Mammalia), with comments on the Palaeanodonta. J. Paleontol., 52(3):658–74.

———. (In press). The Clarkforkian Land-Mammal "Age" and mammalian faunal composition across the Paleocene-Eocene boundary. Mich., Univ., Mus. Paleontol., Papers on Paleontol.

Savage, D. E., and Russell, D. E. 1977. Comments on mammalian paleontologic stratigraphy and geochronology; Eocene stages and mammal ages of Europe and North America. Géobios, Mém. spécial, 1:47–56.

Schorn, H. E. 1971. What is type Paleocene? Amer. J. Sci., 271:402–9.

Smedes, H. W., and Prostka, H. J. 1972. Stratigraphic framework of the Absaroka Volcanic Supergroup in the Yellowstone National Park region. U. S. Geol. Surv., Prof. Paper, 729-C:i–vi, 1–33.

Wood, H. E., 2nd; Seton, H.; and Hares, C. J. 1936. New data on the Eocene of the Wind River Basin, Wyoming. Geol. Soc. Amer., Proc. Vol., 1935:394–95.

MARSUPIAL PALEOBIOGEOGRAPHY

by Larry G. Marshall

Introduction

Marsupials are taxonomically diverse and geographically widespread. Their fossil record is fairly complete, and the interrelationships of most marsupial groups are relatively well established. Despite these positive features, the biogeographic history of marsupials is hotly disputed.

Native marsupials occur today in Australasia (which includes the Australian mainland, Tasmania, New Guinea, and numerous small islands surrounding these larger land masses), in South America, in Central America (including Mexico), and with one recent immigrant species, *Didelphis virginiana*, in North America. The Australasian forms include, among others, dasyurs, phalangers, kangaroos, wombats, koalas, and bandicoots; while opossums, microbiotheres, and caenolestids are found in South America (Kirsch and Calaby, 1977; Tyndale-Biscoe, 1973; Marshall, in press).

Prior to acceptance of plate tectonic theory ("Continental Drift"), biogeographers based interpretations of the distribution of marsupials on the present "fixed" or "static" positions of continents. Their distribution was explained by invoking the presence of now-submerged land bridges between various continents, or by assuming chance dispersals by swimming or on rafts of vegetation. These workers have been called "Stabilists" (Tedford, 1974).

Reconstructions of past continental positions based on geological and geophysical data now indicate that continents have moved significant distances relative to each other. The interpreted movements are explained by plate tectonic theory and evoke a mobile Earth crust. Biogeographers who incorporate plate tectonic theory in attempting to explain present and past distributions of animals and plant groups have been called "Mobilists" (Tedford, 1974).

The principal views of marsupial biogeography can be organized into scenarios which I term models. These are divisable into Stabilist Models of which I recognize four, and Mobilist Models of which I recognize nine. In some cases authors present various alternative models, yet invariably favor one. In such cases I include the cited work and author in the model favored. In other cases, the model championed by an author has changed through the years. As a result, it is not possible in all cases to associate one author with only one model, and some authors are credited with supporting different models at different times. Close attention should thus be given to authorship and to the date of the reference cited.

In order to place the models into proper perspective, I first present consideration of the following topics: (1) biogeographic principles as they relate to marsupial paleobiogeography; (2) the marsupial-placental dichotomy; (3) the marsupial fossil record, with special emphasis on first appearances; and (4) the influence marsupials have had on the historical development of plate tectonic theory.

BIOGEOGRAPHIC PRINCIPLES
AND MARSUPIAL PALEOBIOGEOGRAPHY

Marsupials afford an opportunity to entertain the biogeographic concepts of *dispersal* and *vicariance* (sensu Croizat, Nelson, and Rosen, 1974; Rosen, 1975) to explain distribution patterns. The early history of marsupials is of particular interest in this regard. Did marsupials indeed have a center of origin that can be specified, and if so, where was that center of origin, when did dispersal(s) occur, and in what direction(s)? For a terrestrial group of organisms like the marsupials, dispersal will depend primarily on three factors: (1) the availability of land connections; (2) the nature of physical, ecological and/or climatic barriers to be crossed; and (3) the dispersive abilities of the particular group concerned (Keast, 1977, p. 84).

Marsupial biogeography may alternatively be explained by vicariance, in which their disjunct distribution resulted from subdivision of ancestral biotas in response to changing geography. Of special importance in this regard is acceptance of plate tectonic theory and realization that continents have moved relative to one another through time. Many of these documented movements occurred after marsupials first appeared in the fossil record in the Late Cretaceous, and may have thus been responsible for part or all of their known distribution pattern.

It now seems likely that both dispersal and vicariance account for marsupial distribution patterns. However, there is little consensus on which group or taxa fit which concept. To complicate matters, both concepts can explain the same distribution. For example, McKenna (1975b) has shown that didelphid marsupials dispersed from North America to Europe in the early Eocene via a dry land connection (the DeGeer Route) across the North Atlantic. Since a continuous interbreeding population once existed across the DeGeer Route their subsequent disjunct distribution is explained by vicariance. Thus, this particular vicariance event requires dispersal to account for a significant ancestral range.

It has been demonstrated for other animal groups (e.g., ratite birds—see Cracraft, 1973) that absolute times of fragmentation of continental land masses correlate with relative times of dichotomies of taxonomic groups (as inferred from cladistics) inhabiting these land masses. In such cases, detailed phylogenetic studies are useful in constructing a group's biogeographic history. Indeed, some authors (e.g., C. Patterson, in press) insist that the phylogenetic relationships of a group must be understood before one can accurately reconstruct paleobiogeography.

Groups with overlapping distributions should show similar or comparable dichotomies in their phylogenetic histories if their distribution is explained by vicariance. If dispersal was involved, these same groups need not necessarily show similar and comparable dichotomies in thier phylogenetic histories since the times and directions of dispersal need not be the same for each group. This is especially true if dispersal did not occur via a dry land connection (i.e., filter bridge or corridor), but involved a distinct chance event (i.e., waif or sweepstakes dispersal).

MARSUPIAL-PLACENTAL DICHOTOMY

The basic dichotomy between marsupials (metatherians) and placentals (eutherians) is documented by numerous derived characters unique to one or the other of these groups, establishing that each is monophyletic. Similarly, the common possession by marsupials and placentals of other derived characters indicates that these groups have a common ancestor not shared with other therian groups (Marshall, 1979). Although we probably have not yet discovered representatives of the exact stage of evolution immediately ancestral to marsupials and placentals, there is no fossil evidence to support anything but a monophyletic origin for advanced therians from some post-*Aegialodon* early Cretaceous common ancestor (Lillegraven, 1976, p. 710). The majority of evidence now suggests that placentals did not evolve from marsupials (as was long considered to be the case), but that both were derived concurrently from a common therian ancestor. In a cladistic framework they are regarded as "sister-groups" (Marshall, 1979).

It is now generally agreed that the marsupial-placental dichotomy occurred sometime in the Early Cretaceous and almost certainly no later than middle Cretaceous time (ca. 100 Ma). Air, et al. (1971), obtained an estimate of about 130 Ma for the time of their divergence. This estimate agrees with the date postulated on fossil evidence (Lillegraven, 1974), and the chronological framework of Maxson, Sarich, and Wilson (1975). Based on analysis of rates of amino acid sequence changes in albumins, Maxson, et al. (1975), estimate that South American and Australian marsupials were separated for at least 73 Ma. This estimate is not contradicted by known fossil data.

Vicariance and geographic isolation on a major scale were possibly responsible for the marsupial-placental dichotomy. The Early Cretaceous was a time of exceptional restriction of intercontinental migration of terrestrial faunas due to marine barriers, and probably was a time of development of "island continent" centers of endemism (Lillegraven, 1976; Tedford, 1974). Some workers believe that marsupials and placentals are too alike in their adaptive strategies for them to have developed simultaneously in the same region. If so, it is possible that mutual geographic isolation protected them from the beginning (Hoffstetter, 1970, 1971; McKenna, 1975a).

MARSUPIAL FOSSIL RECORD

The known fossil record of marsupials with emphasis on first appearances is as follows (Figure 21.1):

North America—*Holoclemensia* from the Trinity fauna (Albian in age) of Texas was once believed to be the earliest-known marsupial (Slaughter, 1968a,b). It is now regarded by Butler (1978) as a mammal of metatherian-eutherian grade [as it had originally been by Patterson (1956) and later by Turnbull (1971)] and is placed, with *Pappotherium*, in the family Pappotheriidae, Order Pappotherida, infraclass Tribotheria. The Tribotheria includes mammals with tribosphenic molars that are not classified as marsupials or placentals.

The oldest-known unequivocal marsupials are of early Campanian age (i.e., 70–80 m.y. ago) and are from the upper part of the Milk River Formation, Alberta, Canada. The known taxa include two genera of Didelphidae (*Alphadon, Alberta-*

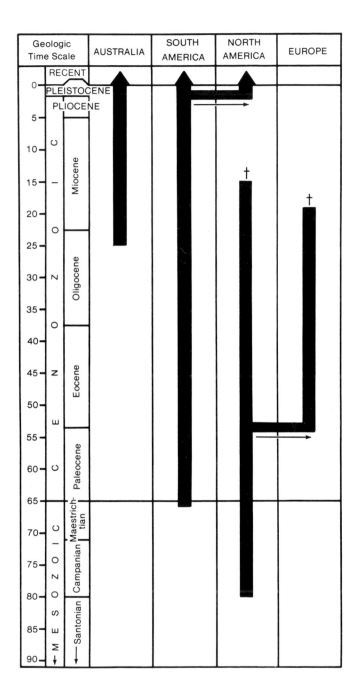

Figure 21.1. Known geographical and chronological distribution of Marsupialia. Arrows indicate direction of known dispersals and crosses indicate termination of fossil record and presumed extinction.

therium), two genera of Pediomyidae (*Pediomys, Aquiladelphis*), and one genus of Stagodontidae (*Eodelphis*) (Fox, 1971, 1979a,b). Members of these families occur throughout Late Cretaceous beds in North America. The family Stagodontidae and probably the family Pediomyidae became extinct at the end of the Cretaceous, although Slaughter (1978) presents data suggesting that pediomyids "may have survived well into the Tertiary along the Texas coast though extinct in the western United States about the close of the Cretaceous." Members of the Didelphidae are found in North America until middle Miocene time when they are last recorded in beds of lower Barstovian age at the Town Bluff locality of Texas (Slaughter, 1978). Marsupials are unknown in North America in beds of late Miocene (Clarendonian, Hemphillian) and Pliocene (Blancan) age. The genus *Didelphis* is first recorded in North America in the Coleman IIA local fauna of Florida, regarded as late Irvingtonian (middle Pleistocene) in age (R. A. Martin, 1974); it is represented on that continent by one living species, *D. virginiana* (Gardner, 1973).

South America—The earliest-known record of fossil marsupials in South America comes from the Laguna Umayo local fauna from the Vilquechico Formation, Peru (Sigé, 1972). Stratigraphic data (Portugal, 1974), the reported occurrence of dinosaur eggs (Sigé, 1968), and late Cretaceous charophytes (Grambast, Martinez, Mattauer, and Thaler, 1967), indicate this earliest South American mammal fauna is probably Late Cretaceous in age. The marsupial fauna includes a didelphid originally described as a species of *Alphadon, A. austrinum* (Sigé, 1972), a genus common in North American late Cretaceous faunas (Clemens, 1968). Recently, Crochet (1978, 1979a,b) recognized *austrinum* as a species of *Peradectes*, a genus which also occurs in beds of early Eocene through middle Oligocene age in North America and early Eocene age in Europe. *Peradectes*, along with *Alphadon, Albertatherium, Nanodelphis*, and possibly *Bobbschaefferia*, were placed in the tribe Peradectini by Crochet.

Sigé (1972) also reported several didelphoids of uncertain affinity from this same fauna, and noted that one may represent a pediomyid. Another specimen (?Didelphidé indét. 1 in Sigé, 1972) Crochet (1979a,b) recognized as a member of the didelphid tribe Didelphini, in which he included all Didelphinae not included in the tribe Peradectini.

Marsupials are next known in the South American fossil record in beds regarded as late Paleocene (Riochican) in age in Patagonia, southern Argentina and Brazil (for reviews see Simpson, 1971; Clemens and Marshall, 1976), and they are a characteristic element in all later fossil and Recent faunas on that continent.

Australasia—The oldest-known marsupials are from the Geilston Travertine near Hobart, Tasmania (Tedford, Banks, Kemp, McDougall, and Sutherland, 1975). The known fauna includes remains of the families Burramyidae, Phalangeridae, and Palorchestidae. Tedford, et al. (1975), demonstrated that formation of this travertine deposit was interrupted by a basalt incursion, and that the basalt and travertine are reasonably contemporaneous in age. Furthermore, the travertine containing the marsupial bones certainly predates the basalt incursion. A sample of this basalt was radioisotopically dated by the Potassium-Argon method and yielded an age of 22.4 ± 0.5 Ma. The Geilston Travertine is thus late Oligocene in age or older. Marsupials are known from beds of at least late Oligocene age through Recent in Tasmania, Pliocene to Recent in New Guinea, and early Miocene to Recent on the Australian mainland (Archer and Bartholomai, 1978).

Europe—In Europe, marsupials are known from beds of early Eocene (Sparna-cian) through middle Miocene (Sarmation) age. All of the twenty-six recognized species are referrable to the family Didelphidae. Three genera are recognized: *Pera-dectes* with three species, placed in the tribe Peradectini; *Amphiperatherium* with twelve species, and *Peratherium* with eleven species, both placed in the tribe Didel-phini (Crochet, 1978, 1979a,b).

Antarctica—Fossil marsupials are not known from either East or from West Antarctica, yet there is sufficient circumstantial evidence to indicate that they almost certainly did occur there in the past (see below). A good place to begin looking for them is in terrestrial beds called the Cross Valley Formation mapped as Paleocene in age on Seymour Island.

Africa, Asia, India—Undoubted fossil marsupials are not known from these continents and there is presently no direct evidence to suggest that they ever oc-curred there. But if they did, they clearly represented evolutionary dead ends as they eventually were in Europe and in North America (Simpson, 1978, p. 323).

MARSUPIAL DISTRIBUTION
AND PLATE TECTONIC THEORY

Many early biogeographers noted the striking similarity of living and fossil spe-cies of animals and plants in South America with those of Australia. Some of these similarities were attributed to parallel or convergent evolution. These types of evolution occur during an adaptive radiation in different regions, when various groups evolve to fill similar roles. A feature of this process is that, as stated by Simpson (1965):

> different lines of descent become adapted in similar ways. They are then likely to develop special functional and structural resemblances to each other. If the two lines involved are similar to begin with (hence near their common ancestry and rather closely related) they continue to be similar as they evolve, and this phenomenon is called parallelism. If they were more distinct to begin with (further from a common ancestry and less closely related), they become more similar as they acquire similar adaptations, and this is called convergence . . . Parallelism and convergence intergrade and are not always sharply distinguishable. Parallelism tends to involve closer resemblances than convergence, but neither process is known to result in identity.

These same similarities have also been attributed to a very close phylogenetic relationship between the taxa in question. For groups occurring in Australia and others in South America, such views were used as evidence to indicate the exist-ence of a former land connection between these continents (e.g., Stabilist Model 3, see below). Tedford (1974) states: "Such biological relationships were accepted as welcome independent evidence of former continental connections by the leading exponents of continental drift, Wegener (1967) and DuToit (1937)."

In the case of marsupials four groups in particular have entered into this debate: the "dog-like" borhyaenids of South America and the "dog-like" *Thylacinus* of Australia, and the "diprotodont" caenolestoids (and polydolopoids) of South Amer-ica and the "diprotodont" phalangeroids of Australia (= Diprotodonta of Ride, 1962). These groups have at various times and by various workers been interpreted as being close phylogenetic allies and hence, those relationships have been used as evidence to support continental drift. Other workers interpreted the resemblance as good examples of parallelism or convergence in evolution.

Case 1—the "dog-like" marsupials

The "Tasmanian Wolf," *Thylacinus cynocephalus*, is the largest Recent predaceous marsupial known, and is about the size of a small wolf (Figure 21.2). Thylacines are known from beds of late Miocene to Holocene age on the Australian continent, from Pliocene to Pleistocene age in New Guinea, and Pleistocene to Recent age in Tasmania. *T. cynocephalus* inhabited Tasmania at the time of the European settlement, but is now believed extinct. The last-known wild specimen was killed in 1930 and the last known live animal died in the Hobart Zoo in 1933.

In South America a fossil group of predaceous marsupials called borhyaenids were "dog-like" in appearance and presumably in ecology as well. They are known from beds of late Paleocene (Riochican) to Pliocene (Montehermosan) in age (Marshall, 1978).

The phylogenetic relationships of borhyaenids and thylacinids has been, and remains, disputed. The first-known borhyaenids were described and named at the end of the last century by the Argentinian paleontologist Florentino Ameghino. A fossil genus from the early Miocene Santa Cruz Formation of Patagonia, southern Argentina, was considered so similar to *Thylacinus* that Ameghino (1891) named it *Prothylacynus* (Figure 21.2). He implied that *Prothylacynus* was ancestral to *Thylacinus*, and placed both in the Family Thylacinidae.

Ameghino later changed his views, and in subsequent papers did not recognize the marsupials as a natural unit. He (e.g., 1906) divided the so-called polyprotodont marsupials into three groups: Pedimana (didelphoids), Dasyura (Australian carnivorous marsupials), and Sparassodonta (large South American carnivorous marsupials). He placed these (along with the Insectivora and Carnivora in a group Sarcobora, which included all the more or less carnivorous mammals. The borhyaenids, divided into various families, were grouped as sparassodonta, considered indirectly related to the Australian canivorous marsupials and through them to the Fissipedia and Pinnipedia.

Sinclair (1906), Gregory (1910), and Wood (1924) placed borhyaenids and *Thylacinus* in the Thylacinidae. Osborn (1910) and Matthew (1906) believed that this and other evidence indicated a link between South America and Australia via Antarctica.

Matthew (1915), Simpson (1941), and Ameghino (1906, and elsewhere as regards the origin of the group) believed borhyaenids were of didelphoid origin and paralleled the Australian thylacinids without any special affinity. Simpson (1941) pointed out that early workers were dealing with relatively late (early Miocene) borhyaenids and comparing these with *Thylacinus*, a highly specialized animal of of relatively recent age. Simpson emphasized that when characters of Paleocene and Eocene borhyaenids are examined, none belong exclusively to the Thylacinidae. Furthermore, some early borhyaenids (e.g., *Arminiheringia*) are more specialized than *Thylacinus* in some features (e.g., closed palate and nasal-lacrimal contact), so that borhyaenids cannot possibly be the ancestors of *Thylacinus*. There seems to be abundant, conclusive evidence that *Thylacinus* is a specialized dasyuroid.

The whole issue is admirably summarized by Simpson (1941, p. 5):

A

B

C

Figure 21.2. *A*, restoration of *Prothylacynus patagonicus*, a borhyaenid marsupial from the early Miocene Santa Cruz Formation of Patagonia, southern Argentina. Restoration is based on the skeletal reconstruction of a single individual given by Sinclair (1905, Plate 2; 1906, Plate 61). *B*, the recently extinct marsupial *Thylacinus cynocephalus* from Tasmania, Australia. *C*, the living wolf, *Canis lupus*, a placental from North America. *B* and *C* illustrations are based on a composite of numerous photographs of living animals. All are drawn to same relative size.

A common ancestry combining the primitive dasyuroid with the primitive didelphoid characters would be more didelphoid, that being the more conservative of the two lines. Such an ancestry for the Borhyaenidae is the only one well supported by the evidence now at hand. Given such a common ancestry, with the same genetic constitution at the beginning of divergence and with the same approximate repertory of mutational possibilities, and given similar environments, it seems not surprising but indeed inevitable that specialized animals evolved independently by adaption for essentially identical ways of life would resemble each other as much as do, for instance . . . [*Lycopsis*] . . . and *Thylacinus*. If they had any more immediate ancestry, they would be expected to resemble each other still more than they do, and in characters less visibly adaptive.

Simpson concludes that some authors (e.g., Scott, 1937) have over-emphasized the resemblance between these forms. It must be remembered that the common ancestor was not a didelphid in the sense of the living forms, but doubtless lacked some special characters of modern opossums and had some features now preserved in dasyurids and not in didelphids. Most importantly, this resemblance is not convergence, but parallelism, which yields a closer similarity than a more distant relationship.

For many years it was agreed that the similarity between borhyaenids and thylacinids resulted from parallel evolution. Recently, however, Archer (1976a,b) reopened the issue and promoted the view that they are closely related. This view has been followed and championed by Kirsch (e.g., 1977, 1979). If Archer is right, his views have important biogeographic consequences, for Australian marsupials could then be regarded polyphyletic (C. Patterson, in press).

I (1977, 1978) favor the view that Australian marsupials are monophyletic and that thylacinids evolved from dasyurid-like ancestors in Australia, while borhyaenids evolved independently from didelphid-like ancestors in South America. I thus follow Simpson and others and regard borhyaenids and thylacinids as representing a classic example of parallelism in evolution.

Case 2—the "diprotodont" marsupials

Phalangeroid marsupials of Australia are characterized by having a pair of large, procumbent lower incisors (Figure 21.3a). This condition has been called *diprotodonty*, as opposed to *polyprotodonty* in which there are a number of more or less subequal teeth. During much of the nineteenth century it was believed that marsupials with diprotodont dentitions were found only in Australia, and that the group was monophyletic. It thus came as quite a surprise when living and fossil marsupials with diprotodont dentitions were reported from South America. The phylogenetic relationships of the Australian and South American diprotodonts perplexed systematists. Views supporting a close phylogenetic relationship were, as in the case of the "dog-like" marsupials, accepted as welcome independent evidence for advocates of Continental Drift.

In 1887, Florentino Ameghino described and named a number of fossil taxa, some of which are now regarded as caenolestids. These taxa were based on specimens collected from the early Miocene Santa Cruz Formation of Patagonia by his brother Carlos. Florentino immediately recognized their marsupial affinity.

The first reference to a living caenolestid was by Tomes (1860), and was based on observation of a single juvenile specimen from Ecuador. It was not then given a scientific name, although reference to "a small and redimentary pouch" clearly indicated that it was a marsupial. In 1863 Tomes named this animal *Hyracodon fuliginosus*, but made no further reference to its marsupial affinity.

In 1895 a second specimen of an extant caenolestid was sent to the British Museum. Thomas (1895) recognized its affinities with Tomes's *Hyracodon fuliginosus*, but as the new specimen was larger and different in appearance he gave it a different specific name, *obscurus*. He placed this specimen along with Tomes's *fuliginosus*, in a new genus, *Caenolestes*, as *Hyracodon* was preoccupied by *Hyracodon* Liedy, 1856, a genus of fossil rhino. *Caenolestes* and some of Ameghino's taxa, with which it has affinity, were placed by Thomas in the family now called the Caenolestidae (Marshall, 1980).

Weber (1904) and Gregory (1910) recognized a tripartite division of Marsupialia. *Caenolestes* and its fossil allies were considered a distinct suborder, Paucituberculata; the other marsupials were included in the suborders Diprotodontia or Polyprotodontia.

Caenolestids are unique among marsupials in possessing diprotodont lower incisors, polyprotodont upper incisors, and quadrate molars. In attempting to establish caenolestid affinity, various workers have emphasized the special importance of one of these features over the other. The principal questions are: (1) are caenolestids indeed as distinct from diprotodonts and polyprotodonts as those groups are from each other; or (2) are caenolestids related by structure of their lower incisors to Australasian diprotodonts; or (3) by their upper incisors to polyprotodonts?

Bensley (1903) concluded that the similarities between *Caenolestes* and Australasian diprotodonts were the result of parallel evolution, and the diprotodont specializations evolved independently in the two groups.

Sinclair (1906, p. 443) and Scott (1937) believed, as did Ameghino and Thomas, that caenolestids were closely related to Australasian diprotodonts. This relationship was used as part of their evidence for existence for a former land connection between South America and Australia. Sinclair was firm in this view but he did concede (1906, p. 443) that the striking similarities in dental structure displayed by these groups "may be explained by convergence." Most recently, Pascual and Herrera (1973) opted for a special relationship between caenolestids and Australian diprotodonts, pointing out that such a relationship is even more likely if we accept the existence of a dispersal route via Antarctica.

Dederer (1909) and Broom (1911) emphasized the polyprotodont affinities of *Caenolestes*, and maintained that apart from the diprotodont specializations, it is a typical polyprotodont in all its cranial characters. Gregory (1910, p. 211) also believed that the skull of *Caenolestes* showed no striking diprotodont characters, and regarded it and its allies as a distinct suborder, "an offshoot of primitive polyprotodonts, which has paralleled the diprotodonts of Australia in certain characters of the dentition."

Based on an exhaustive monographic treatment of *Caenolestes*, Osgood (1921) concluded that caenolestids lay phylogenetically between the Australasian Peramelidae and Phalangeridae. The relationships "are best expressed by classifying . . . [*Caenolestes*] in the suborder Diprotodontia, family Palaeothentidae, subfamily Caenolestinae."

Caenolestid affinities were clarified by Abbie's (1937) study of the marsupial

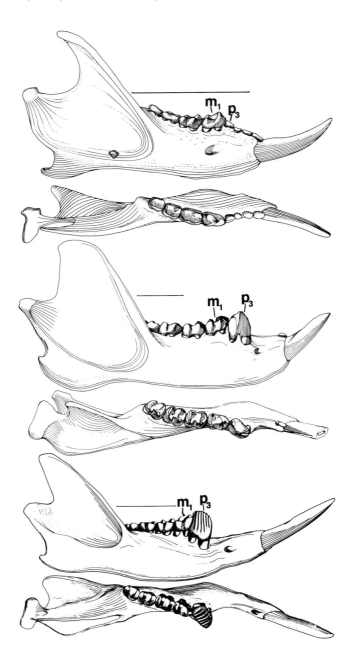

Figure 21.3a. Labial and occlusal views of right mandibular rami of living Australasian "diprotodonts." Top, *Petaurus breviceps,*; middle, *Trichosurus vulpecula;* and bottom, *Hypsiprymnodon moschatus.* Scales = 10 mm.

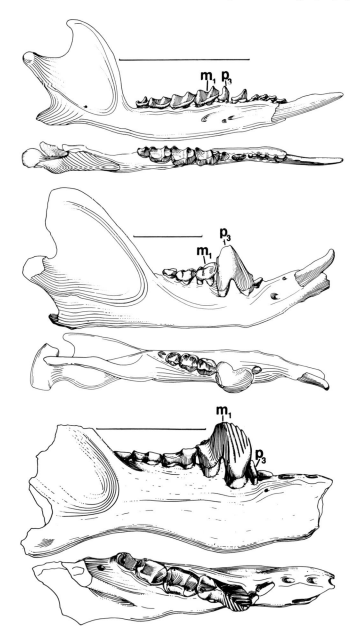

Figure 21.3b. Labial and occlusal views of right madibular rami of living and fossil South American "pseudodiprotodonts." Top, *Caenolestes fuliginosus* (living); middle, *Epidolops ameghinoi* (late Paleocene-Riochican); and bottom, *Abderites meridionalis* (early Miocene-Santacrucian). Scales = 10 mm.

brain. He documented further Elliot Smith's discovery that Australasian diprotodonts have a *fasciculus aberrans* in the anterior commissure of the brain, while polyprotodonts lack this structure. Since caenolestids lack this structure, they must be assigned to the Polyprotodontia.

Ride (1962, p. 299) reviewed data bearing on the identity and homology of the diprotodont incisors in caenolestids and phalangeroids. He concluded that the procumbent tooth in phalangeroids is probably the I_3 (by homology with that of didelphids), while in caenolestids it is the I_1 or I_2. This substantiated the view that the dental similarities between Phalangeroidea and Caenolestoidea are the result of convergence. Ride referred to the procumbent specializations in the Phalangeroidea as *diprotodonty,* compared with *pseudodiprotodonty* in caenolestids.

Further clarification of caenolestid affinity has resulted from study of sperm. Biggers and DeLameter (1965) demonstrated that pairing of spermatozoa in the epididymis occurs in American didelphids and caenolestids tested, but not in Australasian forms. Sperm pairing is a specialized mammalian feature suggesting that all living American marsupials form a monophyletic group. These workers further demonstrated that the morphology of the sperm in *Caenolestes* is quite different from those of the other marsupials examined.

Kirsch (e.g., in Hayman, et al., 1971) compared sera of *Caenolestes obscurus* and *Lestoros inca* with members of all the families of living Marsupialia except the Thylacinidae. He (1977, p. 92) concluded that the two caenolestids are more similar to each other than they are to any other marsupial group. The tests (Hayman, et al., 1971, p. 194–95) indicated that:

> . . . caenolestids have no greater similarity to any Australasian marsupial superfamily examined than to the [American] didelphids; in fact, the data suggest that caenolestids are less like any of the other superfamilies of marsupials than those superfamilies are like each other. Thus serology does not seem to support a separation of living marsupials into Australasian and American stocks, but suggests that the caenolestoids diverged from the principal line of marsupial evolution before the separation of didelphoids and Australasian marsupials. Such a conclusion, however, assumes that serological affinity strictly reflects propinquity of descent, and furthermore, because only two closely similar, modern genera of caenolestids were studied, it leaves little allowance for variation within groups or for aberrant results.

I thus favor the view that South American caenolestids and Australasian phalangeroids independently evolved diprotodont dental specializations from different polyprotodont ancestors on their respective continents. As in the case of the doglike marsupials, I regard these diprotodont marsupial groups as representing a classic example of parallelism in evolution (Marshall, 1980).

Caenolestid-polydolopid affinity

For many years it was conventional practice (e.g., Simpson, 1945) to classify the South American groups Caenolestidae and Polydolopidae in the superfamily Caenolestoidea. This practice was based in part on the belief that the sectorial or *plagiaulacoid* tooth was the first molar. Paula Couto (1952) clearly demonstrated, however, that the polydolopid sectorial tooth is the last premolar and not the first molar as in those caenolestids with plagiaulacoid dentitions (Figure 21.3b). Thus, evolution of sectorial teeth is a convergent (or parallel) feature in these lineages (Pascual and Herrera, 1973). Because there is no convincing evidence to suggest that polydolopids are any closer phylogenetically to caenolestids than they are to didelphids, Clemens and Marshall (1976, p. 9) allocated the Polydolopidae to a separate superfamily, the Polydolopoidea. This left the Caenolestoidea an uncluttered and cohesive monophyletic group (Marshall, 1980).

CONTINENTAL DISTRIBUTIONS: PRESENT AND PAST

Stabilist Models

The fundamental assumption of stabilists is that the arrangement of ocean basins and continents have remained the same during the time covered by the origin, evolution, and dispersal of mammals (i.e., from Mesozoic through Cenozoic time) (Tedford, 1974, p. 110). An additional assumption is that there have been no land bridges other than those that now exist (e.g., the Panamanian land bridge), except for one (Beringia) that connected Asia and North America (Simpson, 1965, p.130). The dispersal patterns proposed by many stabilists are often based on the belief that all major groups of mammals originated in Holarctica (continents of the northern hemisphere) (e.g., Darwin, 1859; Matthew, 1915). According to this model, primitive groups of mammals, such as marsupials, were displaced to continents of the southern hemisphere as a result of adaptive radiations of more advanced groups, such as placentals, on the northern continents (Tedford, 1974, p. 110). This displacement scenario creates what Hershkovitz (1972, p. 316) called the "Sherwin-Williams effect." It is further based on Matthew's "rule-of-thumb" axiom that the age and site of the oldest known representative of a taxon be taken as the time and place of origin of that taxon.

There are two pieces of evidence which now make stabilist models virtually untenable. The first is based on intense recent studies of late Cretaceous and early Cenozoic Asiatic faunas which demonstrate absence of definite marsupials.

> Still more persuasive . . . is the plate tectonic evidence . . . , which indicates that Australia was probably in the vicinity of East Antarctica until some time in the Eocene and did not reach somewhat its present relationship to the East Indies and Asia until approximately the Miocene. . . . If that evidence is accepted, and I believe it should be, the spread of marsupials (in either direction) between Asia and Australia in the Cretaceous or early Cenozoic seem to be impossible (Simpson, 1978, pp. 322–23).

Stabilist Model 1 (Figure 21.4)—The classical, and for a long time the most popular, model of marsupial paleobiogeography was that they arose in North America, an assumption based on the fact that the oldest known marsupials are found on that continent. From there marsupials dispersed in the Late Cretaceous by rafting or island-hopping to South America via an archipelagic chain, or the precursor of the Panamanian land bridge (Simpson, 1950; Patterson and Pascual, 1972), and separately to Asia across a filter bridge via Beringia. From Asia they dispersed by rafting or by island-hopping, through Indonesia to Australasia (e.g., see Matthew, 1915; Simpson, 1961, 1965; Darlington, 1965; Clemens, 1968).

One of the objections to the hypothesis that the immigrant marsupials came

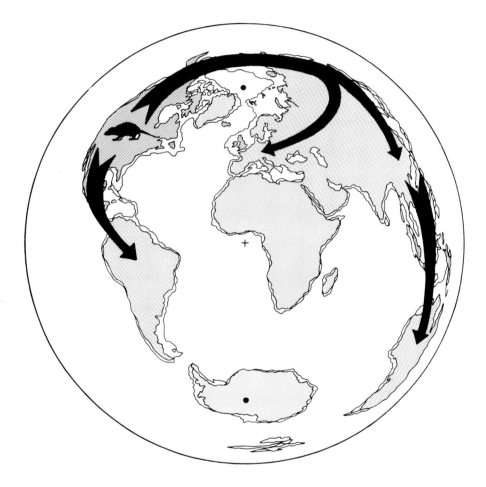

Figure 21.4. Center of origin and early biogeographic history of marsupials as proposed by Stabilist Model 1. Present positions of the continents shown on Lambert's equal-area projection. Based on Map 40 in Smith and Briden (1979).

from Asia has been the absence of remains of marsupials in the Asian fossil record. . . . the Asian Cretaceous and early Cenozoic fossil records are limited and biased in favor of sites well removed from the coastal regions of these times. At present there is no way to verify the suggestion that marsupials were present in the contemporaneous faunas of Asian coastal lowlands. However, the occurrences of Cretaceous and early Cenozoic marsupials in North America, indicate coastal lowlands probably provided a favorable environment for them. Thus, marsupials might have reached the Cretaceous or early Cenozoic coast of southeastern Asia by dispersal across the Bering Straits and then southward along the Pacific strand (Clemens, 1968, p. 15).

Stabilist Model 2 (Figure 21.5)—This model, suggested by Clemens (1968), is similar to Stabilist Model 1, except that marsupials may have reached Europe via a North Atlantic dispersal route. From there a species of *Peratherium* could have extended its range across southern Asia and from there dispersed by a sweepstakes route to Australasia.

Stabilist Model 3 (Figure 21.6)—According to proponents of this model, marsupials originated in South America and dispersed to North America either by island-hopping across the Caribbean or by an early emergent phase of the Panamanian land bridge. From North America they dispersed to Asia via an emergent phase of the Bering land bridge and from there to Europe. Dispersal from South America to Australasia occurred via Antarctica, and involved a now submerged land bridge that united these continents.

Sinclair (1906, p. 444), an early proponent of this model, summarized his views as follows:

> The reality of a former land connection between the Australian region and South America is plainly indicated by several lines of evidence based on the distribution of fishes, land shells, decapod crustaceans, plants, and Tertiary marine molluscs. This land connection is believed to have existed not later than the close of the Cretaceous or beginning of the Tertiary, and it is only by such a connection that the distribution of the Thylacinidae can be explained. The direction, continuity or discontinuity of this land bridge need not enter into the present discussion. So far as the Thylacinidae are concerned, there can be little doubt of their South American origin, judging from the marked adaptive radiation which they attained during the Santa Cruz epoch, but whether the same can be said of marsupials in general is still a matter of question. It is believed, however, that the order may be properly regarded as of southern origin and that the occurrence of opossums in North America and Europe may be explained as the result of migration from the southern hemisphere.

Other workers, including Florentino Ameghino (op. cit.), Osborn (1910), and Anderson (1940), championed or favored this model either in total or in part.

Stabilist Model 4—A number of other early workers advocated the former existence of now-submerged land bridges in the southern hemisphere to account for otherwise inexplicable distribution patterns. They differed from those advocating Stabilist Model 3 in believing that Antarctica was the area of origin for the groups in question. There are many variations of this model. For example, Hedley (1899) noted:

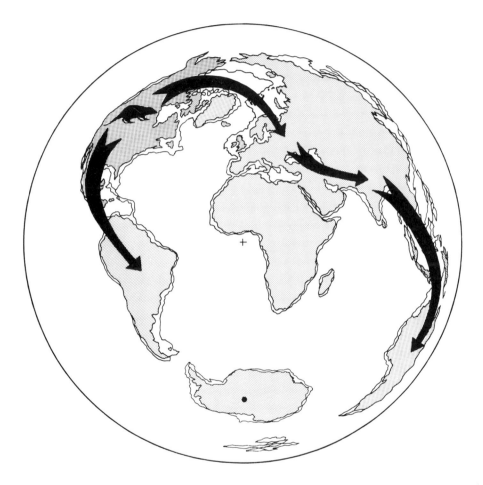

Figure 21.5. Center of origin and early biogeographic history of marsupials as proposed by Stabilist Model 2. Present positions of the continents shown on Lambert's equal-area projection. Based on Map 40 in Smith and Briden (1979).

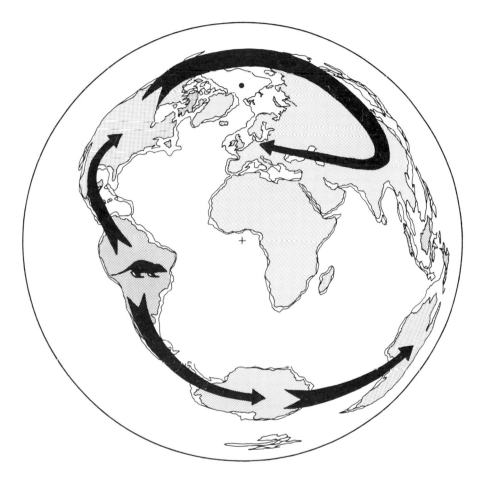

Figure 21.6. Early biogeographic history of marsupials as proposed by Stabilist Model 3. Present positions of the continents shown on Lambert's equal-area projection. Based on Map 40 in Smith and Briden (1979).

A rich fauna of Antarctic origin, which, entering by Tasmania, overran the whole continent [of Australia], crossed Torres Strait into New Guinea, and reached its utmost eastern limit in the Solomons. Characteristic members of it are marsupials, monotremes, cystignathous frogs, venomous snakes, and snails of the order Macroogna.

Longman (1924, p.12) has summarized the pertinent literature as follows:

The assumption of a continent in the Antarctic in early Tertiary times, in order to account for related elements in the faunas and floras of South America, Australia, New Zealand, various Pacific Islands, Madagascar, and South Africa, has proved a veritable Pandora's box to biologists. In order to explain many of the difficulties raised, it is suggested that this circumpolar area had radiating land bridges, inconstant in direction and dimension, apparently reaching out on the one hand to deposit certain types in Pacific archipelagoes; elsewhere stretching to Madagascar to transfer iguanas; extending to New Zealand to allow other forms to gain new land; and independently reaching Tasmania to enable the ancestors of our marsupials to complete a transpolar journey and find a congenial refuge.

Mobilist Models

"One of the most exciting aspects of plate tectonic theory is that it provides a framework for reconstructing the past positions of the continents. Rarely in science can one so easily turn back the hands of time" (Scotese, 1980, p. 93).

With the acceptance of plate tectonic theory in the latter part of the last decade, biogeographers began to actively reinterpret present and past distributions of animal and plant groups based on changing continental configurations (e.g., Colbert, 1973). The dynamic model of the Earth's crust as postulated by plate tectonic theory provided the geological basis for a new concept of paleogeography and a set of physical constraints on biogeographic theory (Tedford, 1974, p. 124). As such, it provided a new basis for speculation about the biogeographic history of living and fossil organisms (del Corro, 1977; Kurtén, 1969).

"Important consequences of the new geography are the reduction in probability for some zoogeographic hypotheses and the increase in probability of others" (Tedford, 1974, p. 124). In short, plate tectonic theory places constraints on biogeographic hypotheses and helps to establish levels of probability among the various models proposed on biological grounds. Since oceanic barriers strongly influence dispersal probabilities of terrestrial organisms, it is important to know where, when, and for how long such barriers existed. The principles of dispersal as formulated by Simpson (e.g., 1965) under the stable continent model were "applied to the dynamic model with logical modifications and extensions (McKenna, 1973)" (Tedford, 1974, p. 124).

Many mobilists used the continental configurations as originally proposed by Wegener (1967). For the Cretaceous, Wegener showed Australia, Antarctica, South America, and North America forming a single continental land mass after Africa and South America were separated by the nascent development of the Atlantic Ocean. This relationship, however, is not totally conformable with present geophysical evidence.

With increase in geophysical data, new and better-documented continental configurations began to emerge. A literature relay ensued in which papers presenting new geophysical data modifying previous views of continental relationships were shortly followed by papers from biogeographers attempting to incorporate these data into new models explaining distribution patterns. The publication rate by biogeographers was staggering, and during the past decade over forty papers were devoted to marsupials alone. Plate tectonic theory has thus had a "bandwagon effect" on biogeography resulting in the production of a plethora of models for marsupial paleobiogeography.

What is most intriguing is that the data set for fossil marsupials has changed little in the past decade. It is the geophysical evidence which has spurred on the model builders, and has strained their abilities to either explain or explain away the negative evidence as dictated by an incomplete fossil record. Fossils give only a minimal age of a group's occupation of an area while maximal ages are at the mercy of a model builder's imagination.

The various mobilist models (except 8, Figure 21.15) are shown on a Lambert's equal-area projection. In all cases the land positive areas of today are shaded and outlines are given of shelf margins. The area of marsupial origin is indicated, where appropriate, by the silhouette of an opossum. Each arrow indicates a single inferred dispersal event and the direction in which the inferred dispersal occurred.

A number of excellent reviews deal with the geological histories of areas critical to discussion of marsupial paleobiogeography and plate tectonic theory (e.g., Cracraft, 1973, 1974, 1975; Keast, 1977; McKenna, in press; Rich, 1975; Tedford, 1974). These attempt to establish the times of initial rifting and geographic separation of various continental land masses, as well as the significance of these separations on potential dispersal of terrestrial vertebrates. In some cases, paleoecological data are considered to see if a particular route at a particular time was or was not climatically amenable for dispersal of warm-blooded terrestrial vertebrates. The map sequence in Figure 21.7 shows approximate times of separation of the different land masses. These continental reconstructions are shown on a Lambert's equal-area projection at the following times: 140 m.y. ago (A), 80 m.y. ago (B), 40 m.y. ago (C), and present day (D) (modified after Smith and Briden, 1979).

The following information is particularly relevant to marsupial paleobiogeography. South America was in at least partial contact with Africa until about 90 Ma; South America and West Antarctica may have been connected, at least periodically, until about 40 Ma (late Eocene); South America probably was not connected with North America during most, if not all, of the Tertiary until the appearance of the Panamanian land bridge about 3.0 Ma; North America was last connected with Europe in the early Eocene; North America and Asia were in direct contact periodically during the Cenozoic; and Australia was in juxtaposition with East Antarctica until the beginning of the Eocene. These events provide latest dates for direct interchange of terrestrial faunas between respective continents.

Mobilist Model 1 (Figure 21.8)—This model promotes the view that marsupials originated in North America in the middle Cretaceous, the site of their first documented adaptive radiation. They dispersed to South America either by island-hopping or waif dispersal in the Late Cretaceous, and by similar means to Australia via Antarctica in the Late Cretaceous or earliest Cenozoic. A dispersal route from South America, across Antarctica to Australia, may have been available until the Eocene (McGowran, 1973; Rich, 1975; Raven and Axelrod, 1972). From North America marsupials dispersed to Europe in the early Eocene via a dry land connec-

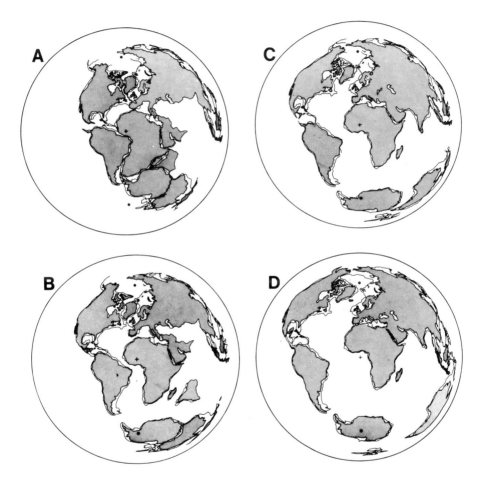

Figure 21.7. Positions of the continents shown on Lambert's equal-area projection at: *A*, 140 m.y. ago (Tithonian–late Jurassic); *B*, 80 m.y. ago (Santonian–Late Cretaceous); *C*, 40 m.y. ago (late Eocene); and *D*, present day. Based on Maps 48, 45, 43, and 40, respectively, in Smith and Briden (1979).

Figure 21.8. Center of origin and early biogeographic history of marsupials as proposed by Mobilist Model 1. Positions of the continents at 80 m.y. ago (Santonian–Late Cretaceous) shown on a Lambert equal-area projection. Based on Map 45 in Smith and Briden (1979).

tion across the North Atlantic (i.e., the DeGeer Route of McKenna, 1975b). This route was at a lower paleolatitude than one which may have occurred between Asia and North America (Beringia) at that time (Szalay and McKenna, 1971).

This model has, in total or in part, been championed or favored by Jardine and McKenzie (1972), Lillegraven (1974), Clemens (1977, p. 61), and P. G. Martin (1977, p. 111). Proponents of this model use the same argument as proponents of Stabilist Models 1 and 2 for inferring North America as the area of origin of marsupials. Thus, this fundamental premise is unaltered in this model, although the routes of dispersal are.

Mobilist Model 2 (Figure 21.9)—According to this model, marsupials originated on what is now the South American part of Gondwana. From there they dispersed to Australia via Antarctica in the Late Cretaceous or early Cenozoic along a volcanic archipelago bordering the Pacific side of Antarctica (Tedford, 1974), and to North America in the Late Cretaceous by island-hopping across the Caribbean. From North America marsupials subsequently dispersed to Europe via a dry land connection across the North Atlantic in the early Eocene. This model has been promoted or favored by Cox (1973), Tedford (1974), and Paula Couto (1974). Crochet (1978, 1979a,b) has recently argued for a South American origin for at least some Marsupialia, with dispersal occurring in two groups (*Peradectes* and Didelphini) to North America around the Cretaceous–Tertiary boundary, and from there to Europe in the latest Paleocene or earliest Eocene. The Australian marsupials are immigrants from South America via Antarctica.

Proponents of this model argue that the spectacular diversity of the late Paleocene (Riochican) marsupial faunas of South America "makes plausible the hypothesis that the didelphids had already experienced a long evolutive [*sic*] history in South America before the beginning of the Cenozoic Era, that is since the Cretaceous period, perhaps the mid-Cretaceous or even earlier" (Paula Couto, 1974, p. 116).

Furthermore, absence of mammal faunas older than that of Laguna Umayo (Late Cretaceous) does not necessarily exclude the possibility that marsupials were there before that time because this absence is based on lack of evidence and is not an established fact (Tedford, 1974).

In opposition to this line of reasoning, Clemens (1977, p. 63) has noted that in the absence of a fossil mammal record in South America, Tedford has:

> developed a hypothetical phylogeny of South American Cretaceous marsupials, with chance dispersals to explain their occurrence in North America.
> . . . Tedford (1974) goes on to argue that the scope of the adaptive radiation of marsupials recorded in the Laguna Umayo and other local faunas of unequivocal Tertiary age, "suggests that marsupials may have been on that continent (South America) for a long time, possibly from early in the Cretaceous." This assessment cannot be disproven. However, considering the scope of the adaptive radiations of the marsupials in Australia (which Tedford suggests reached that continent in the late Cretaceous or Paleocene) or those of Primates or the Condylarthra in the North American Paleocene, his argument loses force. Currently the available evidence favors the hypothesis of origin of marsupials in the Americas, although origin in Antarctica (or even possibly, but unlikely, Australia) cannot be lightly dismissed. Of the working hypotheses I still prefer that of North American origins, which is based on

Figure 21.9. Center of origin and early biogeographic history of marsupials as proposed by Mobilist Model 2. Positions of the continents at 80 m.y. ago (Santonian–Late Cretaceous) shown on a Lambert equal-area projection. Based on Map 45 in Smith and Briden (1979).

a Cretaceous fossil record. Only when comparable samples of the South American Cretaceous fauna become available will there be a basis for choice.

Simpson (1978, p. 323) has noted that "the fact that marsupials occur earlier in North America than any known from elsewhere has been taken as a positive indication, but that is a false positive because no Cretaceous mammals at all are known from Australia or Antarctica and none of equal or greater age from South America." Besides, to infer the antiquity of a given taxon (i.e., marsupials) by comparison of its morphological and/or taxonomic differentiation with that of another taxon (i.e., primates, condylarths) over a similar span of time is unwarranted (Simpson, 1980, p. 199).

Mobilist Model 3 (Figure 21.10)—Kirsch (1979) proposed an Australian origin for Marsupialia, with subsequent dispersal to Europe via Antarctica, South America, and North America. He (1979, p. 116) justifies this model as follows:

> . . . je crois que l'idée d'une origine australienne pour les marsupiaux, si naturelle au vu de leur grande diversité dans cette région, est une idée valable, et même vraisemblable, si l'on rejette l'hypothèse que les plus vieilles formes connues de fossiles indiquent nécessairement l'endroit d'origine d'un groupe et si l'on s'affranchit de l'idée que les marsupiaux son toujours des concurrents démunis devant les placentaires. En outre, une origine australienne des marsupiaux concorde avec un grand nombre de preuves: cela résout le problème posé par les hypothèses d'une dispersion à partir de l'Amérique du Sud au cours de laquelle les placentaires n'auraient pu atteindre l'Australie [Riek, 1970, p. 746]. Elle correspond mieux à la dynamique de dispersion et est plus réaliste étant donné la vraie nature des interactions compétitives. L'argumentation présentée ici est donc une extension logique de celle de Hoffstetter. Elle ne se borne pas à l'examen des données fossiles limitées et présente des arguments biologiques permettant de choisir parmi les origines vraisemblables. Une des prédictions spécifiques de mon hypothèse est que l'on devrait trouver dans le crétacé précoce d'Australie (et naturellement de l'Antarctique) des marsupiaux fossiles primitifs. Cette prédiction est peut-être en train de se réaliser par le petit bout, au sens littéral; il y a quelques années, E. F. Riek, un entomologiste australien, a décrit une puce du crétacé inférieur de QueensLand en Australie. Or, cet insecte semble appartenir à la famille des puces vivant sur les marsupiaux australiens . . .

Mobilist Model 4 (Figure 21.11)—Hoffstetter (1970, 1971, 1972) proposed that the initial divergence of marsupials and placentals was the result of vicariance involving breakup of Pangaea. This model was originally based on Wegener's (1967) continental configurations for the Cretaceous. Marsupials are believed to have originated on the combined land mass of what is now North America, South America, Antarctica, and Australia, with dispersal to Europe occurring in the early Cenozoic via a dry land connection across the North Atlantic. Placentals, on the other hand, originated on the combined land mass of what is now Africa and Eurasia (India was also included here by Wegener), and later spread to North America from Europe via the dry land connection across the North Atlantic and to South America. Hoffstetter (1972) presented a map promoting these same views, but used the continental configurations at the beginning of the Cretaceous as proposed by Dietz and Holden (1970).

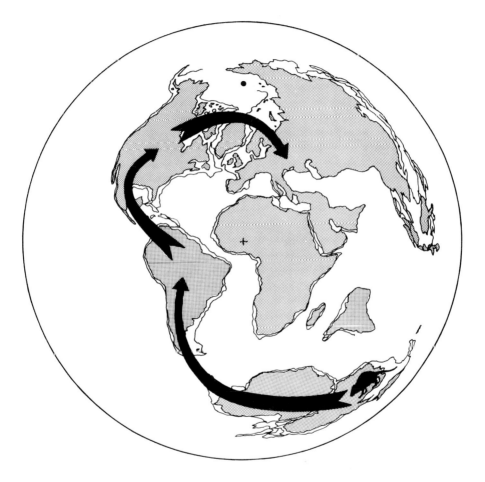

Figure 21.10. Center of origin and early biogeographic history of marsupials as proposed by Mobilist Model 3. Positions of the continents at 80 m.y. ago (Santonian–Late Cretaceous) shown on a Lambert equal-area projection. Based on Map 45 in Smith and Briden (1979).

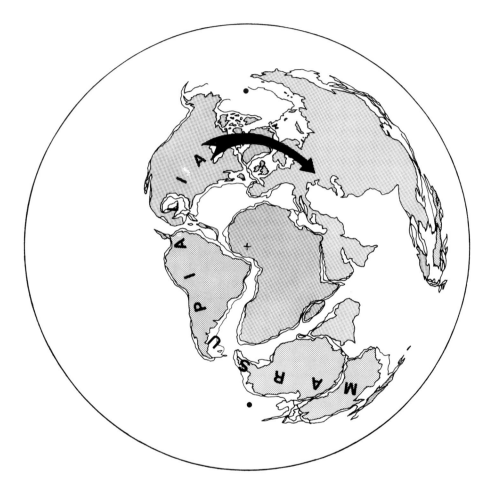

Figure 21.11. Area of origin and early biogeographic history of marsupials as proposed by Mobilist Model 4. Positions of the continents at 120 m.y. ago (Hauterivian–Early Cretaceous) shown on a Lambert equal-area projection. Based on Map 47 in Smith and Briden (1979).

Simpson (written communication, February 23, 1980) offers two points of consideration with regard to this model:

> First, I think it virtually impossible that the Marsupialia originated all over a connected land mass of Australia–Antarctica–South America North America everywhere all at the same time. It is extremely more probable that they originated in only one of those, even if those lands were then fully confluent, and probably only in a part of that one. (I don't know where that was.) Then they spread to all three other parts, and later evolved separately on at least three of the four. Second, I think it unlikely that the route Australia–Antarctica–South America was an uninterrupted land bridge, on which *only* marsupials crossed in only one direction, whatever the direction. I still think such a land bridge extremely improbable.

Mobilist Model 5 (Figure 21.12)—This model is a modified version of Mobilist Model 4. Hoffstetter (1975, 1976) now favors the view that marsupials originated on what is now North and South America. From South America they dispersed to Antarctica across a narrow water barrier and then on to Australia. In other respects this model is identical to Mobilist Model 4. The geographic isolation of the North and South American marsupials in the early Tertiary was due to vicariance.

Rage (1978) argued that the distribution of numerous vertebrate groups in Late Cretaceous beds in North and South America favors the existence of at least an intermittent land bridge between these continents during that time.

This model is similar to that proposed by Keast (1977, p. 85), who opts that marsupials originated in North and/or South America about 100 Ma and dispersed to Australia through a cool-temperate Antarctic bottleneck, probably in the early Tertiary.

Mobilist Model 6 (Figure 21.13)—Thenius (1971) proposed that marsupials probably originated on one or more of the southern continents, but which one or ones he was not certain. "Whether the center of origin of marsupials lay in Antarctica, Africa, or South America cannot be determined at this time" (translated from German) (Thenius, 1971, p. 193).

The unique aspect of this model is that the continents involved were South America, and/or Africa, and/or Antarctica, but not Australia. According to this model, marsupials were once present in Africa, but early became extinct on that continent. They dispersed to Australia from or through what is now Antarctica and to North America from or through what is now South America.

Mobilist Model 7 (Figure 21.14)—This model, championed by Cox (1970), assumes that marsupials existed before placentals and that placentals evolved from marsupials and may thus be regarded as "specialized marsupials" (Cox, 1970, p. 769). This assumption is based largely on Lillegraven's (1969) statement: "It is my opinion that if it were possible to study living specimens of the ancestral therian stock, most observers would not hesitate to assign them to the infraclass Metatheria, order Marsupialia."

This view led Cox (1970, p. 769) to conclude:

> The few relevant facts, then, suggest that marsupials were in existence before the placentals. If so, the existence of Mid-Cretaceous placentals in turn suggests that the marsupials had differentiated by the Lower Cretaceous at the latest. . . . It is also accepted that they may have been able to spread to

Figure 21.12. Area of origin and early biogeographic history of marsupials as proposed by Mobilist Model 5. Positions of the continents at 80 m.y. ago (Santonian–Early Cretaceous) shown on a Lambert equal-area projection. Based on Map 45 in Smith and Briden (1979).

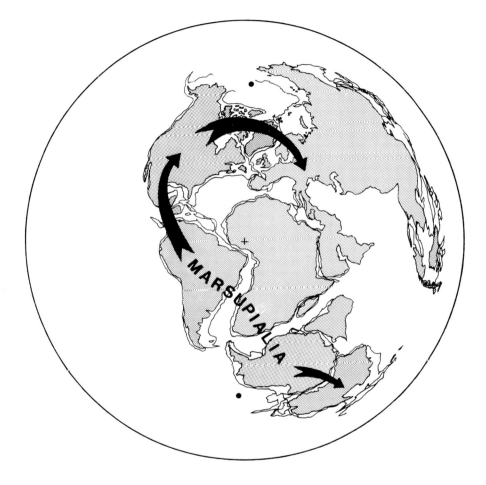

Figure 21.13. Early biogeographic history of marsupials as proposed by Mobilist Model 6. Positions of the continents at 120 m.y. ago (Hauterivian–Early Cretaceous) shown on a Lambert equal-area projection. Based on Map 47 in Smith and Briden (1979).

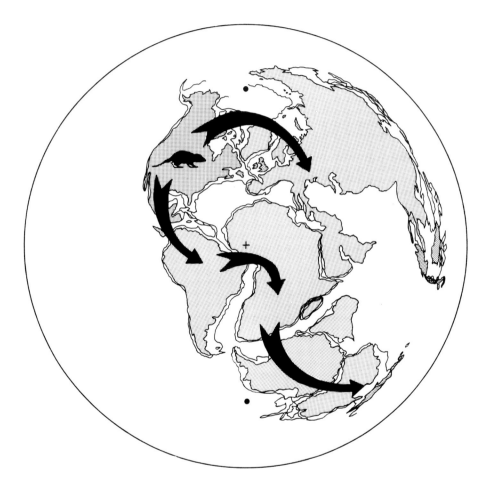

Figure 21.14. Early biogeographic history of marsupials as proposed by Mobilist Model 7. Positions of the continents at 120 m.y. ago (Hauterivian–Early Cretaceous) shown on a Lambert equal-area projection. Based on Map 47 in Smith and Briden (1979).

Africa [from South America] at that time, the problem of deciding whether they could have spread from there to Australia is reduced to that of determining the time by which land access to that continent had been broken.

Cox (1970, p. 769) presents geophysical evidence which he interprets as permitting "the suggestion that marsupials could have migrated from Africa to Australia via Antarctica in the Lower Cretaceous" [ca. 120 m.y. ago].

This dispersal took place before the Antarctica–Australia land mass separated from Africa, while the separation occurred before placentals came into existence, and thus deterred a similar dispersal route for that group. This scenario thus attempts to account for the absence of placentals on the land mass of Antarctica-Australia.

Cox does not discuss the origin of the South American marsupial stock, although considering the references which he cites and lack of statements to the contrary, he seems to accept a North American origin for the group with subsequent dispersal to South America.

This model, like Mobilist Models 8 and 9, suggests the presence of marsupials in lands where they do not occur today and where they are unknown as fossils. Mobilist Model 7 has not been accepted by subsequent workers and was abandoned by its author (Cox, 1973). It was resurrected by Raven and Axelrod (1972), but with little success. The latter authors used the continental reconstructions of Smith, Briden, and Drewry (1973) which suggested that Africa may have remained in contact with Antarctica, Australia, and India until about 100 Ma. Keast (1977, p. 78) recommended rejection of this model because "(1) the land connections were disrupted too early; (2) it is unnecessarily circuitous and complicated; and (3) fossil marsupials are unknown from Africa. Negative evidence is, of course, inconclusive."

Mobilist Model 8 (Figure 21.15)—This model, as in the preceding, assumes that metatherians originated and dispersed earlier than did eutherians, and that metatherians are directly ancestral to eutherians. It further implies that a partial or total "evolutionary relay" occurred in which less progressive groups in a "world fauna" were replaced by more progressive groups. Thus, marsupials once had a world-wide distribution but are presently restricted because of replacement elsewhere by competitively superior placentals.

Fooden (1972) hypothesized that while the continents were united in the single land mass of Pangaea, it was populated by a "world fauna" consisting of Prototheria and Metatheria (Figure 21.15). Pangaea subsequently fragmented into three primary land masses: Laurasia (including North America and Eurasia), West Gondwana (including Africa, South America and Madagascar), and East Gondwana (including Antarctica and Australasia) (Figure 21.15). India was at this time believed to be an island continent. Laurasia and West Gondwana remained in contact by a connection between northwestern Africa and southcentral Laurasia. This land mass contained the "world fauna" which included the Metatheria and their descendants, the Eutheria; while the Prototheria were extinct. East Gondwana was populated by the original "world fauna" which included only Prototheria and Metatheria and was now isolated by marine barriers from other land masses and their faunas. This land mass configuration and distribution of faunas was achieved by the Upper Jurassic or Lower Cretaceous.

By the Middle or Upper Cretaceous, South America became isolated from Africa (Figure 21.15). South America retained the earlier "world fauna" which included

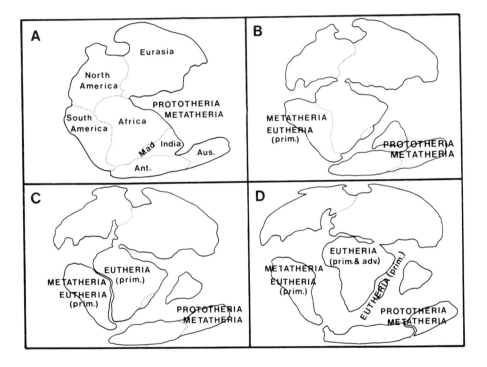

Figure 21.15. Distribution of continents and faunal groups through time as proposed by Mobilist Model 8. Continental configurations are based on Dietz and Holden (1970, Figures 2 through 5): *A*, Permian, 225 m.y. ago; *B*, late Triassic, 180 m.y. ago; *C*, late Jurassic, 135 m.y. ago; *D*, late Mesozoic–Early Cenozoic, 65 m.y. ago. The composition of "world mammal faunas" and "isolated mammal faunas" during successive stages of continental fragmentation are hypothetical. Modified after Figure 1 in Fooden (1972, p. 897). For explanation see text.

metatherians and primitive eutherians, while the new "world fauna" developing on the land mass consisting of Africa, Eurasia, and North America was composed primarily of primitive Eutheria. In the final sequence Madagascar separated from Africa in the Paleocene or lower Eocene, and carried with it an isolated fauna composed totally of primitive eutherians. The "world fauna" by middle Eocene time consisted of primitive and advanced eutherians distributed over the connected land masses of Africa, Eurasia, and North America.

Thus, the distribution of prototherians, metatherians, and primitive eutherians can be seen as representing "successively detached samples of the evolving world mammal fauna as it existed when each of these land masses became faunally isolated from the rest of the world as a result of the progressive fragmentation of Pangaea" (Fooden, 1972, p. 897).

A major objection to this model is that marsupials and placentals apparently arose simultaneously, not successively, and shared a common Early Cretaceous therian ancestor.

Mobilist Model 9 (Figure 21.16)—P. G. Martin (1970, 1975) proposed that marsupial paleobiogeography is best explained by assuming origin on a hypothetical mid-Pacific continent, *Pacifica*. Pacifica was originally situated over the "Darwin Rise," a rift zone in the mid-Pacific. Activity of this rift zone caused Pacifica to split in the Early Jurassic, and the two portions served as Noah's Arks (sensu McKenna, 1972, 1973). One "ark" drifted southwest forming part of New Guinea and the present southern Coral Sea Platform. Upon "docking" with Australia, this portion of Pacifica discharged its marsupial fauna. A second "ark" drifted northeast, and was incorporated into the western edge of North America. The marsupial fauna was discharged upon "docking" and colonized that continent, eventually dispersing to South America. P. G. Martin (1975, p. 217) concluded that this model is compatible with known tectonic histories of the north and east Pacific, "and our present ignorance of the tectonic history of the western Pacific means that the hypothesis cannot be excluded."

This model has come under strong criticism from contemporary workers. It is not given much credence, it is complex, and is based almost totally on no evidence. P. G. Martin has since abandoned this model and most recently noted (1977, p. 111) that:

> the safest hypothesis about marsupial biogeography is that the first significant radiation was in western North America at about 75 m.y.B.P. and that marsupials dispersed from there near the end of the Cretaceous to South America and to eastern North America; later one genus reached Europe. From South America they dispersed, probably by island-hopping, across the Scotia arc and the archipelago of West Antarctica to East Antarctica and so to Australia, arriving no later than 49 m.y.B.P.

Mobilist Model 9 is similar to that proposed by Rusconi (1936) who hypothesized that marsupials originated on a now submerged *terra incognita* in the South Pacific. From there they dispersed to South America and Australia sometime before the Cretaceous (Paula Couto, 1974, p. 118).

DISCUSSION AND CONCLUSIONS

The preceding Stabilist and Mobilist Models are partly or entirely mutually exclu-

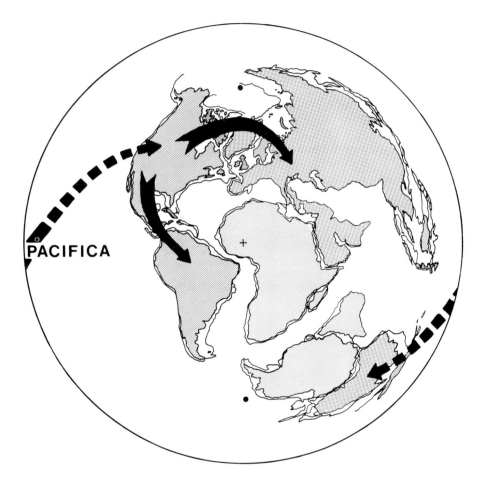

Figure 21.16. Early biogeographic history of marsupials as proposed by Mobilist Model 9. Positions of the continents at 100 m.y. ago (earliest Cenomanian–mid Cretaceous) shown on a Lambert equal-area projection. Based on Map 46 in Smith and Briden (1979).

sive. Models implying different areas of origin dictate different dispersal patterns, and different dispersal patterns have been proposed for the same area of origin. Thus, "despite the enormous burst of geological, geophysical, and paleoecological information. . . , there exists in the literature rampant contradictions and misinterpretations. . . , with many disciplines working to the near exclusion of the others" (Lillegraven, 1974, p. 263).

In all cases, the differences of opinion regarding the early biogeographic history of marsupials stem from attempts to explain where they may (or should) have been. The guiding axiom in the construction of most of these models has been to explain the known facts while adding the smallest number of additional assumptions (Cox, 1973, p. 118). It is thus intriguing that supposedly parsimonious use of virtually the same data set by Mobilists should result in nine different models, of which at least six are still viable.

In reality, "we simply do not know where marsupials evolved" (Simpson, 1978, p. 323), and it is therefore not now possible to establish an area of origin of the group. Consequently, it is also not yet possible to establish what, if any, part of their initial distribution was due to vicariance, or of the dispersal direction(s), if any, taken during their early biogeographic history.

There is, however, little doubt that marsupial distribution was influenced by continental movements. Also, most recent workers agree on one important point —that the early biogeographic history of marsupials involved what are now the continents of North America, South America, Antarctica, and Australia (Simpson, 1978, p. 323). However, the evidence at hand does not favor any one or any combination of these as an "area of origin" over that of any other(s) (Simpson, 1978, p. 323). It is only agreed that marsupials originated somewhere on part or all of the land mass composed of these four continents. These continents were also part of the initial dispersal route taken by marsupials, assuming, of course, that dispersal events were involved and that their initial distribution did not cover all four of these continents.

Knowledge of tectonic events has helped explain aspects of marsupial distribution. This is particularly true of two events in the later biogeographic history of marsupials, both involving dispersal.

First, the occurrence of fossil didelphids in Europe is explained by their dispersal from North America across the North Atlantic in the early Eocene (Figure 21.1). This, the DeGeer Route of McKenna, was represented by a dry land connection between North America and Europe via Ellesmere Island and Greenland. The dispersal route for marsupials to Europe was not via a land connection between North America and Asia (Beringia) as postulated by Stabilist Models 1 and 3.

Second, some authors (e.g., Matthew, 1915) maintained that absence of didelphine marsupials in beds of (late) Miocene and Pliocene age in North America was not conclusive evidence that they were not present. Slaughter (1978) recently reconsidered this view, based on his reported discovery of a didelphine in Barstovian (middle Miocene) deposits in Texas. However, it must be assumed that North American marsupials, autochthonous descendents of Late Cretaceous ancestors in North America, became extinct at the end of Barstovian time. The presence of *Didelphis* in the middle to late Pleistocene and Recent faunas of North America is attributed to immigration of a species of this genus from South America after completion of the Panamanian land bridge sometime after 3.0 Ma (Marshall, et al., 1979). The presence of *Didelphis* in North America thus represents the reappearance of marsupials after their apparent extinction on that continent in middle Miocene times (Figure 22.1).

Acknowledgments

For helpful comments on various phases of this study and/or for reading the manuscript, I thank J. A. W. Kirsch, M. C. McKenna, C. Patterson, G. G. Simpson, R. H. Tedford, and W. D. Turnbull. Colin Patterson kindly allowed me to see a copy of his excellent unpublished paper on paleobiogeography. Figures 21.2 and 21.3 were drawn by Marlene Hill Werner and the rest were drafted by Elizabeth Liebman. Initial stages of this study were made possible by three grants (nos. 1329, 1698, 1943) from the National Geographic Society, Washington, D. C., and its completion was made possible by a grant from the Systematic Biology Program of the National Science Foundation, NSF Grant DEB-7901976. Special thanks to Cambridge University Press for permission to base the continental configurations in Figures 21.4–21.14 and 21.16 on Maps 40 to 48 in *Mesozoic and Cenozoic Paleocontinental Maps* by A. G. Smith and J. C. Briden (1979).

Literature Cited

Abbie, A. A. 1937. Some observations on the major subdivisions of the Marsupialia with special reference to the position of the Peramelidae and Caenolestidae. J. Anat., 71:429–36.

Air, G. M.; Thompson, E. O. P.; Richardson, B. J.; and Sharman, G. B. 1971. Amino acid sequences of kangaroo myoglobin and haemoglobin and the date of marsupial-eutherian divergence. Nature, 229:391–94.

Ameghino, F. 1887. Enumeración sistemática de las especies de mamíferos fósiles coleccionados por Carlos Ameghino en los terrenos eocenos de la Patagonia austral y depositados en el Museo La Plata. La Plata, Mus., Bol., 1:1–26.

———. 1891. Nuevos restos de mamíferos fósiles descubiertos por Carlos Ameghino en el eoceno inferior de la Patagonia austral. Especies nuevas, adiciones y correcciones. Rev. Arg. Hist. Nat., 1:289–328.

———. 1906. Les formations sédimentaires du crétacé supérieur et du tertiaire de Patagonie avec un parallèle entre leurs faunes mamalogiques et celles de l'ancien continent. An. Mus. Nac. Buenos Aires, (3)VIII:1–568.

Anderson, C. 1940. Origin and migration of Australian marsupials. Austral. Sci., 2:102–6.

Archer, M. 1976a. The basicranial region of marsupicarnivores (Marsupialia), interrelationships of carnivorous marsupials, and affinities of the insectivorous marsupial peramelids. Linn. Soc., Zool. J., 59:217–322.

———. 1976b. The dasyurid dentition and its relationships to that of didelphids, thylacinids, borhyaenids (Marsupicarnivora) and peramelids (Peramelina: Marsupialia). Austral. J. Zool., Suppl., 39:1–34.

Archer, M., and Bartholomai, A. 1978. Tertiary mammals of Australia: a synoptic review. Alcheringa, 2:1–19.

Bensley, B. A. 1903. On the evolution of the Australian Marsupialia: with remarks on the relationships of the marsupials in general. Linn. Soc., Trans., ser. 2, 9:83–217.

Biggers, J. D., and DeLamater, E. D. 1965. Marsupial spermatozoa pairing in the epididymis of American forms. Nature. 208:402–4.

Broom, R. 1911. On the affinities of *Caenolestes*. Linn. Soc. N.S.W., Proc., 36:315–20.

Butler, P. M. 1978. A new interpretation of the mammalian teeth of tribosphenic pattern from the Albian of Texas. Breviora, 446:1–27.

Clemens, W. A. 1968. Origins and early evolution of marsupials. Evolution, 22:1–18.

———. 1977. Phylogeny of the marsupials. *In,* B. Stonehouse and D. Gilmore (eds.), The biology of marsupials. Univ. Park Press, London, pp. 51–68.

Clemens, W. A., and Marshall, L. G. 1976. American and European Marsupialia. Fossilium Catalogus. I: Animalia. W. Junk, The Hague, Pars. 123:1–114.

Colbert, E. H. 1973. Wandering lands and animals. New York, Dutton, 323 pp.

Corro, G. del. 1977. Parasitos, marsupiales y deriva continental. Rev. Mus. Arg. Cienc. "Bernardino Rivadavia," Paleontol., 2(3):34–68.

Cox, C. B. 1970. Migrating marsupials and drifting continents. Nature, 226:767–70.

———. 1973. Systematics and plate tectonics in the spread of marsupials. Paleont. Assoc., Spec. Pap. in Paleont., No. 12:113–19.

Cracraft, J. 1973. Continental drift, paleoclimatology, and the evolution and biogeography of birds. J. Zool., 169:455–545.

———. 1974. Continental drift and vertebrate distribution. Ann. Rev. Ecol. Syst., 5:215–61

———. 1975. Mesozoic dispersal of terrestrial faunas around the southern end of the world. Mus. Nat. Hist. Natur., Paris, Mem., n.s., Sér. A., 88:29–54.

Crochet, J. Y. 1978. Les Marsupiaux du Tertiaire D'Europe. Acad. de Montpellier, Univ. des Sciences et Techn. du Larguedoc, These, Vol. 1, pp. 1–360; Vol. 2, figs. I pls.

———. 1979a. Diversité systématique des Didelphidae (Marsupialia) Européens Tertiares. Geobios, 12(3):365–78.

———. 1979b. Donées nouvelles sur l'histoire paléogéographique des Didelphidae (Marsupialia). Acad. Sci., C. R., Ser. D, 288:1457–60.

Croizat, L.; Nelson, G.; and Rosen, D. E. 1974. Centers of origin and related concepts. Syst. Zool., 23:265–80.

Darlington, P. J. 1965. Biogeography of the southern end of the world. Harvard University Press, Cambridge, Massachusetts, 236 pp.

Darwin, C. 1859. On the origin of species by means of natural selection, or the preservation of favoured races in the struggle for life. John Murray, London, 502 pp.

Dederer, P. H. 1909. Comparison of *Caenolestes* with Polyprotodonta and Diprotodonta. Amer. Natur., 43:614–18.

Dietz, R. S., and Holden, J. C. 1970. Reconstruction of Pangaea: breakup and dispersion of continents, Permian to present. J. Geophys. Res., 75:4939 56.

DuToit, A. L. 1937. Our wandering continents: an hypothesis of continental drift. Oliver and Boyd, Edinburgh, 366 pp.

Fooden, J. 1972. Breakup of Pangaea and isolation of relict mammals in Australia, South America and Madagascar. Science, 175:894–98.

Fox, R. C. 1971. Marsupial mammals from the early Campanian, Milk River Formation, Alberta, Canada. *In* D. M. Kermack and K. A. Kermack (eds.), Early mammals. Linn. Soc., Zool. J., Suppl., 50, pp. 145–64.

———. 1979a. Mammals from the Upper Cretaceous Oldman Formation, Alberta. I. *Alphadon* Simpson (Marsupialia). Can. J. Earth Sci., 16:103–13.

———. 1979b. Mammals from the Upper Cretaceous Oldman Formation, Alberta. II. *Pediomys* Marsh (Marsupialia). Can. J. Earth Sci., 16:103–13.

Gardner, A. L. 1973. The systematics of the genus *Didelphis* (Marsupialia: Didelphidae) in North and Middle America. Texas Tech. Univ., The Museum, Spec. Publ., 4:1–81.

Grambast, L.; Martinez, M.; Mattauer, M.; and Thaler, L. 1967. *Peratherium altiplanese, nov. gen., nov. sp.* premier mammifère Mésozoique d'Amérique du Sud. Acad. Sci., C. R., Sér D, 264:707–10.

Gregory, W. K. 1910. The orders of mammals. Amer. Mus. Natur. Hist., Bull., 27:1–524.

Hayman, D. L.; Kirsch, J. A. W.; Martin, P. G.; and Waller, P. F. 1971. Chromosomal and serological studies of the Caenolestidae and their implications for marsupial evolution. Nature, 231:194–95.

Hadley, D. 1899. A zoogeographic scheme for the mid-Pacific. Linn. Soc. N.S.W., Proc., 24:391–417.

Hershkovitz, P. 1972. The Recent mammals of the Neotropical Region. *In*, A. Keast, F. C. Erk and B. Glass (eds.), Evolution, mammals and southern continents. State Univ. N. Y. Press, Albany, N. Y., pp 311–431.

Hoffstetter, R. 1970. L'Histoire biogéographique des marsupiaux et la dichotomie marsupiaux-placentaires. Acad. Sci., C. R., Sér D, 270:3027–30.

———. 1971. Le peuplement mammalien de l'Amerique du Sud. Rôle des continents australs comme centres d'origine, de diversification et de dispersion pour certains groupes mammaliens. An. Acad. brasil. Ciênc., 43 (Suppl.):125–44.

———. 1972. Donées et hypothèses concernant l'origine et l'histoire biogéographique des Marsupiaux. Acad. Sci., C.R., Sér D, 274:2635–38.

———. 1975. Les Marsupiaux et l'histoire de mammifères: aspects phylogéniques et choro-
logiques. Colloque international C.N.R.S., 218:591–610.

———. 1976. Histoire des mammifères et dérive des continents. La Recherche, 64:124–38.

Jardine, N., and McKenzie, D. 1972. Continental drift and the dispersal and evolution of
organisms. Nature, 234:20–24.

Keast, A. 1977. Historical biogeography of the marsupials. *In*, B. Stonehouse and D. Gilmore
(eds), The biology of marsupials. Univ. Park Press, London, pp. 69–95.

Kirsch, J. A. W. 1977. The comparative serology of Marsupialia, and a classification of marsupials. Austral. J. Zool., Suppl. ser., no 52:1–152.

———. 1979. Les marsupiaux. La Recherche, 10(97):108–16.

———, and Calaby, J. H. 1977. The species of living marsupials. *In*, B. Stonehouse and D.
Gilmore (eds.), The biology of marsupials, Univ. Park Press, London, pp. 9–26.

Kurtén, B. 1969. Continental drift and evolution. Sci. Amer., 220:54–63.

Lillegraven, J. A. 1969. Latest Cretaceous mammals of upper part of Edmonton Formation
of Alberta, Canada, and a review of marsupial-placental dichotomy of mammalian evolution. Kans. Univ. Paleontol. Contrib., 50 (Vert. 12):1–22.

———. 1974. Biogeographical considerations of the marsupial-placental dichotomy. Ann.
Rev. Ecol. Syst., 5:263–83.

———. 1976. Biological considerations of the marsupial-placental dichotomy. Evolution,
29:707–22.

Longman, H. A. 1924. The zoogeography of marsupials. With notes on the origin of the
Australian fauna. Queesl. Mus., Mem., 8(1):1–15.

Marshall, L. G. 1977. Cladistic analysis of didelphoid, dasyuroid, borhyaenoid, and thylacinid
(Marsupial) affinity. Syst. Zool., 26:410–25.

———. 1978. Evolution of the Borhyaenidae, extinct South American predaceous marsupials. Calif., Univ., Publ. Geol. Sci., 117:1–89.

———. 1979. Evolution of metatherian and eutherian (mammalian) characters: a review
based on cladistic methodology. Linn. Soc., Zool. J., 66:369–410.

———. 1980. Systematics of the South American marsupial family Caenolestidae. Fieldiana,
Geology, n.s. 5:1–145.

———. In press. The families and genera of Marsupialia. Fieldiana: Geology.

Marshall, L. G.; Butler, R. F.; Drake, R. E.; Curtis, G. H.; and Tedford, R. H. 1979. Calibration
of the Great American Interchange. Science, 204:272–79.

Martin, P. G. 1970. The Darwin Rise hypothesis of the biogeographical dispersal of marsupials. Nature, 225:197–98.

———. 1975. Marsupial biogeography in relation to continental drift. Mus. Nat. Hist. Natur.,
Paris, Mém., Sér. A, 88:216–37.

———. 1977. Marsupial biogeography and plate tectonics. *In*, B. Stonehouse and D. Gilmore
(eds.), The biology of marsupials. Univ. Park Press, London, pp. 97–115.

Martin, R. A. 1974. Fossil mammals from the Coleman IIA Fauna, Sumter County. *In*, S. D.
Webb (ed.), Pleistocene mammals of Florida. Univ. Presses of Florida, Gainesville, pp.
35–99.

Matthew, W. D. 1906. Hypothetical outlines of the continents in Tertiary times. Amer. Mus.
Natur. Hist., Bull., 22:353–83.

———. 1915. Climate and Evolution. N. Y. Acad. Sci., Ann., 24:171–318. Reprinted, E. M.
Schlaikjer (ed.). N. Y. Acad. Sci., Spec. Publs., 1:i–xii, 1–223 (1939).

Maxson, L. R.; Sarich, V. M.; and Wilson, A. C. 1975. Continental drift and the use of
albumin as an evolutionary clock. Nature, 255:394–400.

McGowran, B. 1973. Rifting and drift of Australia and the migration of mammals. Science,
180:759–61.

McKenna, M. C. 1972. Possible biological consequences of plate tectonics. BioScience, 22(9):
519–25.

———. 1973. Sweepstakes, filters, corridors, Noah's arks, and beached Viking funeral ships
in palaeogeography. *In*, D. H. Tarling and S. K. Runcorn (eds.), Implication of continental drift to the earth sciences. Academic Press, London and New York, pp. 21–46.

————. 1975a. Toward a phylogenetic classification of the mammalia. *In*, W. P. Luckett and F. S. Szalay (eds.), Phylogeny of the Primates. Plenum Publ. Corp., New York, pp. 21–46.

————. 1975b. Fossil mammals and early Eocene North Atlantic Land Continuity. Missouri Botanical Garden, Ann., 62(2):335–53.

————. In Press. Early history and biogeography of South America's extinct land mammals. *In*, R. L. Ciochon and A. B. Chiarelli (eds.), Evolutionary biology of the New World monkeys and continental drift. Plenum Publ. Corp., New York.

Osborn, H. F. 1910. The age of mammals in Europe, Asia and North America. MacMillan & Co., New York, 635 pp.

Osgood. W. H. 1921. A monographic study of the American marsupial *Caenolestes*. Field Mus. Nat. Hist., Zool. Ser., 14(1):1–162.

Pascual, R., and Herrera, H. E. 1973. Adiciones al conocimiento de *Pliolestes tripotamicus* Rieg, 1955 (Mammalia, Marsupialia, Caenolestidae) del Plioceno superior de la Argentina. Ameghiniana, 10:36–50.

Patterson, B. 1956. Early Cretaceous mammals and the evolution of mammalian molar teeth. Fieldiana: Geol., 13:1–105.

————, and Pascual, R. 1972. The fossil mammal fauna of South America. *In*, A. Keast, F. C. Erk, and B. Glass (eds.), Evolution, mammals, and southern continents. State Univ. of New York Press, Albany, pp. 274–309.

Patterson, C. In press. Methods of paleobiogeography. *In*, G. Nelson and D. E. Rosen (eds.), Vicariance biogeography: a critique. Columbia University Press, New York.

Paula Couto, C. de. 1952. Fossil mammals from the beginning of the Cenozoic in Brazil. Marsupialia: Polydolopidae and Borhyaenidae. Amer. Mus. Nov., 1559:1–27.

————. 1974. Marsupial dispersion and continental drift. An. Acad. Brasil. Ciênc., 46(1):103–26.

Portugal, J. A. 1974. Mesozoic and Cenozoic stratigraphy and tectonic events of Puno–Santa Lucia Area, Department of Puno, Peru. Amer. Assoc. Petrol. Geol., Bull., 58(6):1982–99.

Rage, J. C. 1978. Une connexion continentale entre Amérique du Nord et Amérique du Sud au Crétacé supérieur? L'example des vertébrés continentaux. Soc. Geol. Fr., C. R., 6:281–85.

Raven, P. H., and Axelrod, D. I. 1972. Plate tectonics and Australasian paleobiogeography. Science, 176:1379–86.

Rich, P. V. 1975. Antarctic dispersal routes, wandering continents, and the origin of Australia's non-passeriform avifauna. Victoria, Nat. Mus., Mem. 36:63–126.

Ride, W. D. L. 1962. On the evolution of Australian marsupials. *In*, G. W. Leeper (ed.), The evolution of living organisms. Melbourne Univ. Press, Melbourne, pp. 281–306.

Riek, E. F. 1970. Lower Cretaceous fleas. Nature, 227:746–47.

Rosen, D. E. 1975. A vicariance model of Caribbean biogeography. Syst. Zool., 24:431–64.

Rusconi, C. 1936. La supuesta afinidad de *Argyrolagus* con los Typotheria. Bol. Acad. Nac. Cien. Córdoba, 33:173–82.

Scotese, C. R. 1980. Mesozoic and Cenozoic paleocontinental maps, by A. G. Smith and J. C. Briden. Amer. J. Sci., 280(1):93–96.

Scott. W. B. 1937. A history of land animals in the Western Hemisphere. Revised Edition, New York, The MacMillan Co., 786 pp.

Sigé, B. 1968. Dents de micromammifères et fragments de coquilles d'oeufs de dinosauriens dans le fauna de vertébraés du Crétacé supérieur de Lagune Umayo (Andes peruviénnes). Acad. Sci., C. R., Sér. D, 267:1495–98.

————. 1972. La faunule de mammifères du Crétacé supérieur de Lagune Umayo (Andes peruviénnes). Mus. Nat. Hist. Natur., Paris, Bull., Sér. 3, 19(19):375–409.

Simpson, G. G. 1941. The affinities of the Borhyaenidae. Amer. Mus. Nov., 1118:1–6.

————. 1945. The principles of classification and a classification of mammals. Amer. Mus. Natur. Hist., Bull., 85:1–350.

————. 1950. History of the fauna of Latin America. Amer. Sci., 38(3):361–89.

————. 1961. Historical zoogeography of Australian mammals. Evolution, 15:431–46.

————. 1965. The geography of evolution: collected essays. Chilton Books, New York, 249 pp.

————. 1971. The evolution of marsupials in South America. An. Acad. Brasil. Ciênc., 43: 103–18.

————. 1978. Early mammals in South America: Fact, controversy, and mystery. Amer. Phil. Soc., Proc., 122(5):318–28.

————. 1980. Splendid isolation: The curious history of South American mammals. Yale University Press, New Haven, 266 pp.

Sinclair, W. J. 1905. The marsupial fauna of the Santa Cruz beds. Amer. Phil. Soc., Proc., 44:73–81.

————. 1906. Mammalia of the Santa Cruz beds: Marsupialia. Princeton Univ., Rept., Exped. Patagonia, 4(3):333–460.

Slaughter, B. H. 1968a. Earliest known marsupials. Science, 162:254–55.

————. 1968b. *Holoclemensia* instead of *Clemensia.* Science, 162:1306.

————. 1978. Occurrence of didelphine marsupials from the Eocene and Miocene of Texas Gulf coastal plain. J. Paleontol., 52(3):744–46.

Smith, A. G.; Briden, J. C.; and Drewry, G. E. 1973. Phanerozoic world maps. *In,* N. F. Hughes (ed.), Organisms and continents through time. Palaeontol. Assoc. London, Spec. Paper Palaeontol., 12, pp. 1–43.

Smith, A. G., and Briden, J. C. 1979. Mesozoic and Cenozoic paleocontinental maps. Cambridge Earth Science Series, Cambridge University Press, Cambridge, 63 pp.

Szalay, F. S., and McKenna, M. C. 1971. Beginning of the Age of Mammals in Asia: the late Paleocene Gashato fauna, Mongolia. Amer. Mus. Natur. Hist., Bull., 144:271–317.

Tedford, R. H. 1974. Marsupials and the new paleogeography. *In,* C. A. Ross (ed.), Paleogeographic provinces and provinciality. Soc. Econ. Paleontol. Miner., Spec. Publ. 21, pp. 109–26.

Tedford, R. H.; Banks, M. R.; Kemp, N. R.; McDougall, I.; and Sutherland, F. L. 1975. Recognition of the oldest known fossil marsupials from Australia. Nature, 255(5504):141–42.

Thenius, E. 1971. Zum gegenwärtigen Verbreitungsbild der Säugethiere und seiner Deutung in erdgeschichtlicher Sicht. Natur Mus., 101(3):185–96.

Thomas, O. 1895. Description of four small mammals from South America, including one belonging to the peculiar marsupial genus *"Hyracodon"* Tomes. Ann. Mag. Natur. Hist., 16:367.

Tomes, R. F. 1860. Notes on a second collection of mammalia made by Mr. Fraser in the Republic of Ecuador. Zool. Soc. London, Proc., pp. 211–21.

————. 1863. Notice of a new American form of marsupial. Zool. Soc. London, Proc., pp. 50–51.

Turnbull, W. D. 1971. The Trinity therians: their bearing on evolution in marsupials and other therians. *In,* A. A. Dahlberg (ed.), Dental morphology and evolution. Univ. Chicago Press, Chicago, pp. 151–79.

Tyndale-Biscoe, H. 1973. Life of Marsupials. Edward Arnold Ltd., London, 254 pp.

Weber, M. 1904. Die Säugethiere. Jena. 866 pp.

Wegener, A. 1967. The origin of continents and oceans. Methuen, London, 248 pp. (new translation of 1929 edition).

Wood, H. E. 1924. The position of the "sparassodonts": with notes on the relationships and history of the Marsupialia. Amer. Mus. Natur. Hist., Bull. 51:77–101.

WOODRATS AND PICTURESQUE JUNIPERS

by Terry A. Vaughan

Introduction

The character of a natural scene is often determined in large part by the growth forms of plants. The spreading, umbrella-like crown of *Acacia tortilis* provides the floral signature of the East African savanna, just as the twisted and semi-prostrate white-barked pine (*Pinus albicaulus*) is the symbol of a Sierra Nevadan subalpine landscape. In the southwestern United States juniper trees are the dominant plants over wide areas of foothills and mesas, and a gnarled juniper heightens the dramatic impact of many of the best photographs of the Southwest. Skeletons of long-dead branches trace angular patterns against the sky, and spiny leaves cluster awkwardly at the bases of still-living branches. What has determined the growth form of such junipers?

The growth forms of leaves and entire plants are not only under genetic control but can be modified by weather, soils, competition with other plants for water, light, or nutrients, and by the depredations of herbivores. Because plants provide the most abundant of all food sources, herbivory is the major feeding strategy of animals, and herbivores strongly influence plant form.

This paper describes the impact that selective feeding on junipers by Stephens' woodrats (*Neotoma stephensi*) (Figure 22.1) has on the growth form of some junipers, and considers the hypothesis that the feeding activity of this woodrat is a selective force affecting the course of evolution of some populations of juniper.

Methods and Study Area

This paper is in part an outgrowth of a study in progress on the feeding ecology and reproductive cycle of the Stephens' woodrat. This study began in November of 1978; the primary study area is 32 km northeast of Flagstaff, Coconino County, Arizona. (This locality will hereinafter be referred to as the Flagstaff study area.) Dietary information is based on microscopic examinations of 317 samples of fecal material deposited by 106 live-trapped woodrats. Monthly sample sizes are given in Figure 22.2. The methods of dietary analysis that I used are those of Hansen and Flinders (1969).

Observations on damage to junipers by woodrats were made from September of 1967 until February of 1980 at Wupatki National Monument and many other localities in Coconino County, and in the Cerbat and Black mountains, Mohave County, Arizona. Degrees of damage to junipers were estimated by counting the numbers of clipped and unclipped twigs over 3 mm in diameter. For junipers under 2 m high, counts were made on four branches, one in each cardinal direc-

Figure 22.1. Stephens' woodrat in a picturesque juniper.

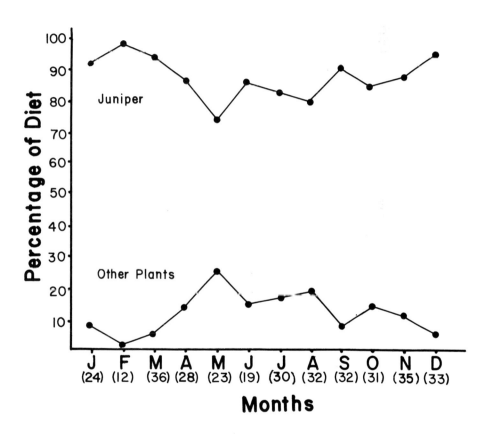

Figure 22.2. Summary of the diet of the Stephens' woodrat, showing the importance throughout the year of one-seed juniper. All of the monthly samples but one were taken in 1979. The January sample is from 1980. The figures in parentheses beneath the months are sample sizes.

tion from the main trunk. For junipers over 2 m high counts were made on twelve branches, four each from the top third, the middle third, and the bottom third of the tree.

Data on heights of junipers and the coverage of the foliage canopy were taken in a north-south trending plot 6 m wide and roughly ½ km long at the Flagstaff study area. The height and width of the canopy of every juniper (a total of seventy-five) with foliage extending into the plot were measured. Canopy width was measured once for each tree, and was measured alternately in a north-south and an east-west direction.

Juniper density was estimated by counting junipers in sixteen sample plots within 4 ha of the 5-ha area at the Flagstaff study area where woodrats were live trapped. Each of the 4 ha was divided into four ¼-ha parts, and each of these was divided into quarters (each 1/16 ha in area). The junipers within one randomly chosen, 1/16-ha part of each ¼ ha were counted. The mean number of junipers for all of the 1/16-ha plots was multiplied by sixteen to give a rough estimate of the number of junipers/ha.

Results

Stephens' woodrat and its dependence on juniper. Over wide areas of northern Arizona, Nevada, Utah, and southern Colorado, junipers (one-seed juniper, *Juniperus monosperma*, Utah juniper, *J. osteosperma*, or California juniper, *J. californica*), in association with piñon pines (*Pinus edulis* or *P. monophylla*), form a rather open woodland. Junipers are generally more tolerant of dryness than are piñons (Short and McCulloch, 1977), however, and in many areas are the dominant plants and the only conifers in a well-defined vegetational belt immediately below the piñon-juniper woodland. Precipitation is infrequent and unpredictable in juniper woodland, and most small mammals must be independent of drinking water. Some, such as the woodrats, solve their water-balance problems by eating succulent food (Schmidt-Nielsen and Schmidt-Nielsen, 1952; Lee, 1963; MacMillen, 1964).

Throughout its geographic range, the Stephens' woodrat lives in association with junipers and depends on this plant for food and water. This woodrat is a dietary specialist: considering the entire year, over 87 percent of the diet is one-seed juniper at the Flagstaff study area. The intensity of use of juniper varies seasonally, with the highest use in winter (95 percent of the diet) and the lowest in summer (83 percent; see Figure 22.2). Juniper occurred in 100 percent of the samples of fecal material that I examined, indicating that Stephens' woodrats eat juniper consistently, probably each day.

Compared to most other species of woodrats, the Stephens' woodrat occupies a rather limited geographic range and has narrow ecological tolerances. The range is centered in the juniper woodlands of northern Arizona, but extends eastward into juniper areas in western New Mexico and westward to some desert mountains in western Arizona (Hoffmeister and de la Torre, 1960). Near the eastern limit of its range, in the San Juan Basin of New Mexico, Harris (1963) found this woodrat living in association with juniper. Similarly, in the Black and Cerbat mountains at the western margin of its range, I have invariably found it living with juniper.

Juniper is used by Stephens' woodrats for shelter as well as for food. Of sixty dens examined at the Flagstaff study area, thirty-eight (63 percent) were beneath living, or in a few cases dead, junipers (Table 22.1). Many, perhaps most, of the

TABLE 22.1

Summary of den sites of Stephens' woodrats at a juniper-woodland study area
32 km NE Flagstaff, Coconino County, Arizona.

Den Site	Number	Percentage of Total
Base of live juniper	31	52
Rock outcrop	19	32
Rocks under juniper	5	8
Base of dead juniper	2	3
Under piñon logs		5
Total	60	

TABLE 22.2

Summary of constituents in the leaves of two species of *Juniperus*. Samples of *J. monosperma* were taken at the locality given in Table 22.1; those of *J. osteosperma* were taken 5 km E of Flagstaff. Figures are based on fresh (wet) weights of leaves.

Constituents		*J. monosperma*			*J. osteosperma*
		Oct. 1979	Nov. 1979	Feb. 1980	Feb. 1980
Protein	$\overline{X}$	3.57	3.33	3.22	3.33
	N	16	12	6	4
	S.D.	0.42	0.30	0.31	0.26
Fat	$\overline{X}$	2.50	2.85	1.57	1.98
	N	11	9	6	4
	S.D.	1.30	0.37	0.15	0.26
Carbo-hydrates	$\overline{X}$		36.15	36.8	37.23
	N		8	6	4
	S.D.		1.19	0.44	0.83
Ash	$\overline{X}$		2.23	1.32	1.63
	N		8	6	4
	S.D.		0.51	0.10	0.29
Water	$\overline{X}$	54.83	55.08	57.08	55.83
	N	11	12	6	4
	S.D.	1.81	1.14	0.65	0.79

trees that shelter woodrat dens are used for food by the resident woodrat. Dens not beneath junipers are never far from them. The distances from twenty-one such dens to their nearest living junipers averaged but 5.1 m; extremes were 0.9 and 10.8 m.

Parenthetical comments should be made concerning the use of juniper by species of woodrats other than *N. stephensi*. I am discussing the Stephens' woodrat-juniper association because this woodrat specializes on juniper, but other species are known to eat juniper. Finley (1958), for example, reported that *N. albigula, N. lepida,* and *N. mexicana* eat juniper in Colorado, and one of his photographs (Plate 36, Figure 1) shows a one-seed juniper heavily damaged by *N. albigula*. In Arizona also, several species of woodrats take juniper, at least locally. Some fecal samples of *N. albigula* from the Black Mountains that I have examined microscopically contained juniper together with cactus, and Hoffmeister and Durham (1971) found fragments of juniper (perhaps discarded food) in some dens of *N. lepida* from the Arizona Strip (the exteme northwestern corner of Arizona).

Contributions of junipers to Stephens' woodrats. Juniper is primarily important to Stephens' woodrats as a consistently available source of food and water. One-seed juniper at the Flagstaff study area averages about 55 percent water by weight in winter and is the only plant except for the uncommon piñon pine that remains green and succulant through all seasons. The leaves of this juniper contain fairly low amounts of protein and fat, but are high incarbohydrates (Table 22.2). As indicated by the woodrats' nearly total dependence on this plant for food in winter, it must contain a reasonably balanced array of the nutrients that these animals need.

Although juniper is seemingly a marginal food in terms of nutrients, its abundance, size, and growth form favors efficient foraging. In the Flagstaff study area junipers occur at a rate of 111/ha, but are not evenly dispersed. (The mean number of junipers in the sixteen 1/16-ha plots studied is 6.94, the range, 0–15, and the S.D., 4.37.) The average area of ground covered per tree is 14.9 m^2 (N = 75; S.D. = 2.2). Approximately 16.6 percent of the ground is covered by junipers. Because the mean density of woodrats is only some 2/ha (T. A. Vaughan, unpublished data) and the animals occupy sites near junipers, each individual seemingly has a virtually limitless supply of juniper for food; indeed, a single juniper could provide a lifetime supply of foliage for a woodrat.

In the case of the Stephens' woodrat and most other rodents, pressure from predators has been a selective force affecting choice of food, foraging routes and refuge sites, styles of locomotion, and the evolution of capacious cheek pouches as means of transporting food. One-seed junipers characteristically lack trunks; the branches often spread widely from ground level and form an intricate maze of terminal branches that are a barrier to predators. (Other species of junipers occasionally have this growth form.) Woodrats are rapid and agile climbers, and are difficult to detect or capture when feeding in such junipers.

The spreading growth form of many junipers also provides an advantageous thermal microenvironment. Radiation of heat to the night sky is slower beneath a juniper or in the canopy of foliage than in the open (Figure 22.3), and the large, prostrate limbs offer shelter from wind. Although the thermal advantage gained by a woodrat within the canopy or at the base of a juniper is slight, it may be of importance energetically to a mammal with as narrow a thermoneutral zone as

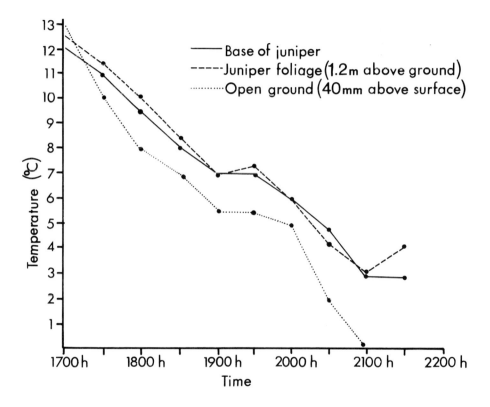

Figure 22.3. An example of patterns of temperature change at the base of a California juniper (the probe was 25 mm above the ground at the mouth of the entrance to a woodrat den), in the foliage of the juniper, and on open ground 10 m from any juniper. These temperatures were taken on 25 January 1980 in the Black Mountains, 5.5 km S and 2.0 km W of Grasshopper Junction, Mohave County, Arizona.

TABLE 22.3

Comparisons of damage to junipers near dens of Stephens' woodrats. The locality is that given in Table 22.1. The junipers at dens 1 and 3 were less than 2 m apart; those at den 2 were 15 m apart. Heights of trees are in parentheses.

	Distance from den	Total no. twigs clipped / Total no. twigs counted	Percentage clipped
Den 1			
Juniper A (2.0 m)	1.6 m	$\frac{90}{183}$	49.2
Juniper B (2.2 m)	1.3 m	$\frac{17}{169}$	10.1
Den 2			
Juniper A (2.8 m)	13.0 m	$\frac{475}{581}$	81.8
Juniper B (4.2 m)	5.0 m	$\frac{0}{739}$	0.0
Den 3			
Juniper A (3.0 m)	7.0 m	$\frac{144}{232}$	62.1
Juniper B (3.0 m)	7.0 m	$\frac{57}{251}$	22.7

TABLE 22.4

Summary of damage to one-seed junipers by Stephens' woodrats in Wupatki National Monument. Damage to a group of trees or to a single tree is expressed as the percentage of the twigs found clipped out of the total number of twigs (clipped or unclipped) that were counted. Numbers of trees studied are in parentheses after height classes.

Height class	Total no. twigs clipped / Total no. twigs counted	Average damage	Greatest damage to a tree	Least damage to a tree
Less than 1.0 m (9)	$\frac{749}{963}$	77.8%	99.0%	60.4%
1.0–2.0 m (14)	$\frac{1193}{2064}$	57.8%	95.3%	14.9%
More than 2.0 m (7)	$\frac{1485}{2365}$	62.8%	84.4%	40.1%

that of the Stephens' woodrat. This zone is between approximately 32 and 26 degrees C (John L. Fisher, unpublished data).

Lest the foregoing comments picture junipers as providing Stephens' woodrats with an idyllic (for a rodent) life, the secondary compounds of juniper leaves should be mentioned. Many secondary compounds of plants provide defenses against herbivores by acting as toxins or by reducing digestibility (Rhoads and Cates, 1976; Harborne, 1972). Juniper leaves contain an array of terpenoids and other secondary compounds (Adams, 1969) that may serve a defensive function. Although nothing is known of the specific effects of these compounds on woodrats, some observations and lines of evidence are suggestive. Stephens' woodrats strongly prefer some juniper trees over others for food (Table 22.3); such preferences may reflect differences in palatability among junipers due to different amounts of secondary compounds. Also, densities, growth rate, and reproductive rates of this woodrat are remarkably low in juniper woodland (T. A. Vaughan, unpublished data). For example, densities of Stephens' woodrats are about 2/ha for much of the year in optimal habitat at the Flagstaff study area, whereas for two woodrats that eat mostly cactus and shrub leaves (*N. lepida* and *N. albigula*) densities of 38/ha (Brown and Lieberman, 1972) and 49/ha (Spencer and Spencer, 1941), respectively, were recorded. The low densities of Stephens' woodrats may be a result of metabolic-energetic problems associated with eating juniper.

Patterns of woodrat damage to junipers. Stephens' woodrats damage some junipers by feeding heavily on them and thereby altering their growth forms. The conspicuousness of this alteration is enhanced by the woodrats' selective pattern of feeding. The following aspects of this selectivity will be considered in the following paragraphs: (1) woodrats typically feed on certain junipers repeatedly and leave other individuals untouched; (2) woodrats selectively clip twigs of a specific size; (3) in the case of some trees, feeding is concentrated on one section of the tree; (4) some trees are fed on by many generations of woodrats; (5) small junipers sustain the heaviest woodrat damage.

In juniper woodland, where there are many junipers from which to choose, Stephens' woodrats are selective. As indicated by the percentages of twigs clipped, a woodrat often feeds repeatedly on one juniper and leaves a neighbor completely or relatively untouched (Table 22.3). Clearly the woodrats distinguish individual differences (in secondary compounds?) between the junipers. Only under exceptional conditions, at localities where junipers are widely scattered, is every juniper heavily pruned.

Woodrats remove the ends of twigs by clipping them from 50 to 150 mm from their tips. At this point a twig is usually from 3 to 4 mm in diameter. Woodrats feed persistently enough on some junipers to remove a high percentage of such twigs (Table 22.4). Laboratory feeding trials demonstrate that a Stephens' woodrat eats an amount of juniper leaves each night to equal an average of 12.5 percent of its body weight, or from 12.5 to 25 g (wet weight). In addition, wild woodrats discard quantities of uneaten juniper foliage around their dens. An average woodrat probably discards as much juniper each night as it eats, and must therefore clip some 38 g each night. At this rate, a tree fed on consistently by a woodrat would lose nearly 24 kg of foliage a year, an amount probably greater than the annual twig production of a juniper less than 2 m in height. Although such losses can perhaps be readily sustained year after year by a large tree, one less than 2 m in height may be

Figure 22.4. A small California juniper, 2.4 m in height, that is heavily damaged by Stephens' woodrats. Of the 257 twigs that were examined, 160 (62 percent) were clipped by woodrats. Photo taken at the locality given for Figure 22.3.

Figure 22.5. A, a California juniper, 1.1 m in height, from which most of the twigs have been clipped by Stephens' woodrats. Photo taken at the locality given for Figure 22.3. B, a one-seed juniper, 2.6 m in height, from which most of the apical meristematic tissue has been clipped by Stephens' woodrats. The bundles of foliage are largely adventitious growth. Photograph taken at Wupatki National Monument, Coconino County, Arizona.

severely stressed. Trees of this height class that are heavily clipped by woodrats seldom produce cones, even during years when other trees of similar size growing in similar soils bear abundant cone crops.

The woodrats' effect on the form of a tree partly depends on which twigs are clipped. Perhaps because the ends of branches bend sharply under a woodrat's weight, an individual often clips twigs to within 20 or 30 cm of the tip of a main branch, which is left untouched. This habit produces branch silhouettes of the sort shown in Figure 22.4. Where juniper trees are widely scattered, however, woodrats may clip all of the tips of each branch over much of the tree (Figure 22.5). In extreme cases, repeated clipping apparently contributes to the death of one section of a juniper.

At the Flagstaff study area rock outcrops or spreading junipers suitable as den sites are at a premium, and many such sites are therefore in almost continuous use by woodrats over many years. Some one-seed junipers probably live for 500 years (Preston, 1968), and might shelter a den for 100 years or more; but some dens in rock crevices or cliffs are occupied vastly longer. Plant remains from middens associated with woodrat dens have been radiocarbon dated at ages of over 21,000 years (Van Devender and Spaulding, 1979), and some of these dens are currently in use by woodrats (A. Phillips, personal communication). Because of the more-or-less continuous long-term use of a den site by woodrats, an "edible" juniper near a den could be fed upon for hundreds of years by woodrats. These dens are in suitable locations for long term use but have not necessarily been surrounded throughout their long histories by the same plant community. The point is that an individual juniper may be fed upon for its entire life by successive generations of woodrats. Long-term use and selective patterns of feeding strongly alter the growth forms of such junipers, and may cause the death of some.

Woodrats feed especially heavily on small junipers (see Figure 22.5). Considering the data in Table 22.4, the difference between the degree of damage to trees less than 1 m and to those over 1 m in height is highly significant ($X^2 = 114.21$; $P \ 0.001$). The woodrats' preference for small trees, as indicated by differences in damage per unit of time, are far greater than Table 22.4 suggests, however, for damage to small (relatively young) junipers accumulated over a shorter time than did damage to large (relatively old) ones. At the Flagstaff study area the distribution of juniper sizes is strongly skewed toward large trees. Of the seventy-five junipers within a study plot, fifty-nine were over 3 m in height, fourteen were between 2 and 3 m, and two were less than 2 m. None were less than 1 m.

Responses of junipers to damage. The anti-herbivore chemical defenses of junipers are poorly understood. Nothing is known of the effects that these compounds have on woodrats, nor of the defensive chemical responses of a tree suffering repeated damage. The following remarks deal entirely with easily observed responses of junipers to damage. These responses seem to have defensive functions, may strikingly alter the appearance of a juniper, and may be costly energetically.

Woodrats remove a high percentage of the twigs from some junipers (Table 22.4), and thereby destroy much of the apical meristematic growth. Seemingly in response to such clipping, some branches produce clusters of adventitious growth. (Adventitious growth is defined by Esau, 1977, as growth that is initiated from more or less mature tissues without connection with apical meristem.) These clusters generally occur well back from the tip of the branch. Trees that are extensively damaged produce largely adventitious growth, and the "normal" form of the plant is lost. Repeated clippings of apical meristematic tissue apparently contributes to the death of some large branches or entire sections of a tree.

A second obvious response to woodrat damage is the production of spiny juvenile growth. The leaves of this growth terminate in stiff sharp spines (see Figure 22.1), and occur on seedlings and on heavily clipped branches of junipers of all sizes. Conspicuous clusters of juvenile growth also develop around the bases of small, heavily-damaged junipers. Juvenile growth apparently deters woodrats. As an example, a seedling one-seed juniper 24 cm high, which had seventeen branches with "mature" leaves and five basal branches with juvenile leaves, had lost all of the ends of the mature branches to woodrats; none of the juvenile growth had been clipped. Similarly, I have not observed clipped twigs of juvenile growth on mature trees.

Most of my experience has been with one-seed juniper, but other junipers have similar defenses. Juvenile growth occurs on woodrat-damaged Utah junipers and California junipers in the Cerbat Mountains and Black Mountains respectively.

Discussion

Herbivores seem to have limitless food for the taking. Rather than passively accepting animal depredations, however, plants have evolved a wide array of defenses in the form of toxin-bearing thorns, spines, and hairs, as well as toxins and/or digestive retardants in the tissues. Although plants and herbivores continue to thrive, neither has emerged from the contest unscathed. The following discussions and speculations deal primarily with the impact of the Stephens' woodrat on Juniper.

Among North American mammals, only the Stephens' woodrat and the red tree vole (*Phenacomys longicaudus*) feed primarily on conifer leaves. The latter lives in humid coniferous forests of coastal northern California and Oregon and eats the needles of Douglas fir (*Pseudotsuga menziesii*). The life histories of both of these rodents depart from the norm for rodents; this departure may be a price they pay for their diets. As mentioned previously, Stephens' woodrats occur often at low densities and have low reproductive rates and low growth rates. (These points will be considered in detail in a later paper.) Similarly, red tree voles never attain high densities, as do many other voles, litter size is low, and the growth of young is unusually slow (Hamilton, 1962).

During the 11,000-year period from the early part of the last Wisconsinan glacial maximum, some 22,000 years before present (B.P.), until 11,000 years B.P., when the ice sheets were retreating, piñon-oak-juniper woodlands occupied broad areas (at elevations from about 550 to 1525 m) that are now dominated by desert scrub plant communities (Van Devender and Spaulding, 1979). As climatic conditions shifted toward increasing aridity and probably colder winters, the woodlands retreated to higher elevations, and after 8,000 years B.P. the present climatic and vegetational zones of the Southwest were established. The duration of the association between Stephens' woodrats and juniper has not been fully documented, but clearly extends back into the period when woodlands were widespread in the southwest. Records of jaw fragments of Stephen's woodrats associated with juniper have been found in woodrat middens dated at about 15,000 years B.P. (Van Devender, et al., 1977).

Today, in some mountain ranges of the Southwest, relict stands of juniper are last vestiges of these woodlands. Some of these insular patches of juniper are inhabited by Stephens' woodrats, which, like the junipers, have probably been isolated there for some 8,000 years. Such a juniper-woodrat "island" occurs at the westernmost periphery of the range of the woodrat, in the Black Mountains of northwestern Arizona. Here California junipers are scattered at low densities over rocky slopes. Although the density of woodrats is also low (less than 6/ha), at some sites they have fed on at least 30 percent of the junipers.

In parts of Wupatki National Monument, in north-central Arizona, one-seed junipers reach the lower limit of their local distribution, and the last groups of junipers are often scattered along rocky walls of arroyos. Sharing these places are Stephens' woodrats, and in some such sites virtually every juniper is severely pruned.

In both of these settings woodrat damage to junipers is so frequent that during geological or even ecological time it could affect the junipers in several ways. In the case of the Wupatki junipers, severely pruned trees rarely produce seeds, and, because small junipers (less than 1 m in height) are unusually heavily damaged, their survival is apparently low. Under these conditions, the lower limit of the range of the juniper could be set as much by pressure from woodrats as by the effects of hostile climatic conditions. Of greater importance, the isolation of small populations of junipers that are under heavy pressure from woodrats provides a favorable setting for differentiation of the junipers. Accelerated evolution of secondary compounds serving a defensive function could be expected if only the least edible junipers survived as seedlings, and if as adults they had a strong reproductive advantage. The marked preference that the woodrats show for small trees suggests that the conspicuous damage to large trees may be of little selective importance relative to the less obvious damage to seedlings. In addition, the combination of woodrat damage and suboptimal climatic conditions may reduce longevity in junipers living in montane isolation, and could alter rates of evolution.

At the Flagstaff study area woodrats may contribute to the paucity of small trees and the uniformity in size of the junipers. Seedling survival there may depend on both an optimal annual weather cycle and a very low woodrat population. The probability that these conditions may occur simultaneously is extremely low.

I suggest, then, that the Stephens' woodrat not only contributes today to the picturesque growth form of some junipers, but that at least since late Pleistocene it has been an important selective force "guiding" the evolution of junipers in parts of the Southwest.

Acknowledgments

For assistance in the field I would like to thank M. M. Bateman and L. R. Pyc. R. M. Warner prepared slides for dietary analyses, cared for captive woodrats, and assisted with feeding trials. C. N. Slobodchikoff and T. G. Whitham made valuable critical comments on early drafts of this paper.

Literature Cited

Adams, R. P. 1969. Chemosystematic and numerical studies in natural populations of *Juniperus*. Ph.D. Dissertation, University of Texas, Austin.

Brown, J. H., and Lieberman, G. A. 1972. Woodrats and cholla: dependence of a small mammal population on the density of cacti. Ecology, 53:310–13.

Esau, K. 1977. Anatomy of seed plants. Second edition. John Wiley and Sons, New York, 550 pp.

Finley, R. B. 1958. The wood rats of Colorado: distribution and ecology. Kans., Univ., Mus. Natur. Hist., Publ., 10:213–552.

Hamilton, W. J. III. 1962. Reproductive adaptations of the red tree mouse. J. Mammal., 43:486–504.

Hansen, R. M., and Flinders, J. T. 1969. Food habits of American hares. Colo. St. Univ., Range Sci. Dept., Sci. Ser. No. 1, 18 pp.

Harborne, J. B. (ed.). 1972. Phytochemical ecology. Academic Press, New York, 272 pp.

Harris, A. 1963. Ecological distribution of some mammals in the San Juan Basin, New Mexico. Museum of New Mexico Papers in Anthropol., 8:1–63.

Hoffmeister, D. F., and Durham, F. E. 1971. Mammals of the Arizona Strip including Grand Canyon National Monument. Mus. North. Ariz., Tech. Ser., No. 11, 44 pp.

———, and de la Torre, L. 1960. A revision of the wood rat *Neotoma stephensi*. J. Mammal. 41:476–91.

Lee, A. K. 1963. The adaptations to arid environments in wood rats of the genus *Neotoma*. University California Publ. Zool., 64:57–96.

MacMillen, R. M. 1964. Population ecology, water relations, and social behavior of a Southern California semidesert rodent fauna. Calif., Univ. Publ. Zool., 71:1–66.

Preston, R. J., Jr. 1968. Rocky Mountain trees, a handbook of the native species with plates and distribution maps. Dover Publ., Inc., New York, 285 pp.

Rhoads, D. F. and Cates, R. G. 1976. Toward a general theory of plant anti-herbivore chemistry, pp. 168–213. *In* J. W. Wallace and R. L. Mansell (eds.), Biochemical interactions between plants and insects. Recent Advances in Phytochemistry 10. Plenum Press, New York, 425 pp.

Schmidt-Nielsen, K., and Schmidt-Nielsen, B. 1952. Water metabolism of desert mammals. Physiol. Rev., 32:135–166.

Short, H. L., and McCulloch, C. Y. 1977. Managing piñon-juniper ranges for wildlife. USDA Forest Service, General Technical Report RM-47. 10 pp.

Spencer, D. A., and Spencer, A. L. 1941. Food habits of the white-throated woodrat in Arizona. J. Mammal., 22:280–84.

Van Devender, T. R., and Spaulding, W. G. 1979. Development of vegetation and climate in the southwestern United States. Science, 204:701–10.

Van Devender, T. R., Phillips, A. M. III; and Mead, J. I. 1977. Late Pleistocene reptiles and small mammals from the lower Grand Canyon of Arizona. Southwest. Natur., 22:49–66.

INDEX

PAPER: Paloma matte text; cloth bound, linen finish; paper, Carolina cover
TYPOGRAPHY: Meridien by Tiger Typographics, Flagstaff, Arizona
LITHOGRAPHY: Classic Printers, Prescott, Arizona
BINDERY: Roswell Bookbinding, Phoenix, Arizona
DESIGN: Stanley Stillion, John F. Stetter, Stephen Trimble
PRODUCTION: Earl Hatfield, Stephen Trimble, Sandy Davidson, Robin T. Squire
EDITORIAL: Louis L. Jacobs, Stephen Trimble, John F. Stetter